Evolutionary Biology

VOLUME 8

Evolutionary Biology

Biology

VOLUME 8

Edited by

THEODOSIUS DOBZHANSKY
Department of Genetics
University of California
Davis, California

MAX K. HECHT
Queens College
Flushing, New York

and

WILLIAM C. STEERE
New York Botanical Garden
Bronx, New York

PLENUM PRESS ● NEW YORK AND LONDON

The Library of Congress cataloged the first volume of this title as follows:

Evolutionary biology. v. 1–   1967–
New York, Appleton-Century-Crofts.

v. illus. 24 cm. annual.

Editors: 1967–   T. Dobzhansky and others.

1. Evolution — Period.   2. Biology — Period.   ɪ. Dobzhansky,
Theodosius Grigorievich, 1900–

QH366.A1E9   575'.005   67—11961

Library of Congress   ₍67n7₎

QH366
A1
E9
v.8

Library of Congress Catalog Card Number 67-11961
ISBN 0-306-35408-X

©1975 Plenum Press, New York
A Division of Plenum Publishing Corporation
227 West 17th Street, New York, N.Y. 10011

United Kingdom edition published by Plenum Press, London
A Division of Plenum Publishing Company, Ltd.
Davis House (4th Floor), 8 Scrubs Lane, Harlesden, London, NW10 6SE, England

Contributors

FRANCISCO J. AYALA • *Department of Genetics, University of California, Davis, California*

P. R. GRANT • *Department of Biology, McGill University, Montreal, Quebec, Canada*

ALESSANDRO MORESCALCHI • *Institute of Histology and Embryology, University of Naples, Naples, Italy*

JEFFREY R. POWELL • *Department of Biology, Yale University, New Haven, Connecticut*

V. N. SOYFER • *All-Union Institute of Applied Molecular Biology and Genetics, Moscow, U.S.S.R.*

Contents

5. Chromosome Evolution in the Caudate Amphibia

Alessandro Morescalchi

1

Genetic Differentiation During the Speciation Process

FRANCISCO J. AYALA

Department of Genetics
University of California
Davis, California 95616

INTRODUCTION

The evolution of populations through time occurs by changes in the constitution of their gene pools, that is, by changes in the frequencies of genes and genotypes. Genetic variation arises as a result of the processes of mutation (including chromosomal changes), recombination, and gene flow from other populations. Gene frequencies may also change because of genetic drift and natural selection. Genetic drift is due to sampling errors from generation to generation in populations consisting of finite numbers of individuals. Mutation, recombination, gene flow, and genetic drift are random processes with respect to adaptation. They bring about genetic changes independently of whether such changes are beneficial or harmful to the populations in which they occur. The only adaptive process of evolutionary change is natural selection, which increases the frequencies of genetic variants which are useful to their carriers as adaptations to the environment.

The process of evolution has two dimensions: (1) anagenesis or phyletic evolution and (2) cladogenesis or splitting. Changes occurring in a given population, or group of populations, as time goes on are anagenetic or phyletic evolution. As a rule, they result in increased adaptation to the environment, and often reflect changes in the physical or biotic conditions of the environment. Cladogenesis occurs when a phylogenetic lineage splits into two or more independently evolving lineages. The great diversity of the living world is the result of cladogenetic evolution, which results in adaptation to a greater variety of niches, or ways of life.

The most decisive among cladogenetic processes is speciation, the process by which one species splits into two or more species. In outbreeding sexual organisms, a species is an array of Mendelian populations among which gene exchange can occur without impediments other than geographical separation, but reproductively isolated from any other such arrays. A mutation or other genetic change originating in a single individual of a species may spread to all members of the species by natural selection or genetic drift. Owing to reproductive isolation, such a change cannot be passed on to different species. Speciation is, then, a highly significant stage of evolutionary differentiation: species are discrete and independent evolutionary units.

The question of how much genetic differentiation occurs during the process of speciation is, therefore, one of the most important, and in a sense the cardinal problem of evolutionary genetics. In this chapter I shall discuss this question and summarize whatever information exists to provide an answer.

In recent years, the application of certain techniques of molecular genetics has made it possible to quantify, at least to a first approximation, the amount of genetic differentiation among populations, reproductively isolated or not. It may be worth pointing out that estimates of genetic differentiation among populations measure the amount of change involved not only in cladogenesis but also in anagenetic evolution. Assume that it is found that two species, S_1 and S_2, differ at a certain proportion, x, of their genes. If we ignore (or have a way to correct for) parallel and convergent changes, as well as the fact that a gene may have changed more than once, x is an estimate of the amount of anagenetic change that has taken place in the phyletic line going from their common ancestral population to S_1, *plus* the anagenetic change in the phyletic line going from the common ancestor to S_2. To a first approximation, then, one may assume that approximately half of x changed in each of the two phyletic lines. Comparison with other species may allow more precise estimation of how much genetic change occurred in each of the two phyletic lineages.

MODES OF SPECIATION

The number of species named and described is somewhere between 1.5 and 2 million (Dobzhansky, 1972); the number of existing species may be twice as large or larger. The concept of species advanced above, as arrays of

Mendelian populations reproductively isolated from other arrays, does not apply to organisms that are exclusively asexual, parthenogenetic, or self-fertilizing. Arrays of organisms phenotypically or ecologically similar to each other, and different from other arrays can, nevertheless, be identified in these organisms, and the arrays may be identified and named as "species"; genetic differentiation within and between the arrays can be measured.

Among outcrossing sexually reproducing organisms, the most common mode of speciation requires, for its inception, that populations become geographically isolated. If there is little or no gene flow between them, populations may gradually become genetically differentiated, particularly as a result of their adaptation to different environments. Different races arise by this process, which is a necessary albeit not sufficient condition for species differentiation. It is not sufficient, because the process of geographic or race differentiation is reversible. Not every race is a future species; without reproductive isolation differentiated populations may converge or fuse by gene exchange. This is precisely what is happening in the human species.

What processes bring about the development of reproductive isolating mechanisms? Two, not mutually exclusive, answers have been suggested. One answer is that reproductive isolation is a by-product of the accumulation of genetic differences between geographically separated diverging populations. There is little doubt that if the process of genetic divergence continues long enough, geographically separated populations might become so different as to be unable to interbreed if the opportunity did eventually arise.

The second answer considers the conditions under which the development of reproductive isolation might be accelerated. If two populations geographically isolated for some time come again into geographic contact, two outcomes are possible: (1) a single gene pool comes about, because the populations hybridize readily without significant loss of fitness (or because one of the populations is eliminated by the other through ecological competition); and (2) two species ultimately arise because natural selection favors the development of reproductive isolation between the populations. If matings between individuals of different populations leave progenies with reduced fertility or viability, natural selection would favor genetic variants promoting matings between individuals of the same population. Reproductive isolation, and therefore speciation may ensue. Which one of the two possible outcomes just mentioned will, in fact, occur in a particular instance depends, of course, on the degree of genetic differentiation achieved by the populations previous to regaining sympatry. There is ample evidence showing that natural selection favors the development of reproductive isolation among genetically differentiated populations when these exchange genes by hybridization (Mayr, 1963; Dobzhansky, 1970).

The question, "how much genetic differentiation is concomitant to the process of speciation?" needs, therefore, to be divided into two separate questions. The first concerns the amount of genetic differentiation during the first stage of speciation, when allopatric populations of the same species become differentiated to the point that they are likely to evolve into different species if they come into geographic contact. The second question refers to the amount of genetic differentiation during the second stage of speciation, when reproductive isolation is completed by natural selection between genetically differentiated populations that have come into geographic contact.

Carson (1970, 1971a) has recently advanced a model that would make possible the development of new species in a relatively short time, without involving natural selection directly favoring the development of reproductive isolation between populations exchanging genes. This could occur when a single gravid female (or very few individuals) colonize an isolated propitious territory, where its descendants find few or no competitors. The founder individual(s) would carry only a small fraction of the total genetic variation of the parental population. Given favorable conditions, its descendants would increase in number until the carrying capacity of the environment becomes saturated, at which point intraspecific competition would occur. Carson argues that the initial inbreeding followed by rapid expansion in numbers would generate numerous new genotypes that would not have appeared or would not have survived in the parental population. Natural selection acting on these genotypes might eventually give rise to reproductively isolated "neospecies."

Carson claims that speciation according to his model may have played a key role in the proliferation of drosophilid species in the archipelago of Hawaii. More than 500 species of drosophilids are endemic to Hawaii, all descended from one or two colonizer species, introduced in the archipelago by accidental long-distance transport across the ocean. A great majority of species are endemic not only to the archipelago but to single islands (or to adjacent islands which were connected in recent geological time). According to Carson, the evidence from cytogenetic, morphological, ecological, and biogeographic studies suggests that each species is descended from a single gravid female. Groups of species, sometimes from different islands, have "homosequential" polytene chromosomes, i.e., no inversions or other chromosomal rearrangements have occurred in the origin of one from another species, or in their origin from a common ancestral one. Carson believes that "homosequential" species may represent instances of rapid allopatric speciation without major genetic differentiation.

New species may arise in a single generation by polyploidy, whether through multiplication of the chromosomes of a single species (auto-polyploidy) or through doubling of the chromosomal complement in the hybrid between two previously existing species (allopolyploidy). Species formation through polyploidy is a common phenomenon in plants; perhaps more than one-third of all living species of plants have arisen by polyploidy, although the proportion varies greatly in different plant groups. Polyploidy is a much rarer phenomenon in the animal kingdom, but polyploid species occur among hermaphroditic animals such as earthworms and planarians, and also among animals with parthenogenetic females, including some beetles, moths, sow bugs, shrimps, fishes, and salamanders. Whether it is an allopolyploid or an autopolyploid, a polyploid possesses, at least initially, all the genes present in its ancestors and no new ones. In this sense, polyploidy represents speciation without differentiation of individual genes.

Modes of rapid, or "saltational," speciation not involving polyploidy are known to occur in plants. *Clarkia franciscana* is a narrow endemic growing near San Francisco, California, within the distribution area of *C. rubicunda;* *C. amoena* replaces *C. franciscana* just north of San Francisco Bay, and is found northward to British Columbia. The three species are morphologically very similar, but they are reproductively isolated because of meiotic irregularities in the hybrids owing to gross differences in chromosomal structure. Lewis and Raven (1958) have proposed that *C. franciscana* originated relatively recently from *C. rubicunda* as a consequence of rapid reorganization of the chromosomes brought about by inbreeding among surviving individuals of a population of *C. rubicunda* which had been subjected to a severe environmental stress. However, findings of substantial genic differentiation between *C. rubicunda* and *C. franciscana* have brought into question whether *C. franciscana* did indeed evolve from *C. rubicunda* by rapid, saltational speciation (Gottlieb, 1973*a*). On the other hand, findings of considerable genic similarity between two other species of *Clarkia,* *C. lingulata,* and *C. biloba* (Gottlieb, 1974) are consistent with the hypothesis that the former originated from the latter in a rapid series of events involving various chromosomal rearrangements (Lewis and Roberts, 1956). These two species differ by a translocation, several paracentric inversions, and an extra chromosome in *C. lingulata* which is homologous to parts of two chromosomes of *C. biloba*.

An instance of apparent saltational sympatric speciation has been discovered by Gottlieb (1973*b*). The new diploid species, designated "Malheurensis," is thought to have arisen in a geographically peripheral population of the annual plant, *Stephanomeria exigua* ssp. *coronaria*. The new species may have emerged, perhaps from a single progenitor individual, by a change in breeding system; *S. exigua* ssp. *coronaria* is an obligate outcrosser, while

"Malheurensis" is highly self-pollinating. The reproductive isolation resulting from differences in breeding systems is reinforced by chromosomal structural differences. The contention that "Malheurensis" arose rapidly and recently is supported by the fact that virtually all alleles found in "Malheurensis" at 13 loci coding for enzymes are apparently identical to alleles found in its progenitor *S. exigua* ssp. *coronaria*.

White (1973, for a review) has proposed a model of "stasipatric" speciation to account for the evolution of some species and semispecies of flightless Australian grasshoppers of the subfamily Morabinae, particularly in the superspecies *Moraba viatica* and *M. scurra*. Semispecies differing by chromosomal translocations occupy adjacent territories. According to White, a translocation establishes itself at first in a small colony by genetic drift. If members of this colony possess high fitness, they may subsequently spread and displace the ancestral form from a certain area. The ancestral and the derived form may coexist contigually, their individuality maintained by the low fitness of the hybrids found in the contact zones, since the hybrids are translocation heterozygotes. Saltational speciation analogous to the type of speciation postulated by White, may have been involved in the evolution of closely related species differing in chromosome numbers in two rodent groups, the mole rats of the complex *Spalax ehrenbergi* in Israel (Nevo and Shaw, 1972), and the pocket gophers of the complex *Thomomys talpoides* in the southern Rockies of the United States (Nevo *et al.*, 1974).

THE METHODS

De Vries, Goldschmidt, and others believed that a new species may arise by a single mutation of a special kind, variously called "macromutation," "systemic mutation," and the like. Goldschmidt claimed that a systemic mutation may generate not only a new species, but also a new genus and even a family. These claims are not tenable today, although new species may emerge by polyploidy, or by some other form of saltational speciation.

As early as 1919, Morgan argued that species differ from each other "not by a single Mendelian difference, but by a number of small differences" (Morgan, 1919). Studies of segregation in progenies of interspecific hybrids in plants, and more rarely in animals, as well as other studies provided gradually increasing evidence of multiple gene differences between closely related species (for a brief historical review, see Dobzhansky, 1970). Some authors, however, argued that speciation often involves few genetic changes. The controversy around the amount of genetic differentiation involved in the process of speciation has continued until the present time.

It may appear paradoxic that the question of how much genetic differentiation occurs in the process of speciation has not received an early answer, since the question is of paramount importance in evolutionary genetics. This apparent paradox becomes understandable when one considers the kind of information required to answer the question. To find out the proportion of the genes at which two populations or species differ, one must be able to study genes that are an unbiased sample of the total genome of the organisms; i.e., the genes studied must be on the average neither more similar nor more different in the two populations than the rest of the genome. Yet, a classical Mendelian geneticist ascertains that a character difference is determined by a gene by studying segregation in the progenies of individuals differing phenotypically in the character. The applicability of the study of segregation is severely limited in determining species differences by the inviability or sterility of most interspecific hybrids. Even when segregation can be observed in the progenies of hybrids, this method discovers only genes with different allelic forms in the two populations compared; but how many or what proportion of the genes are identical in the two populations cannot be ascertained. Still one more difficulty arises because phenotypic, or behavioral traits are many steps removed from the genes, and often result from the interaction of many genes.

Molecular genetics and certain technical developments have made it possible in recent years to answer questions concerning the amount of genetic differentiation between populations, at least to a first approximation. The basic conceptual breakthrough was the discovery of the relationship between genes and the proteins coded by them. Enzymes and other proteins are the translational products of genes; variation in enzymes and proteins substantiates allelic variation in the genes coding for them. The existence of a gene can now be ascertained even when the gene is not variable. The important point is that genes can be selected for study without *a priori* knowledge of whether or not they are variable within or between populations. Genes can be chosen for study that represent a random sample of the genome with respect to their variation in the populations to be compared. The results obtained from the study of a moderate number of gene loci can, therefore, be extrapolated to the whole genome.

The techniques used in recent years to measure genetic differences between species include sequencing the amino acids of proteins, microcomplement fixation, DNA–DNA and DNA–RNA hybridization, and gel electrophoresis. When comparisons are made between closely related species, gel electrophoresis provides more information by far than any of the other techniques for a given amount of work and cost. The technique essentially consists of placing tissue homogenates in a homogeneous matrix (gel) which is then placed in an electric field for some time, usually a few hours. Enzymes and other proteins migrate in the gel at rates that are a function of their net electric charge and their size. After the electric current is stopped, the gel is subject to a staining solution for proteins or for a specific enzyme. The positions where the proteins or enzymes have migrated thus become visible in the gels. Assume that tissues from two different individuals are electrophoresed and assayed for a specific enzyme, and assume further that the enzyme of one individual migrates faster than that of the other. It follows that the two enzymes are different at least by one amino acid substitution, and therefore that they are coded for by different genes, i.e., by different alleles if the enzyme is coded by the same gene locus in both individuals. On the other hand, if the enzymes of the two individuals migrate equally, the inference is made that they are coded by identical genes. The study of a score or more enzymes in two populations permits a reasonable estimate of the proportion of genes that are different in the populations.

The use of gel electrophoresis to measure genetic variation is subject to several potential sources of bias (Lewontin and Hubby, 1966; Ayala *et al.*, 1970). One important difficulty is that not all allelic variation is detectable by electrophoresis; enzymes with identical electrophoretic mobilities may nevertheless be different in amino acid sequence. Thus, the amount of genetic variation is underestimated, although it is not possible at present to tell by how much. Another difficulty is that only structural genes coding for soluble proteins can be studied. Regulatory genes or those coding for nonsoluble proteins may behave differently than genes coding for soluble proteins, with respect to the questions that interest us. A third difficulty which I shall mention is that when we observe two enzymes with different electrophoretic mobilities, we know that they have different amino acid sequences, but we do not know by how many amino acid substitutions they differ. In practice, however, this is the least serious difficulty of the three mentioned. If the two species compared have identical enzymes at least at a moderate proportion of loci, some fairly reasonable assumptions allow estimating what proportion of the nonidentical enzymes differ by one, two, three, etc., amino acid substitutions.

Electrophoretic study of a variety of enzymes in individuals of two populations provides information about genotype and gene frequencies in the two populations. Many methods have been developed to quantify and express in a single parameter the amount of genetic differentiation between two populations. The statistics most commonly used in electrophoretic studies are genetic identity, *I*, and genetic distance, *D*, proposed by Nei

(1972). Let X and Y be two different populations, and K a given locus. The normalized probability that two alleles, one from each population, are identical is

$$I_K = \frac{\sum x_i y_i}{(\sum x_i^2 \sum y_i^2)^{1/2}} \qquad (1)$$

where x_i and y_i are the frequencies of the ith allele in populations X and Y. The mean *genetic identity* over all loci studied is given by

$$I = \frac{I_{xy}}{(I_x \cdot I_y)^{1/2}} \qquad (2)$$

where I_{xy}, I_x, and I_y are the arithmetic means, over all loci, of $\sum x_i y_i$, $\sum x_i^2$, and $\sum y_i^2$, respectively.

The average genetic differentiation, or *genetic distance*, per locus between the two populations is estimated by

$$D = -\log_e I \qquad (3)$$

If it is assumed that codon substitutions within a locus occur independently from each other, and that the number of allelic substitutions per locus follows a Poisson distribution, D can be interpreted as a measure of the average number of electrophoretically detectable allelic substitutions per locus which have accumulated since the two populations separated from a common ancestral one.

A PARADIGM OF THE PROCESS OF SPECIATION: THE *WILLISTONI* GROUP OF *DROSOPHILA*

The most common mode of speciation among sexually reproducing organisms is the process of geographic speciation discussed above. The process of geographic speciation occurs in two main stages. During the first stage, genetic differentiation occurs between geographically isolated populations. The second stage occurs when sufficiently differentiated populations come into geographic contact, and natural selection results in the development of sexual and other mechanisms of reproductive isolation between the populations so that two species ultimately emerge.

The most extensive and complete study of genetic differentiation during the process of geographic speciation has been carried out in the *Drosophila willistoni* group of species (summary in Ayala *et al.*, 1974a). The *willistoni* group of *Drosophila* consists of at least fifteen closely related species endemic to the tropics of the New World. Six species are siblings, mor-

phologically nearly indistinguishable, although the species of individual
males can be identified by slight but diagnostically reliable differences in
their genitalia. Two sibling species, *D. insularis* and *D. pavlovskiana*, are
narrow endemics; the former in some islands of the Lesser Antilles, and the
latter in Guyana. Four other siblings, namely, *D. willistoni, D. equinoxialis,
D. tropicalis,* and *D. paulistorum,* have wide and largely overlapping
geographic distributions through Central America, the Caribbean, and
much of continental South America (Fig. 1).

Some sibling species consist of at least two subspecies. Populations of
D. willistoni west of the Andes near Lima, Peru, belong to the subspecies *D.
w. quechua,* while east of the Andes and elsewhere the subspecies is *D. w.
willistoni.* Incipient reproductive isolation in the form of partial hybrid
sterility exists between these two subspecies. Laboratory crosses of *D. w.*

FIG. 1.   Geographic distribution of four sibling species of the *Drosophila willistoni* group.

willistoni females with *D. w. quechua* males yield fertile males and females. Crosses between *D. w. quechua* females and *D. w. willistoni* males from continental South America east of the Andes produce fertile females but sterile males. Laboratory tests show no evidence of ethological (sexual) isolation between the subspecies (Ayala and Tracey, 1973).

Two subspecies are also known in *D. equinoxialis: D. e. caribbensis* in Central America, north of Panama, and in the Caribbean islands; and *D. e. equinoxialis* in eastern Panama and continental South America. Crosses between the two subspecies yield fertile females but sterile males, independently of the subspecies of the female parent. Like in *D. willistoni*, there is no evidence of sexual isolation between the subspecies of *D. equinoxialis* (Ayala *et al.*, 1974*b*).

Evolutionary divergence beyond the taxonomic category of subspecies, but without complete achievement of speciation, exists in a third sibling species, *D. paulistorum*. This "species" consists of at least six semispecies, or incipient species, named Centroamerican, Transitional, Andean–Brazilian, Amazonian, Orinocan, and Interior (Spassky *et al.*, 1971). Laboratory crosses between the semispecies generally yield fertile females but sterile males. Reproductive isolation between some semispecies is essentially complete, so that two and even three semispecies coexist sympatrically in many localities (Fig. 2). Sexual isolation is essentially complete between some semispecies, particularly when sympatric populations are tested. Gene flow among the semispecies is nevertheless possible, particularly through populations of the Transitional semispecies.

Drosophila willistoni and its relatives are, thus, excellent materials to study the amount of genetic differentiation in the process of geographic speciation. Five increasingly divergent levels of evolutionary divergence or cladogenesis can be recognized:

1. Between geographic populations of the same taxon.

2. Between subspecies. These are allopatric populations that exhibit incipient reproductive isolation in the form of partial hybrid sterility. If populations of two subspecies were to come into geographic contact, intersubspecific matings would leave fewer fertile descendants than intrasubspecific matings. Therefore, natural selection would favor the development of reproductive isolating mechanisms between the subspecies. Whether two species would ultimately result, or one subspecies be absorbed (by introgression) or eliminated (by competition), we of course do not know. The subspecies are allopatric populations in the first stage of the speciation process.

3. Between the semispecies of D. paulistorum. The process of speciation is being completed between the semispecies. Sexual isolation is being superimposed over the preexisting hybrid sterility and is nearly complete in

FIG. 2.   Geographic distribution of the semispecies of the *Drosophila paulistorum* complex.

many cases; some semispecies are sympatric in many localities. The semi-species are populations in the second stage of the speciation process.

   4. Between sibling species.   In spite of their morphological similarity, the sibling species are completely reproductively isolated. Study of genetic differentiation between them will show how much genetic differentiation may occur after speciation without noticeable morphological diversification.

   5. Between morphologically distinguishable species of the same group.   D. nebulosa is a close relative of *D. willistoni* and its siblings, but can be easily distinguished from them by external morphology. Comparison of *D. nebulosa* with the siblings will show how much genetic differentiation occurs at this level of evolutionary divergence.

   Using electrophoretic techniques 36 gene loci coding for enzymes have been studied in each of the sibling species of the *D. willistoni* group and in the morphologically differentiated *D. nebulosa*. The genotypes of large numbers of individuals (from several hundred to several thousand) have been ascertained at each locus in each species, except for the two narrow endemics, *D. insularis* and *D. pavlovskiana*, of which only a few genomes were sampled. The numbers of genomes (equal to twice the number of individuals for autosomal loci) sampled and the amount of genetic variation found in each of the five widely distributed species are given in Table I. These species are genetically very polymorphic. On the average, the proportions of polymorphic loci per population are 51% or 72%, depending on the criterion of polymorphism used. The average frequency of heterozygous loci per individual is 17.7%. This amount of genetic polymorphism is not unusually high for an invertebrate species. Selander and Kaufman (1973) have calculated the average heterozygosity for 24 species of invertebrates as 15.1%.

The amounts of genetic differentiation at each of the five increasing levels of evolutionary divergence found in the *D. willistoni* group are as follows (Ayala *et al.*, 1974*a*):

1. Local populations. Laboratory tests indicate that in the *D. willistoni* group there is no reproductive isolation between geographic populations of the same taxon. Local populations are also genetically very little differentiated, as shown in Table II. Geographic populations of the Transitional semispecies of *D. paulistorum* are genetically more differentiated ($\bar{D} = 0.101 \pm 0.028$) than those of any other taxon (overall average, $\bar{D} = 0.028 \pm 0.006$). This is hardly surprising since the Transitional semispecies consists of a very heterogeneous group of populations, perhaps representing incipient stages of speciation (Spassky *et al.*, 1971).

The distribution of loci with respect to genetic identity is shown in Fig. 3. At most loci, local populations have essentially identical allelic frequencies ($I \geq 0.95$); some populations, however, have substantially different genetic constitutions at an occasional locus.

2. Subspecies. The two subspecies of *D. willistoni*, as well as the two subspecies of *D. equinoxialis,* are populations in the first stage of the process of geographic speciation. They are pairs of allopatric populations which have diverged from each other to the point where some hybrid progenies are sterile; natural selection would favor the development of complete reproductive isolation between them if they were to become sympatric. Table III shows the amount of genetic differentiation between subspecies.

Several relevant facts should be emphasized. First, the amount of genetic differentiation between the subspecies is quite substantial. On the average, 23 electrophoretically detectable allelic substitutions for every 100 loci have taken place during the formation of the subspecies. The genetic distance is about ten times larger between the subspecies than between populations of the same species (see Table II).

Second, the genetic distance is greater between the two subspecies of *D. equinoxialis* than between the two subspecies of *D. willistoni*. This agrees well with the degree of evolutionary divergence between each pair of

TABLE I. Summary of Genetic Variation at 36 Loci in Natural Populations of Five Closely Related Species of *Drosophila*

Species	Genomes sampled per locus	Polymorphic loci per population[a]		Heterozygous loci per individual
		(1)	(2)	
D. willistoni	4,983 ± 636	0.738 ± 0.064	0.491 ± 0.072	0.179 ± 0.037
D. tropicalis	1,731 ± 229	0.647 ± 0.067	0.474 ± 0.076	0.152 ± 0.031
D. equinoxialis	2,356 ± 238	0.792 ± 0.061	0.540 ± 0.075	0.165 ± 0.030
D. paulistorum	1,277 ± 258	0.742 ± 0.062	0.511 ± 0.071	0.194 ± 0.027
D. nebulosa	412 ± 43	0.692 ± 0.071	0.530 ± 0.069	0.195 ± 0.035
All species	10,759 ± 341	0.722 ± 0.025	0.509 ± 0.012	0.177 ± 0.008

[a] A locus is considered polymorphic: (1) when the frequency of the second most common allele is at least 0.01; (2) when the frequency of the most common allele ≤ 0.95.

TABLE II. Average Genetic Identity, $\bar{I}$, and Genetic Distance, $\bar{D}$, between Local Populations of Various Taxa of the *Drosophila willistoni* Group

Taxon	$\bar{I}$	$\bar{D}$
D. w. willistoni	0.986 ± 0.005	0.014 ± 0.005
D. tropicalis	0.960 ± 0.005	0.041 ± 0.005
D. e. equinoxialis	0.977 ± 0.003	0.023 ± 0.003
D. e. caribbensis	0.975 ± 0.004	0.026 ± 0.004
D. paulistorum		
Transitional	0.906 ± 0.026	0.101 ± 0.028
Andean-Brazilian	0.969 ± 0.007	0.031 ± 0.006
Interior	0.968 ± 0.007	0.033 ± 0.007
Amazonian	0.960 ± 0.005	0.042 ± 0.005
Average	0.951 ± 0.015	0.052 ± 0.017
D. nebulosa	0.986 ± 0.004	0.014 ± 0.004
Average among species	0.972 ± 0.006	0.028 ± 0.006

subspecies as measured by the sterility of intersubspecific hybrids. All hybrid males are sterile between subspecies of *D. equinoxialis*, while only those having *D. w. quechua* mothers are sterile in *D. willistoni*.

Third, the pattern of the genetic differentiation deserves consideration. The degree of genetic divergence achieved between the subspecies is not due to a moderate amount of genetic differentiation at each of many or all loci. Rather, populations of different subspecies have remained essentially identical at most loci, while they are nearly completely different at a few loci. This pattern is illustrated in Fig. 4, where the numbers of loci within each range of values of genetic identity are given. The allelic frequencies are essentially identical ($I \geq 0.95$) in the subspecies at 63% of the loci, while nearly complete genetic differentiation ($I < 0.15$) occurs at 18% of the loci.

3. Semispecies. The semispecies of *D. paulistorum* are populations in the second stage of geographic speciation. Populations which had become genetically differentiated while geographically isolated have come into geographic contact. Natural selection is expected to favor the development of sexual isolation and other isolating mechanisms. This has been confirmed by Ehrman (see Dobzhansky, 1970, p. 370), who measured in the laboratory sexual isolation between sympatric and between allopatric strains of different semispecies. The average coefficient of isolation is greater between sympatric strains (0.85) than between allopatric strains (0.67).

The average genetic identity and genetic distance between strains of different semispecies are: $\bar{I} = 0.798 \pm 0.026$, $\bar{D} = 0.226 \pm 0.033$. These values

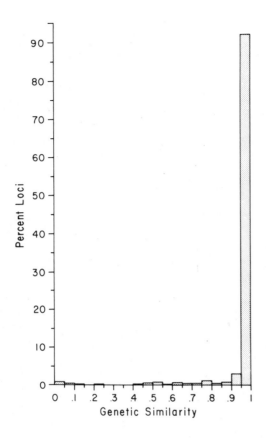

FIG. 3. Histogram of the distribution of genetic identity among gene loci for comparisons between local populations of species of the *D. willistoni* group.

are not significantly different from those observed between subspecies ($\bar{I}$ = 0.795 ± 0.013, $\bar{D}$ = 0.230 ± 0.016; see Table III). It appears that the development of sexual isolation does not require the change of a large fraction of the genes in addition to those differentiating subspecies.

The distribution of loci relative to genetic identity is depicted in Fig. 5. This distribution is similar to that for subspecies comparisons (Fig. 4), except that between the semispecies there is a greater proportion of loci with intermediate degrees of differentiation.

4. Sibling species. Table IV is a matrix of genetic identities (above the diagonal) and genetic distances (below the diagonal) between the sibling species of the *D. willistoni* group. The comparisons involving the two narrowly distributed *D. insularis* and *D. pavlovskiana* are not as reliable as the others since the samples of those two species were small. If such com-

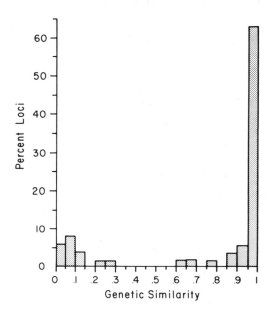

FIG. 4. Histogram of the distribution of genetic identity among gene loci for comparisons between subspecies of the *Drosophila willistoni* group.

parisons are excluded, the mean genetic identity between sibling species is $\bar{I}$ = 0.563 ± 0.023, and the mean genetic distance is $\bar{D}$ = 0.581 ± 0.039.

The siblings of *D. willistoni* are closely related species, distinguishable morphologically mainly by some slight but reliable differences in the male genitalia. They have largely overlapping geographic distributions (see Fig. 1), share their food resources, and are quite similar also in other ecological attributes (Ayala, 1970). Yet we estimate that, on the average, about 58.1 electrophoretically detectable allelic substitutions for every 100 loci have occurred in each pair of siblings since their divergence from a common ancestor. Slight morphological and ecological differentiation cannot be taken as evidence of little genetic differentiation. Dobzhansky (1956) has suggested that the evolution of external morphology has reached in the genus

TABLE III. Genetic Identity, *I*, and Genetic Distance, *D*, between Subspecies of the *Drosophila willistoni* Group

Subspecies	*I*	*D*
D. w. willistoni–D. w. quechua	0.808	0.214
D. e. equinoxialis–D. e. caribbensis	0.782	0.246
Average	0.795 ± 0.013	0.230 ± 0.016

TABLE IV. Genetic Identity (above Diagonal) and Genetic Distance[a] (below Diagonal) between Sibling Species of the Drosophila willistoni Group[b]

	D. willistoni	D. tropicalis	D. equinoxialis	D. paulistorum	D. pavlovskiana	D. insularis
D. willistoni	—	0.663	0.522	0.594	0.596	0.344
D. tropicalis	0.413	—	0.514	0.545	0.496	0.414
D. equinoxialis	0.656	0.665	—	0.540	0.537	0.336
D. paulistorum	0.524	0.609	0.621	—	0.795	0.299
D. pavlovskiana	0.518	0.701	0.633	0.232	—	0.280
D. insularis	1.070	0.883	1.091	1.208	1.273	—

[a] Average genetic identity = 0.475 ± 0.037; average genetic distance = 0.740 ± 0.078.
[b] When subspecies or semispecies occur, the average values between each one of them and the other species are given.

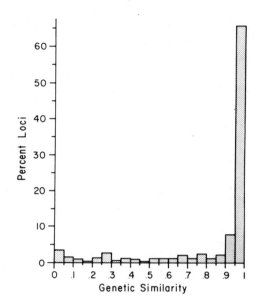

FIG. 5. Histogram of the distribution of genetic identity among gene loci for comparisons between semispecies of *Drosophila paulistorum*.

Drosophila a high degree of perfection and that the adaptive evolution in this genus proceeds largely through physiological channels; the present results corroborate that hypothesis.

The distribution of genetic identities among loci shown in Fig. 6 is strongly bimodal. At most loci, any two sibling species have either virtually identical ($I \geq 0.95$), or completely different ($I < 0.05$) genetic constitutions. Only a minority of the loci have I values in the wide range between 0.05 and 0.95.

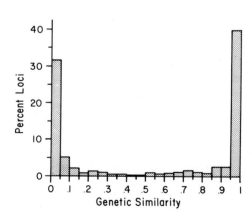

FIG. 6. Histogram of the distribution of genetic identity among gene loci for comparisons between sibling species of the *Drosophila willistoni* group.

5. *Morphologically different species.* Species have independently evolving gene pools. After speciation is completed, species are likely to continue diverging genetically. We have seen that sibling species are genetically quite different; the differences are even greater between *D. nebulosa* and the siblings (Table V). *D. nebulosa* is a close relative of the sibling species although morphologically easily distinguishable from them. On the average, about one electrophoretically detectable allelic substitution per locus ($\bar{D}$ = 1.056 ± 0.068) has taken place between *D. nebulosa* and the siblings since they diverged from a common ancestral population. The distribution of loci relative to genetic identity (Fig. 7) is, as in the comparisons between siblings, strongly bimodal, but the mode near $I = 0$ is greater and the mode near $I = 1$ is smaller than for the siblings.

The results from the study of genetic differentiation in the *D. willistoni* group are summarized in Table VI. The data provide preliminary answers to the question of how much genetic differentiation occurs during the two main stages of the process of geographic speciation. The first stage is represented by the subspecies of *D. willistoni* and of *D. equinoxialis*. Incipient reproductive isolation in the form of partial hybrid sterility exists between the subspecies. If populations of two subspecies were to exchange genes, individuals mating with members of their own subspecies would have greater fitness than individuals mating with members of the other subspecies. If there are genes modifying the probability of intrasubspecific matings, allelic variants increasing such probability would be favored by selection; therefore natural selection would impel the development of re-

TABLE V. Genetic Identity, *I*, and Genetic Distance, *D*, between Morphologically Different Species of the *Drosophila willistoni* Group[a]

Species compared to *D. nebulosa*	*I*	*D*
D. willistoni	0.267	1.325
D. tropicalis	0.391	0.939
D. equinoxialis	0.426	0.854
D. paulistorum	0.367	1.006
D. pavlovskiana	0.344	1.066
D. insularis	0.319	1.144
Average	0.352 ± 0.023	1.056 ± 0.068

[a] When subspecies or semispecies occur, the average value between each one of them and *D. nebulosa* is given.

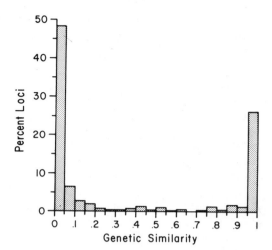

FIG. 7. Histogram of the distribution of genetic identity among gene loci for comparisons between morphologically distinguishable species of the *Drosophila willistoni* group.

productive isolation between the subspecies. The subspecies show a substantial degree of genetic divergence. On the average, 23 electrophoretically detectable allelic substitutions for every 100 loci have occurred in a pair of subspecies since their divergence from a common ancestral population. These results indicate that much genetic differentiation takes place during the first stage of geographic speciation.

Little additional genetic differentiation, however, seems to take place during the second stage of geographic speciation, when reproductive isolation is being completed. The average genetic distance between the semispecies of *D. paulistorum* is not significantly different from the average

TABLE VI. Average Genetic Identity, $\bar{I}$, and Genetic Distance, $\bar{D}$, between Taxa of Various Levels of Evolutionary Divergence in the *Drosophila willistoni* Group

Taxonomic level	$\bar{I}$	$\bar{D}$
Local populations	0.970 ± 0.006	0.031 ± 0.007
Subspecies	0.795 ± 0.013	0.230 ± 0.016
Semispecies[a]	0.798 ± 0.026	0.226 ± 0.033
Sibling species[a]	0.563 ± 0.023	0.581 ± 0.039
Nonsibling species	0.352 ± 0.023	1.056 ± 0.068

[a] These I and D values are based only on well-sampled semispecies or sibling species.

genetic distance between subspecies. The semispecies of *D. paulistorum* are populations in the second stage of speciation; reproductive isolation has in fact been completed in the places where some semispecies are sympatric. Sexual (ethological) isolation is the main mechanism maintaining reproductive isolation between species of *Drosophila*. Perhaps sexual isolation may come about by changing only a few genes that affect courtship and mating behavior.

The observation that the development of reproductive isolation *per se* may require changes in only a small fraction of all genes has interesting implications for saltational modes of speciation. Assume that incipient reproductive isolation has arisen between two populations by chromosomal rearrangements, or (in plants) by changes of the mating system, with little or no allelic change in individual genes. Reproductive isolation, and therefore speciation, could be completed without much additional genic change. Speciation could then be achieved with little differentiation at the single gene level. As we shall see below, this suggestion is corroborated by studies of species emerged by saltational speciation.

An alternative explanation may, however, be advanced for the observation of little additional genetic differentiation during the second stage of speciation, namely, that a substantial fraction of genes may evolve at this stage, but they are not the kinds of genes that are studied by electrophoretic methods. The genes studied by electrophoresis code for enzymes involved, for the most part, in the basic cell metabolism. Conceivably, such genes might not be the genes that affect courtship and mating behavior.

After reproductive isolation is completed, species evolve independently. As expected, they continue to diverge genetically. The average genetic distance between sibling species is more than twice as large as that between subspecies and semispecies. Morphologically distinct species are genetically much more different than sibling species.

Matrices of genetic distances between taxa may be used to construct dendrograms of genetic relationships. If the rate of genetic change is approximately the same in all taxa of a group of closely related species, such dendrograms would also indicate phylogenetic relationships. Figure 8 depicts a dendrogram for the taxa of the *D. willistoni* group obtained by applying a modified Wagner method (Farris, 1972) to the matrix of genetic distances. The dendrogram has been drawn in such a way that the vertical distances between neighboring taxa is roughly proportional to their genetic distances. It is significant that the dendrogram in Fig. 8 indicates phylogenetic relationships which are virtually identical with those suggested by Spassky *et al.* (1971) on the basis of other kind of information, namely, studies of chromosomal polymorphisms, sexual behavior, morphological biometry, ecology, and geographic distribution.

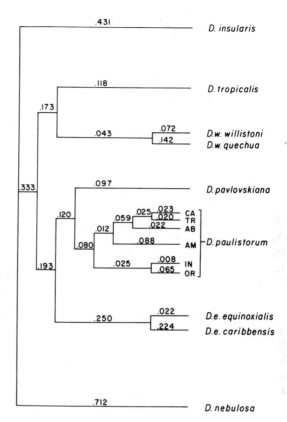

FIG. 8. Dendrogram of the probable phylogenetic relationships of the taxa of the *Drosophila willistoni* group. The vertical distances between neighboring taxa are roughly proportional to their genetic distance calculated from the allelic frequencies at 36 gene loci coding for enzymes. The numbers are the genetic distances between consecutive nodes, or between a node and a taxon.

GENETIC DIFFERENTIATION IN OTHER *DROSOPHILA* SPECIES

In recent years several studies have appeared of genetic differentiation between *Drosophila* populations at various levels of evolutionary divergence. I shall first briefly review this literature, and then present some previously unpublished results from a study of Hawaiian species carried out in my laboratory.

Zouros (1973) has studied genetic differentiation at 16 gene loci coding for enzymes in four species of the *mulleri* subgroup of the *repleta* group. The species *D. arizonensis* and *D. mojavensis* are siblings; the species *D. mulleri* and *D. aldrichi* are also siblings, and moreover have homosequential salivary gland chromosomes, i.e., the sequences of bands are identical in the polytene chromosome of these two species. *D. mojavensis* consists of two allopatric subspecies, *D. m. mojavensis* and *D. m. baja*. In *D. m. baja* it is possible to distinguish at least two allopatric "subraces," I and II, which differ in their inversion polymorphisms (Johnson, 1973) as well as in the allelic frequencies at some enzyme loci.

Zouros found very little genetic differentiation among local populations of the same taxon (average genetic distance, $\bar{D} = 0.001$), but somewhat greater differentiation among the two subraces of $D.\ m.\ baja$ ($\bar{D} = 0.025$). The average genetic distance, $\bar{D}$, is 0.130 ± 0.011 between the two $D.\ mojavensis$ subspecies, 0.203 ± 0.045 between sibling species, and 0.292 ± 0.022 between nonsibling species (see Table VII).

It is instructive to compare these results with those obtained in the $D.\ willistoni$ group. The average genetic distance between the two subspecies of $D.\ mojavensis$ is 0.130, while between subspecies of the $D.\ willistoni$ group is 0.230 (Table III). However, in terms of reproductive isolation, subspecies in the two groups represent different degrees of evolutionary divergence. The two $mojavensis$ subspecies are distinguishable by their chromosomal inversion polymorphisms and by some morphological differences, but they produce fertile hybrids of both sexes when crossed in the laboratory. On the other hand, hybrid males between the two subspecies of $D.\ willistoni$ are sterile when the mothers of $D.\ w.\ quechua$ ($D = 0.214$ between these subspecies); and all hybrid males are sterile in crosses between $D.\ e.\ equinoxialis$ and $D.\ e.\ caribbensis$ ($D = 0.246$).

The degree of reproductive isolation between $D.\ mojavensis$ and $D.\ arizonensis$ is comparable to that found between semispecies of $D.\ paulistorum$. Laboratory crosses between $D.\ mojavensis$ females and $D.\ arizonensis$ males yield fertile hybrids of both sexes; the reciprocal crosses produce fertile females but sterile males. $D.\ arizonensis$ is largely sympatric with $D.\ m.\ baja$, but completely allopatric with $D.\ m.\ mojavensis$. No hybrids between $D.\ arizonensis$ and its sympatric $D.\ m.\ baja$ are found in nature; few or none are produced in the laboratory when males and females of both species are present (Mettler and Nagle, 1966). On the contrary, fertile and vigorous hybrids are yielded when males and females of $D.\ arizonensis$ and its allopatric $D.\ m.\ mojavensis$ are placed together in laboratory populations. The genetic distance between $D.\ arizonensis$ and $D.\ m.\ baja$ is 0.279, between $D.\ arizonensis$ and $D.\ m.\ mojavensis$ is 0.207, and the average of both comparisons is 0.243. This value is fairly similar to 0.226, the average genetic distance between semispecies of $D.\ paulistorum$. On the other hand, $D.\ mulleri$ and $D.\ aldrichi$ are fully reproductively isolated; they maintain their identities over their largely overlapping geographic distributions, yet their genetic distance is only 0.124. (In the laboratory, $D.\ mulleri$ females crossed to $D.\ aldrichi$ males produce sterile hybrids of both sexes; the reciprocal crosses do not yield any progenies.) As pointed out above, $D.$

TABLE VII. Summary of Genetic Identities among Populations of *Drosophila* at Different Stages of Evolutionary Divergence (Calculated According to Nei's, 1972, Method)[a]

Species group or cluster	Local populations	Subspecies	Semispecies	Siblings and very closely related species	Species in different subgroups or groups	Source
willistoni	0.970 ± 0.006	0.795 ± 0.013	0.798 ± 0.026	0.563 ± 0.023	0.352 ± 0.023	Ayala *et al.*, 1974a
mulleri	0.999 ± 0.001	0.878 ± 0.009	—	0.777 ± 0.022	—	Zouros, 1973
bipectinata	—	0.907 ± 0.10	—	0.788	—	Calculated from Yang *et al.*, 1972
pseudoobscura	0.993	0.824	—	—	—	Ayala and Dobzhansky, 1974
obscura	—	—	—	0.466 ± 0.031	0.204 ± 0.004	Lakovaara *et al.*, 1972 and personal communication
Hawaiian	—	—	—	0.539 ± 0.054	0.164 ± 0.019	This paper

[a] The studies of Hubby and Throckmorton (1968) and Nair *et al.* (1971) are not included since Nei's statistic of genetic identity cannot be calculated from the published data.

mulleri and *D. aldrichi* have homosequential banding patterns in their salivary gland chromosomes.

Yang *et al.* (1972) have studied genetic differentiation at 23 enzyme loci in four species of the *D. bipectinata* complex (*melanogaster* group) endemic to Southeast Asia, northern Australia and intervening islands. The four species, morphologically and chromosomally very similar, are *D. bipectinata, D. parabipectinata, D. malerkotliana,* and *D. pseudoananassae.* The two latter species each consist of two allopatric subspecies: *D. m. malerkotliana* and *D. m. pallens,* and *D. p. pseudoananassae* and *D. p. nigrens.* The statistics, *I* and *D,* can be calculated according to Nei's method, using the allelic frequencies given by Yang *et al.* (1972). The average genetic distance between subspecies is $\bar{D}$ = 0.094, and between species is $\bar{D}$ = 0.236. There is, however, no evidence of reproductive isolation between the subspecies; laboratory crosses between individuals of different subspecies produce as many fertile progeny as crosses between strains of the same subspecies. On the other hand, the degree of reproductive isolation between these species is similar to that found between semispecies of *D. paulistorum.* Crosses between species of the *bipectinata* complex yield sterile males but fertile female hybrids, although only between 1% and 20% interspecific single-pair crosses involving *D. pseudoananassae* produce progeny. The average genetic distance between the *bipectinata* species (0.236) is very similar to that between *D. paulistorum* semispecies (0.226).

Two subspecies with a degree of reproductive isolation comparable to that between subspecies in the *D. willistoni* group are *D. pseudoobscura pseudoobscura* and *D. p. bogotana. D. p. pseudoobscura* is a common and ecologically versatile species living in western United States, western Canada, Mexico, and Guatemala. *D. p. bogotana* inhabits a small territory in the highlands above Bogotá, Colombia. The genetic distance between *D. p. pseudoobscura* and *D. p. bogotana,* estimated on the basis of allelic frequencies at 44 enzyme loci, is *D* = 0.194 (Ayala and Dobzhansky, 1974). This value is fairly similar to the genetic distance estimated between *D. w. willistoni* and *D. w. quechua,* *D* = 0.214 (Table III).

The analogies between *D. w. willistoni* and its subspecies *quechua,* and between *D. p. pseudoobscura* and its subspecies *bogotana* are striking. The major nominate subspecies are in both cases very successful and widely distributed populations; the minor subspecies, *quechua* and *bogotana,* are confined to relatively small areas, and are isolated from the main bodies of their species by geographic barriers. In the case of *quechua* the barriers are the high Andes and the extremely arid coastal zone of Peru; *D. willistoni* does not live at high elevations or in parched deserts. The isolation of *bogotana* is due to the intervening lowlands; while *D. p. pseudoobscura* is abundant on the Pacific Coast of the United States, it withdraws to higher elevations in Mexico and Guatemala, where it reaches the southernmost point of its distribution. In both cases intersubspecific matings in one direction yield fertile females and males, but sterile males are produced in the other direction, namely, when the female parent is the minor, narrowly distributed and geographically isolated subspecies. In neither case is any trace of ethological isolation detectable between the subspecies.

Nair *et al.* (1971) have studied allelic variation at 24 loci in six species of the *D. mesophragmatica* group, namely, *D. pavani, D. gaucha, D. viracochi, D. mesophragmatica, D. gasici,* and *D. brncici.* These species are morphologically very similar, and some of them are siblings; they exhibit varying degrees of reproductive isolation, with some interspecific crosses yielding abundant although sterile hybrids. The species live along the Andes, some of them at altitudes over 3000 m. Nair *et al.* have estimated genetic distances between the six species using Rogers' (1972) index of genetic similarity. The average genetic similarity for all pairwise comparisons between species is 0.501 $\pm$ 0.035. Nair *et al.* have not published the allelic frequencies in the *mesophragmatica* species, and thus it is not possible to calculate Nei's I and D for their data; Rogers' index, however, generally gives somewhat lower similarity values than Nei's index. The average genetic similarity between the species of the *mesophragmatica* group appears to be of the same magnitude as that between the sibling species of the *D. willistoni* group ($\bar{I} = 0.563$, $\bar{D} = 0.581$; see Table VI).

Hubby and Throckmorton (1968) have studied about 18 loci coding for proteins and enzymes in 27 species of *Drosophila*. The species included nine groups of three species, two of them siblings and the third a closely related but morphologically distinguishable species. On the average, sibling species shared proteins with identical mobility 50% of the time, and morphologically distinct species of the same triad shared proteins with identical mobility 18% of the time. Although Nei's indices cannot be calculated, and thus precise comparison is not possible, the results of Hubby and Throckmorton indicate genetic divergences of the same magnitude as those observed in the *D. willistoni* group.

Lakovaaara, Saura, and their colleagues (1972 and personal communication) have studied electrophoretic variation at 21 loci in 20 species of the *D. obscura* group. Some species are endemic in Europe, others in America, and two in Japan. The group can be divided into five separate clusters, each with three or four closely related species, some of them siblings, plus three species not belonging to any cluster. Between species of the same cluster, the average genetic identity is $\bar{I} = 0.466 \pm 0.031$, and the average genetic distance is $\bar{D} = 0.764 \pm 0.042$; between species of different clusters $\bar{I} = 0.204 \pm 0.004$, and $\bar{D} = 1.589 \pm 0.025$.

The endemic drosophilid fauna of the Hawaiian archipelago is unique in many respects. Some 500 species, mostly of the genus *Drosophila* but some belonging to other closely related genera, have been described—about one-third of the total number of species known for the entire world. The total number of drosophilid species endemic to Hawaii may actually exceed 700. There is far greater morphological and ecological diversity among *Drosophila* species in Hawaii than anywhere else in the world. As is the case elsewhere, Hawaiian *Drosophila* can be clustered into species groups, each of which has resulted from a major burst of speciation. Virtually every group has one or more species on each of the islands that support an endemic *Drosophila* fauna, although the number of species varies greatly from group to group. The picture-winged species group consists of more than 100 species (Carson, 1971*b*); at the other extreme, the *crassifemur* species group consists of six species (Kaneshiro, 1969); while other groups such as the "modified tarsi," "modified mouthparts," and "white-tip scutellum" are intermediate in numbers of species.

Two issues arise with respect to Hawaiian *Drosophila*. The first one is connected with the enormous proliferation of speciation events that have taken place in Hawaii during a geologically short time span. As pointed out above, Carson (1970, 1971*a*) has suggested that in Hawaii speciation may have often occurred rapidly as a consequence of the reorganization of the gene pool of an inbred population descended from a single gravid female. If Carson is correct, Hawaiian species of the same group should have very similar configurations of allelic frequencies. The second issue arises owing to the different rates of speciation in different groups. The question is whether or not more species emerge in some groups than in others precisely because the former are evolving genetically at a faster rate. There are about 20 times as many picture-winged species as *crassifemur* species, and twice as many picture-winged species as white-tips—all living in essentially the same geographical areas. Has there been more genetic change per unit time in the picture-winged than in the white-tip species, and more in these two than in the *crassifemur* group? I shall discuss the second issue raised in a later section of this chapter. Here, I shall review some data relevant to the first question.

Ayala, Spieth, and Kaneshiro (unpublished data) have undertaken a study of genetically determined enzyme variation in several groups of Hawaiian *Drosophila*. The table in the Appendix gives the allelic frequencies observed at 31 loci in seven species belonging to three different groups—picture-winged, white-tipped, and *crassifemur*; a summary of the amount of genetic variation found in these species is also given. All individuals of the seven species were collected in a single locality, Waikamoi, Maui, during two consecutive days. Like most other species in the genus, Hawaiian *Drosophila* are genetically very polymorphic. The average proportion of heterozygous loci per individual for the seven Hawaiian species (0.179 ± 0.016; see Appendix) is very similar to that observed in species of the *D. willistoni* group (0.177 ± 0.008; see Table I). The average proportion of polymorphic loci per population (using the 95% criterion of polymorphism) is 0.594 ± 0.031 compared to 0.509 ± 0.012 in the *D. willistoni* group.

Table VIII gives the matrix of genetic identities and genetic distances for all pairwise comparisons between eight *Drosophila* species, including the seven species listed in the Appendix. The same 31 enzyme loci were studied in every species. Three species (*D. adiostola, D. truncipenna,* and *D. hamifera*) belong to the *adiostola* subgroup of picture-winged species; *D. planitibia* belongs to a different subgroup of picture-wings. The average genetic identity among the three *adiostola* species is $\bar{I}$ = 0.569 ± 0.051, virtually identical to that among the *D. willistoni* siblings ($\bar{I}$ = 0.563 ± 0.023). Between *D. planitibia* and the three *adiostola* species, $\bar{I}$ = 0.348 ± 0.041, very similar to the value observed between morphologically different species of the *D. willistoni* group, $\bar{I}$ = 0.352 ± 0.023.

The genetic identities among the picture-winged species conform fairly well with their chromosomal affinities (Clayton *et al.*, 1972). *D. hamifera* and *D. truncipenna* differ by a single inversion in the X chromosome; their genetic identity is I = 0.657. *D. adiostola* differs from *D. hamifera* by eight inversions, and from *D. truncipenna* by nine; the corresponding I's are 0.571 and 0.479. *D. planitibia* differs from the three species by about 25 inversions; between *D. planitibia* and the three *adiostola* species, $\bar{I}$ = 0.348.

Table VIII includes three white-tipped species (also called fungus-feeders): *D. nigra, D. nigella,* and *D. dolichotarsis*. Their average genetic identity is $\bar{I}$ = 0.509 ± 0.105, somewhat lower than that among the three adiostola species. *D. nigra* and *D. nigella* are more similar to each other ($\bar{I}$ = 0.718) than either one is to *D. dolichotarsis* (I = 0.427 and 0.382, respectively).

The mean genetic distances between species of different groups are as follows: pictured-wings *vs.* white-tips, $\bar{I}$ = 0.147 ± 0.011; white-tips *vs. crassifemur,* $\bar{I}$ = 0.137 ± 0.032; picture-wings *vs. crassifemur,* $\bar{I}$ = 0.100 ± 0.016. The white-tipped species are more closely related to either the picture-winged species or to *D. crassifemur* than the latter are to each other. In general, Hawaiian *Drosophila* belonging to different groups are genetically somewhat more diversified ($\bar{I}$ = 0.135 ± 0.010) than species belonging to different clusters of the *D. obscura* group ($\bar{I}$ = 0.204).

Although there is a positive correlation between genetic and chromosomal differentiation in Hawaiian *Drosophila*, it appears that lack of chromosomal differentiation does not indicate that none or very little genetic divergence has occurred. As pointed out above, *D. hamifera* and *D. truncipenna* differ by only one chromosomal inversion; yet, their genetic

TABLE VIII. Genetic Identity (above Diagonal) and Genetic Distance (below Diagonal) between Eight Species of Hawaiian Drosophila[a]

	D. planitibia	D. adiostola	D. truncipenna	D. hamifera	D. nigra	D. nigella	D. dolichotarsis	D. crassifemur
Picture-winged species:								
D. planitibia	—	0.426	0.329	0.289	0.066	0.106	0.169	0.135
D. adiostola	0.854	—	0.479	0.571	0.176	0.173	0.209	0.095
D. truncipenna	1.113	0.737	—	0.657	0.122	0.124	0.137	0.060
D. hamifera	1.241	0.560	0.421	—	0.145	0.176	0.163	0.108
White-tipped species:								
D. nigra	2.712	1.738	2.101	1.934	—	0.718	0.427	0.086
D. nigella	2.244	1.753	2.090	1.739	0.331	—	0.382	0.127
D. dolichotarsis	1.781	1.565	1.990	1.816	0.850	0.964	—	0.197
Crassifemur-group species:								
D. crassifemur	2.006	2.358	2.816	2.230	2.450	2.068	1.623	—

[a] The same 31 loci coding for enzymes were studied in every species. Average values are as follows: between species of the *adiostola* subgroup $I = 0.569 \pm 0.051$, $D = 0.573 \pm 0.091$; between *D. planitibia* and the three *adiostola* species $I = 0.348 \pm 0.041$, $D = 1.069 \pm 0.114$; between white-tipped species $I = 0.509 \pm 0.105$, $D = 0.715 \pm 0.195$; between picture-wings and white-tips $I = 0.147 \pm 0.011$, $D = 1.955 \pm 0.089$; between white-tips and *D. crassifemur* $I = 0.137 \pm 0.032$, $D = 2.047 \pm 0.239$; between picture-wings and *D. crassifemur* $I = 0.100 \pm 0.016$, $D = 2.352 \pm 0.171$.

distance is $D = 0.421$, i.e., about 42 allelic substitutions for every 100 loci have occurred in the separate evolution of these two species. Rockwood *et al.* (1971) studied allelic variation at between four and eight enzyme loci in each of at least 44 Hawaiian drosophilids, including several groups of homosequential species. Although the number of loci surveyed by these authors is too small to allow reliable estimates of genetic similarity, they found substantial differentiation between homosequential species. They report that the mean "proportion of the gene pool which is unique between each homosequential species pair" is 54.7%, with a range between 26.3% and 80.0%.

GENETIC DIFFERENTIATION IN INVERTEBRATES OTHER THAN *DROSOPHILA*

Genetic variation using electrophoretic techniques has been measured in a variety of invertebrates belonging to several phyla. In invertebrates other than *Drosophila*, nearly 20 or more loci have been studied in a number of marine organisms, including a horseshoe crab, *Limulus polyphemus* (Selander *et al.*, 1970); a krill, *Euphausia superba* (Ayala, *et al.*, 1975c); a giant clam, *Tridacna maxima* (Ayala *et al.*, 1973); a phoronid, *Phoronopsis viridis* (Ayala *et al.*, 1974c); two brachiopods, *Liothyrella notorcadensis* (Ayala *et al.*, 1975a) and *Frieleia halli* (Valentine and Ayala, 1975); one ophiuroid, *Ophiomusium lymani* (Ayala and Valentine, 1974); and six species of asteroids: *Asterias forbesi* and *A. vulgaris* (Shopf and Murphy, 1973), *Nearchaster aciculosus, Pteraster jordani, Diplopteraster multipes,* and *Mixoderma sacculatum* (Ayala *et al.*, 1975b).

The average proportion of heterozygous loci per individual ranges from 1.1% in *Asterias vulgaris* to about 22% in *Tridacna maxima*. Considerable diversity in the degree of polymorphism exists among organisms of a given phylum. The two asteroids, *Asterias vulgaris* and *A. forbesi,* have, respectively, 1.1 and 2.1 heterozygous loci per individual (Shopf and Murphy, 1973), while the average in four other asteroids is 17.6% (Ayala *et al.,* 1975b). One brachiopod from subtidal waters in Antarctica, *Liothyrella notorcadensis*, has an average heterozygosity of 3.9% (Ayala *et al.,* 1975a) while a brachiopod from the deep sea at intermediate latitudes, *Frieleia halli,* has 16.9% heterozygosity (Valentine and Ayala, 1975).

Data about genetic differentiation during early stages of evolutionary divergence are scanty for invertebrates other than *Drosophila*. Local populations of a given species of the horseshoe crab, *Limulus polyphemus*, extending from Massachusetts to Florida, have an average genetic identity of

$\bar{I}$ = 0.99 (calculated from Selander *et al.*, 1970). Three populations of *Phoronopsis viridis*, geographically only a few hundred meters apart but morphologically and ecologically distinguishable, are also genetically very similar, $\bar{I}$ = 0.996 ± 0.0004 (Ayala *et al.*, 1974c). Two populations of *Tridacna maxima* separated by 2500 miles, one from Eniwetok Atoll and the other from the Great Barrier Reef, are nevertheless fairly similar genetically, $\bar{I}$ = 0.968 (Campbell, *et al.*, 1975).

Two closely related starfish species, *Asterias forbesi* and *A. vulgaris* have been compared genetically at 27 loci by Schopf and Murphy (1973). The two species are closely related phylogenetically, but can be distinguished morphologically by a set of seven traits no one of which is completely reliable by itself. The geographic distribution extends from the Gulf of Mexico to Maine for *A. forbesi*, and from Cape Hatteras to Labrador for *A. vulgaris*. In the region of overlap between Cape Hatteras and Maine, *A. forbesi* lives mostly in the warmer waters of inshore bays, while *A. vulgaris* generally occurs in colder offshore waters. Nevertheless, the two species coexist sympatrically in offshore areas south of the Cape Cod and in a few harbors north of the Cape. Hybrids are found where the two species are sympatric, but apparently there is no massive introgression since the hybrids coexist with individuals morphologically typical of one or the other species. According to Mayr (1963), *A. forbesi* and *A. vulgaris* are semispecies that are not yet completely reproductively isolated.

As pointed out above, *A. forbesi* and *A. vulgaris* have little genetic polymorphism (average heterozygosities 2.1% and 1.1%, respectively). The two species are fixed for the same alleles at 16 (59.3%) monomorphic loci, and have fairly similar frequencies at two polymorphic loci; the other nine (33.3%) loci are species diagnostic. The genetic identity between *A. forbesi* and *A. vulgaris* is I = 0.672; their genetic distance is D = 0.397. On the average, about 40 allelic substitutions per 100 loci are estimated to have occurred in the evolution of *A. forbesi* and *A. vulgaris* from a common ancestral population. The degree of genetic similarity between these two species is similar to that observed between siblings or closely related species of *Drosophila* (see Table VII).

Ayala *et al.* (1975b) have studied allelic variation at 24 loci coding for enzymes in four deep-sea asteroid species collected off-shore at southern California at 1244 m below sea level. *Pteraster jordani* and *Diplopteraster multipes* belong to two confamilial genera; their genetic identity is I = 0.257 (D = 1.358), similar to that observed between species in different groups of *Drosophila* (see Table VII). *P. jordani* and *D. multipes* belong to the order Spinulosida; *Nearchaster aciculosus* belongs to the order Paxillosidosa, and *Myxoderma sacculatum* to the order Forcipulatida. The average genetic identities between species in different orders are: Spinulosida-Paxillosidosa, 0.039; Spinulosida-Forcipulatida, 0.004; Paxillosidosa-Forcipulatida, 0.002. These observations agree with other information concerning the phylogeny of these organisms, since the orders Spinulosida and Paxillosidosa diverged from each other after their common ancestor diverged from the ancestor of the Forcipulatida. However, the techniques of gel electrophoresis have little reliability when the species compared are quite different genetically. With electrophoretic data, estimates of the number of allelic substitutions per locus are based on the proportion of loci at which *no* allelic substitutions have occurred (the zero class of a Poisson distribution). When the zero class is empty no estimates of allelic substitutions per locus can be made; when the zero class is very small, the estimates are unreliable. Moreover, when genetically very different species are compared, it is likely that allozymes with apparently identical electrophoretic mobilities may in fact differ in their amino acid composition.

Two species of brachiopods, *Liothyrella notorcadensis* from Antarctica, and *Frieleia halli* from the deep sea off southern California, have been studied at 34 and 18 electrophoretic loci, respectively (Ayala *et al.,* 1975a; Valentine and Ayala, 1975). Based on the electrophoretic patterns, 12 loci are thought to be homologous in both species, but only one allele (*Mdh-3*^{100}) has the same electrophoretic mobility in both species. The frequency of this allele is 1.00 in *Liothyrella,* and 0.04 in *Frieleia.* The genetic identity between these two species is estimated as $I = 0.004$ ($D = 5.62$); but as just pointed out this estimate has little reliability. A summary of genetic identities in invertebrates other than *Drosophila* is given in Table IX.

TABLE IX. Summary of Genetic Identities (Nei's *I*) between Populations at Different Stages of Evolutionary Divergence in Several Invertebrates

Organisms	Local populations	Closely related species	Genera	Source
Limulus polyphemus	0.99	—	—	Selander *et al.,* 1970
Phoronopsis viridis	0.996	—	—	Ayala *et al.,* 1974d
Tridacna maxima	0.968	—	—	Campbell *et al.,* 1975
Asterias	—	0.672	—	Schopf and Murphy, 1973
Pteraster–Diplopteraster	—	—	0.257	Ayala *et al.,* 1975b

GENETIC DIFFERENTIATION IN FISHES

The genus *Lepomis* of the sunfish family Centrarchidae consists of 11 species distributed throughout much of the eastern and central United States. The species status of all 11 species is well established—the species retain their identity throughout most of their distributions even where they are sympatric (adults of different species are easily distinguishable morphologically). Nevertheless F_1 hybrids between 21 different pairs of species (out of 55 theoretically possible hybrid combinations) have been found in nature, and introgression may occasionally occur (Hubbs, 1955); artificially raised hybrids are frequently fertile.

Avise and Smith (1974a,b) have studied allozyme variation at 14 loci in every one of 10 of the 11 species of *Lepomis*. Two morphologically distinct allopatric subspecies of *L. macrochirus* have been intensively studied, including 17 local populations of *L. m. macrochirus*, eight populations *L. m. purpurescens*, and 22 geographically intermediate populations where hybridization and introgression between the subspecies are taking place. Little genetic differentiation exists between populations of a given subspecies (average similarity, 0.97). "Pure" populations of different subspecies are virtually fixed for different alleles at each of two loci. Using Rogers' (1972) index, the average genetic similarity between the subspecies is estimated as 0.85. This indicates a substantial degree of genetic differentiation, comparable to that observed among subspecies of *Drosophila*. Nevertheless, the subspecies can produce fertile hybrids, as evidenced by the occurrence of extensive introgression in the intergrade zone, which apparently represents a secondary contact of allopatrically evolved subspecies.

The average genetic similarity between the *Lepomis* species, calculated using Rogers' index is 0.53. Avise and Ayala (1975) using the data of Avise and Smith (1974b) have calculated genetic similarities between species of *Lepomis* following Nei's method. The average genetic identity between the species is $\bar{I} = 0.544 \pm 0.016$, and the average genetic distance is $\bar{D} = 0.627 \pm 0.029$. The two most similar species are *L. marginatus* and *L. megalotis* ($I = 0.76$), which are morphologically most similar and were considered by Fowler (1945) to be subspecies of a single species, although their specific status now seems well established (Reeves and Moore, 1951).

In spite of their ability to hybridize both in nature and in the laboratory, species of the genus *Lepomis* are genetically quite different—about 63 allelic substitutions per 100 loci have occurred, on the average, between any pair of species. The degree of genetic differentiation between *Lepomis* species is similar to that observed between the siblings of *D. willistoni*, or between morphologically different but closely related species of other groups of *Drosophila*.

TABLE X. Summary of Genetic Similarities between Populations at Different Stages of Evolutionary Divergence in Several Groups of Fishes

Organisms	Local populations	Subspecies	Species and closely related genera	Genera	Source
Lepomis, sunfish	0.97[a]	0.85[a]	0.544 ± 0.016		Avise and Smith, 1974a,b
Salmonidae	—	—	0.456 ± 0.032[a]		Utter et al., 1973
Cyprinodon, pupfish	—	—	0.894 ± 0.016[a]		Turner, 1974
California minnows	0.99	—	0.592 ± 0.029		Avise and Ayala, 1975
Sciaenidae	—	—	—	0.170 ± 0.027[a]	Shaw, 1970

[a] Similarities calculated using methods other than Nei's (1972).

Utter *et al.* (1973) have studied allozyme differentiation at 19 to 23 loci in six species of the Pacific salmon (genus *Oncorhynchus*) and two species of the closely related trout (genus *Salmo*). Using a method that yields results numerically similar to the methods of Nei or Rogers, the average genetic similarity between all eight species is 0.456 ± 0.032. The two trout species, *Salmo gairdneri* and *S. clarkii* are the most similar pair, with a similarity of 0.90. The average similarity between the six salmon species is 0.453 ± 0.042, and between them and the two trout species is 0.422 ± 0.040. These two genera are genetically not very different, a conclusion also supported by other evidence.

Turner (1974) has studied allozyme variation at 31–38 loci in five closely related species of pupfish, genus *Cyprinodon*, endemic in the Death Valley (three species) or adjacent hydrologic basins of western North America (two species). The species are morphologically, behaviorally, and ecologically quite different. Genetically, however, the species are not very different. The average genetic similarity for all pairwise comparisons between the species is 0.894 ± 0.016, with a range from 0.808 to 0.968. The extensive morphological, ecological, and behavioral differentiation between these pupfish species contrasts with their limited genetic differentiation, although the species status of some forms has been brought into question.

Avise and Ayala (1975) have sampled 24 enzyme loci in nine species classified into nine different genera of California minnows (family Cyprinidae). About 16 species and 10 genera of California minnows are recognized, although at least four genera are probably of recent monophyletic origin. Two local populations were sampled in one species, *Ptychocheilus grandis* and several in two other species, *Hesperoleucus symmetricus* and *Lavinia exilicauda*. Little genetic differentiation exists between local populations, $I \simeq 0.99$. The average genetic identity between the nine species, calculated according to Nei's method, is $\bar{I} = 0.592 \pm 0.029$, and the average genetic distance is $\bar{D} = 0.568 \pm 0.052$. These values are similar to those observed among the siblings of *D. willistoni* ($\bar{I} = 0.563 \pm 0.023$), but the California minnows are far from a homogeneous group. Among the four most closely related genera (*Hesperoleucus, Lavinia, Mylopharodon,* and *Ptychocheilus*) $\bar{I} = 0.871 \pm 0.021$, comparable to that observed between some *Drosophila* subspecies; while between the most differentiated genus, *Notemigonus*, and the other taxa, $\bar{I} = 0.385 \pm 0.015$, similar to that observed between morphologically different species of the *willistoni* or other *Drosophila* groups.

The most genetically similar species are *Hesperoleucus symmetricus* and *Lavinia exilicauda*, with $I = 0.946$ and $D = 0.055$, i.e., only between five and six electrophoretically detectable allelic substitutions per 100 loci have occurred in the evolution of these two species. The two species have virtually identical genetic constitutions at 23 out of 24 loci, but are highly differentiated at a single locus, *Pt-2*, which is indeed species diagnostic for most populations. The overall genetic differentiation observed between *H. symmetricus* and *L. exilicauda* is no greater than what is often found between local populations of some species, but is extremely low for interspecific comparisons. Nevertheless, *H. symmetricus* and *L. exilicauda* are different ecologically and morphologically and retain their identity where they are sympatric, although some hybridization and perhaps introgression takes place; reproductive isolation between these two species is maintained by prezygotic mechanisms (Avise *et al.*, 1975). The close genetic similarity between *H. symmetricus* and *L. exilicauda* suggests that speciation and adaptive differentiation may sometimes occur with changes in only a relatively small fraction of the genome.

The distribution of genetic similarities among loci for pairwise comparisons between the nine species of California minnows is strongly bimodal. Pairs of species are, on the average, essentially identical ($I > 0.95$) at nearly 60% of the loci, and completely different ($I < 0.05$) at about 30% of the loci. Few loci have genetic identities in the broad range between $I = 0.05$ and 0.95.

Six sympatric species belonging to six different genera of the family Sciaenidae have been studied at 16 enzyme loci by Shaw (1970). The numbers of individuals per species were very small (between two and twelve); genetic similarities were calculated as the average numbers of enzyme bands with identical migration in any two species. The average similarity between all pairwise comparisons is 0.170 ± 0.027, with a range from zero (between *Cynoscion arenarius* and *Menticirrhus americanus*) to 0.333 (between *Leiostomus xanthurus* and *Stellifer lanceolatus*). The genetic similarity between the latter two species is similar to that observed in the California minnows between the most differentiated genus, *Notemigonous*, and the other species ($\bar{I} = 0.385 \pm 0.015$).

GENETIC DIFFERENTIATION IN SALAMANDERS

Hedgecock (1974) and Hedgecock and Ayala (1974) have made an extensive study of genetically controlled enzyme variation in salamanders of the genus *Taricha*. This genus consists of three species, *T. granulosa, T. rivularis,* and *T. torosa.* The latter species consists of two described allopatric subspecies, *T. torosa torosa* and *T. t. sierrae.* Under the name *T. t. torosa* there are two undescribed allopatric subspecies which may be referred to as "northern" and "southern" *T. t. torosa.* The genus lives in the

Coast Ranges of western North America, from San Diego in southern California to southern Alaska. The subspecies *T. t. sierrae* lives in the foothills of Sierra Nevada in northern California; an isolated population of *T. granulosa* has been found in Idaho. Hedgecock (personal communication) has also studied a species, *Notophthalmus viridescens,* belonging to a genus closely related to *Taricha*. The genus *Notophthalmus* lives in the eastern United States; the population of *N. viridescens* surveyed by Hedgecock was collected in Scioto County, Ohio.

Four levels of increasing evolutionary divergence have been studied in these salamanders: (1) local populations of the same taxon; (2) subspecies of *T. torosa*; (3) species of the genus *Taricha*; (4) the genera *Taricha* and *Notophthalmus*.

A total of 42 gene loci coding for enzymes and other proteins has been studied in the genus *Taricha*; at least 34 loci in every taxon. More than 100 individuals have been assayed in each taxon with a total of 1111 individuals for the genus. Several local populations have been sampled in each species. The amount of genetic variation in *Taricha* populations is within the typical range of what is found in vertebrates, although somewhat greater than the average (see Selander and Kaufman, 1973). On the average, an individual is heterozygous at 0.096 ± 0.003 loci in *T. granulosa*, 0.064 ± 0.002 in *T. rivularis*, and 0.053 ± 0.019 in *T. torosa*. The proportion of polymorphic loci per population ranges from 0.195 in *T. rivularis*, through 0.250 in *T. torosa*, to 0.441 in *T. granulosa*. (A locus is considered polymorphic when the frequency of the most common allele is no greater than 0.99.)

Table XI gives the average genetic identities and distances (Nei's *I* and *D*) between taxa at increasing stages of evolutionary divergence. Geographic populations of the same taxon are, on the average, quite similar ($\bar{I} = 0.984 \pm 0.004$), but the 12 populations studied of *T. rivularis* are genetically more

TABLE XI. Average Genetic Identity, $\bar{I}$, and Genetic Distance, $\bar{D}$, between Taxa at Various Levels of Evolutionary Divergence in the Genus *Taricha*, and between *Taricha* and a Species of a Closely Related Genus, *Notophthalmus viridescens*

Taxonomic level	$\bar{I}$	$\bar{D}$
Local populations	0.984 ± 0.004	0.017 ± 0.004
Subspecies	0.836 ± 0.020	0.181 ± 0.024
Species	0.745 ± 0.004	0.296 ± 0.006
Genera	0.306 ± 0.013	1.187 ± 0.044

similar to each other ($\bar{I} = 0.993 \pm 0.001$) than the local populations of a given subspecies of *T. torosa* ($\bar{I} = 0.957 \pm 0.026$). A greater degree of genetic divergence has occurred among the four local populations of *T. granulosa* surveyed ($\bar{I} = 0.896 \pm 0.024$). Some authors (Bishop, 1941; Myers, 1942) have suggested on the basis of other evidence that two or three subspecies of *T. granulosa* should be distinguished. Hedgecock's results do not support these claims since the variation follows a north–south cline without sharp discontinuities.

The degree of genetic differentiation between the subspecies of *T. torosa* ($\bar{I} = 0.836 + 0.020$) is of the same magnitude as that observed in *Drosophila* (Table VII). Considerable variation, however, exists among the possible comparisons. Northern *T. t. torosa* is genetically approximately equidistant from *T. t. sierrae* ($\bar{I} = 0.853 \pm 0.018; \bar{D} = 0.160 \pm 0.022$) and southern *T. t. torosa* ($\bar{I} = 0.878 \pm 0.088; \bar{D} = 0.130 \pm 0.009$), but the latter two are nearly twice as different ($\bar{I} = 0.760 \pm 0.026; \bar{D} = 0.274 \pm 0.035$).

The three species of *Taricha* are genetically more similar ($\bar{I} = 0.745 \pm 0.004$) than the siblings of the *D. willistoni* group ($\bar{I} = 0.563$), but not quite as similar as the sibling species of the *mulleri* or the *bipectinata* groups. On the average, 29.6 allelic substitutions for every 100 loci have occurred in the separate evolution of a pair of *Taricha* species. Moreover, about 21% of all loci are diagnostic between the species, i.e., they allow correct identification of the species of a single individual with a probability no smaller than 0.99. Yet, experimentally obtained interspecific hybrids are fertile and produce fully viable F_2 progenies (Twitty 1961, 1964).

Apparently, evolution in the genus *Taricha* has proceeded without development of any detectable postmating isolating mechanisms. Reproductive isolation is nevertheless effectively complete, even in streams where two species breed simultaneously (Twitty, 1961). This is confirmed by the electrophoretic studies. The high proportion of diagnostic loci allows detection of hybrid individuals. Only two hybrids (both between *T. granulosa* and *T. t. torosa*) have been found among more than one thousand individuals studied, and both in streams severely disturbed by human activity. Hybridization apparently occurs rarely in nature and there is no evidence of introgression. Reproductive isolation between species of *Taricha* is achieved ethologically (Davis and Twitty, 1964).

The distribution of loci with respect to genetic similarity is typically bimodal, i.e., populations of different taxa are either very similar or totally different at any one locus. Whether different subspecies or different species are compared, genetic similarity is either smaller than 0.05 or greater than 0.95 in about 80% of the loci; although the proportion of loci with $I > 0.95$ is of course greater for intersubspecific than for interspecific comparisons.

The average genetic identity between populations of *Taricha* and

TABLE XII. Summary of Genetic Similarities between Populations at Different Stages of Evolutionary Divergence in Three Groups of Lizards

Organisms	Local populations	Subspecies	Species	Source
Anolis (Bimini species)[a]	0.965 ± 0.011	0.689 ± 0.001	0.208 ± 0.012	Webster *et al.*, 1973
Anolis (*roquet* group)[b]	0.996	—	0.667 ± 0.048	Yang *et al.*, 1974
Sceloporus grammicus[a]	0.887 ± 0.024	0.787 ± 0.032	—	Hall and Selander, 1973

[a] Estimates obtained using Rogers' index of genetic similarity.
[b] Estimates obtained using Nei's index of genetic identity.

Notophthalmus viridescens is $\bar{I} = 0.306 \pm 0.013$. *Taricha* and *No-tophthalmus* are closely related genera, but more than one allelic substitution per locus ($\bar{D} = 1.187 \pm 0.044$) is estimated to have occurred between populations of the two genera since they had a common ancestor. The degree of genetic divergence between *Taricha* and *Notophthalmus* is of comparable magnitude to that observed among species belonging to different subgroups of *Drosophila*.

GENETIC DIFFERENTIATION IN LIZARDS

Webster *et al.* (1973) and Yang *et al.* (1974) have measured genetic variation in various species of the iguanid lizard *Anolis*, a Neotropical genus with more than 200 species. Three levels of evolutionary divergence are involved in these studies: local populations, subspecies, and closely related species.

Webster *et al.* (1973) assayed 25 to 29 gene loci coding for soluble proteins in four species *A. angusticeps, A. carolinensis lerneri, A. sagrei ordinatus,* and *A. distichus* collected in South Bimini, a small island (five square miles) in the Bahamas. Three samples of *A. carolinensis carolinensis* from Texas, Louisiana, and Florida, and one sample of *A. sagrei sagrei* from Jamaica were also studied. Genetic similarity was estimated using Rogers' (1972) measure on the basis of 23 loci assayed in every species.

The average genetic similarity among the three local populations of *A. c. carolinensis* from Texas, Louisiana, and Florida, is 0.965 ± 0.011. The average similarity between the subspecies *A. c. carolinensis* and *A. c. lerneri,* based on three comparisons is 0.689 ± 0.001; the similarity between the subspecies *A. s. sagrei* and *A. s. ordinatus,* based on one comparison, is greater, 0.816. If all intersubspecific comparisons are averaged, the similarity between subspecies becomes 0.720 ± 0.032. The average similarity for all 21 possible comparisons between populations of different species is 0.208 ± 0.012 (range from 0.136 to 0.288). The distribution of loci with respect to genetic similarity is typically bimodal. At six loci three or four species have very similar genetic constitutions, while at 13 loci no allele occurs with high frequency in more than one species. At the other four loci any two species are also either quite similar or very different.

Yang *et al.* studied nine species of the *Anolis roquet* group from the Windward group of the Lesser Antilles, including two islands, Bonaire and Blanquilla, far to the west of the main chain. Two populations of *A. aeneus* from different islands, Grenada and Bequia, were sampled; only one population was studied for each of the other eight species. Up to 26 loci coding for enzymes and other proteins were assayed in some populations, but Nei's statistics of genetic similarity and distance were estimated for 22 loci studied in every population.

The two populations of *A. aeneus* from Bequia and Grenada are very similar ($I = 0.996$); these two islands were connected about 15,000 years ago when the sea level was some 100 meters lower than today. The average genetic identity for all interspecific comparisons is $\bar{I} = 0.667 \pm 0.048$ ($\bar{D} = 0.405 \pm 0.030$), but there is a broad range. *A. roquet* from Martinique and *A. extremus* from Barbados are the most similar species ($I = 0.987$), and these two are quite similar to *A. aeneus* ($\bar{I} = 0.885$). This finding agrees with results from chromosomal and other studies that indicate that these three species, and particularly the first two, are phylogenetically closely related. The two most different species are *A. richardi* from Grenada and *A. blanquillanus* from Blanquilla ($I = 0.378$).

The degree of genetic differentiation between local populations, subspecies, and closely related species of *Anolis* lizards is, overall, similar to that found in other groups of organisms, although species of the *roquet* group are more closely interrelated than the species from South Bimini. (Note, however, that the numerical values given for the two groups are not directly comparable, since they are based on different indices, that of Rogers resulting in lower similarities). Nevertheless, two allopatric species exist (*A. roquet* and *A. extremus*) which are genetically very similar to each other, even more similar than the minnow fishes *Hesperoleucus* and *Lavinia* ($D = 0.013$ and 0.055 for the *Anolis* and the minnow pair, respectively).

Hall and Selander (1973) have studied allozyme variation at 19 loci in two chromosomally differentiated populations of another iguanid lizard, *Sceloporus grammicus* from Mexico. One chromosomal type (P1) is polymorphic for a centric fission of chromosome 1, but otherwise has the standard karyotype configuration. The other chromosomal type (F6) has fixed a centric fission of chromosome 6, but has the standard karyotype configuration in chromosome 1 and elsewhere. Populations of the two chromosomal types are largely allopatric but contiguous, with a zone of contact and hybridization in the Valley of Mexico. Of 153 individuals sampled in the hybrid zone, 13 were presumed to be F_1 hybrids, and 46 probable backcrosses to one or the other parental type. Using Rogers' coefficient, the average genetic similarity between the two chromosomal types is 0.787 ± 0.032, while between populations of the F6 type it is 0.887 ± 0.024. The degree of genetic differentiation between the P1 and F6 chromosomal types is similar to that observed between subspecies or semispecies of other organisms, including lizards. Indeed, P1 and F6 populations should be considered at least subspecies and perhaps incipient species. Postmating isolating mechanisms between the P1 and F6 populations exist in the form of fitness reduction in the progenies of hybrid crosses, although apparently no premating isolating mechanisms have developed in spite of an estimated 7000 years of geographic contact between the two chromosomal types.

The level of genetic differentiation between *Anolis* subspecies or between the P1 and F6 chromosomal types of *S. grammicus* is comparable to that observed between subspecies of the side-blotched lizard, *Uta stansburiana*, an iguanid species of the western United States and northern Mexico (McKinney *et al.*, 1972). A sample of 18 loci have been studied in 17 local populations of *U. stansburiana*, but disagreement among authors about the boundaries between the subspecies does not warrant numerical calculations of genetic similarities between subspecies.

GENETIC DIFFERENTIATION IN MAMMALS

Selander *et al.* (1969) (see also Hunt and Selander, 1973) have studied genetic differentiation between two subspecies of the house mouse, *Mus musculus musculus* and *M. m. domesticus*. A sample of 41 protein loci were studied in 99 mice from six different localities in Denmark. The average genetic similarity between populations of different subspecies is 0.769 ± 0.001, using Rogers' coefficient (0.84 using Nei's coefficient). The average similarity between local populations of the same subspecies is 0.90 ± 0.01.

M. m. musculus and *M. m. domesticus* are morphologically distinguishable allopatric subspecies with geographic distributions meeting along a boundary running north to south through central Germany and Austria, and east to west through central Jutland in Denmark. The two subspecies hybridize at present, and may have hybridized continuously for about 5000 years, along the narrow zone where they meet. Introgression along the hybrid zone in the Jutland peninsula is mostly unidirectional; *domesticus* alleles are incorporated in *musculus*, but not vice versa. There is no evidence of assortative mating in the zone of hybridization, supporting laboratory evidence of free interbreeding among the subspecies. Hunt and Selander (1973) suggest that the separation of the subspecies may be maintained by a combination of environmental factors and postmating isolating mechanisms. Climatic and other environmental differences may favor the alleles of one subspecies on one side of the hybrid zone, and those of other subspecies on the other side; hybrids and backcross individuals may have low fitness owing to disruption of the coadapted gene complexes of the parental forms. Be that as it may, the degree of evolutionary divergence

TABLE XIII.   Summary of Genetic Similarities between Populations at Various Stages of Evolutionary Divergence in Several Groups of Mammals[a]

Organisms	Local populations	Subspecies	Species	Genera	Source
Mus musculus	0.90	0.769	—	—	Selander *et al.*, 1969
Thomomys	0.933	—	0.845	—	Patton *et al.*, 1972
Geomys	—	—	0.757	—	Selander *et al.*, 1974
Sigmodon	0.983	—	0.763	—	Johnson *et al.*, 1972
Dipodomys	0.96	—	0.61	0.16[b]	Johnson and Selander, 1971
Peromyscus boylii	—	0.793	0.693	—	Avise *et al.*, 1974*b*
Peromyscus	—	—	0.648	—	Avise *et al.*, 1974*a*
Lasiurus	—	—	0.569	—	Shaw, 1970
Homo sapiens- Pan troglodytes	—	—	—	0.538	King and Wilson, 1975

[a] The estimates are obtained using Rogers' index, S, except in the cases of *Geomys* and *Lasiurus* where similarities are calculated as the proportion of predominant alleles shared by pairs of species.
[b] Camparison between *Dipodomys* and *Perognathus*.

between *musculus* and *domesticus* is intermediate between subspecies and semispecies; along the hybrid zone, *M. m. musculus* and *M. m. domesticus* have entered the second stage of geographic speciation.

Pocket gophers are subterranean rodents often geographically differentiated presumably due to their low vagility and consequent lack of gene flow. Patton *et al.* (1972) have studied 27 protein loci in *Thomomys bottae* and *T. umbrinus*, two largely allopatric populations which hybridize at a single point of geographic contact in the Patagonia Mountains of Southern Arizona. Using Rogers' coefficient, the average genetic similarity is 0.933 among five local populations of *T. bottae*; between these and a population of *T. umbrinus* the average similarity is 0.845. Hybrids are often formed in the zone of contact between the two species, but there is no introgression. Hybrid males are at least partially sterile and backcross individuals have very low fitness.

Selander *et al.* (1974) have studied enzyme variation at 23 loci in four species of cotton rats of the genus *Geomys*. The species are morphologically distinguishable, have largely allopatric geographic distributions, and differ in their karyotypes. *G. tropicalis,* a species confined to a small area of some 300 square miles, and separated by some 250 km from its nearest neighbor of this genus, is the most differentiated karyotypically ($2n$ = 38 biarmed chromosomes *versus* 68 to 74 mostly acrocentric chromosomes in the other three species). The average genetic similarity among the four species (calculated as the proportion of predominant alleles shared by pairs of species) is 0.757 ± 0.015. *G. tropicalis* is apparently somewhat less similar (0.746 ± 0.007) to the other three species than these are to each other (0.768 ± 0.031).

Sigmodon hispidus and *S. arizonae* are two allopatric species of cotton rats whose geographic ranges are 17 miles apart at their nearest point. Hybrids have never been found in nature and cannot readily be obtained in the laboratory owing to behavioral incompatibilities between the two forms. The two species are morphologically similar but have very different karyotypes—the diploid number of chromosomes is 52 for *S. hispidus,* and 22 for *S. arizonae.* Based on a sample of 22 protein loci, Rogers' mean coefficient of genetic similarity between the species is 0.763; between local populations the mean coefficient of similarity 0.983 (Johnson *et al.,* 1972).

Johnson and Selander (1971) have studied protein variation at 18 gene loci coding for proteins in 11 species of kangaroo rats, genus *Dipodomys,* and in one species of a closely related genus, the pocket mouse *Perognathus hispidus.* Using Rogers' coefficient the mean genetic similarities at various stages of evolutionary divergence are as follows: between local populations of the same species 0.96; between species of *Dipodomys,* 0.61 (0.89 between *D. panamintinus* and *D. heermani,* the two most similar species); between *Dipodomys* and *Perognathus,* 0.16.

Avise *et al.* (1974a,b) have made an extensive study of genetically controlled protein variation in several species groups of the genus *Peromyscus.* A sample of 21 loci were sampled in four species of the *P. boylii* group, including three subspecies of *P. pectoralis.* Rogers' mean coefficient of similarity is 0.793 ± 0.026 between pairs of subspecies, and 0.691 ± 0.015 between pairs of species of this group. Some 24 protein loci were studied in an additional set of 13 species belonging to three different subgenera of *Peromyscus;* the mean genetic similarity for all interspecific pairwise comparisons is 0.648 ± 0.017.

Shaw (1970) has studied 14 enzyme loci in small samples (one to six individuals per species) in three sympatric species of bats, genus *Lasiurus.* Estimates of genetic similarity were calculated as the average proportion of electrophoretic bands in common between pairs of species. The average similarity among the three species is 0.569 ± 0.019. Two species, *L. seminolus* and *L. borealis,* have sometimes been thought to be color variants of the same species, yet their genetic similarity (0.572) is about equal to that observed between those two species and *L. cinerus* (0.567), a species well differentiated from the other two.

King and Wilson (1975) have studied genetically determined variation at 44 loci in man, *Homo sapiens,* and chimpanzee, *Pan troglodytes.* Chimpanzee and man are conspicuously different in morphology, ecology, and behavior, and are classified in two different families, the Pongidae and the Hominidae, respectively. Yet their similarities and evolutionary relatedness are also notable: chimpanzees are the closest living relatives of man. This close relationship is confirmed by genetic studies. The genetic identity between man and chimpanzee is $I = 0.54$, their genetic distance is $D = 0.62$; on the average, about 62 electrophoretically detectable allelic substitutions per 100 loci have occurred in the evolution of man and chimpanzee from a common ancestor. This degree of genetic differentiation is similar to that observed between the sibling species of *Drosophila willistoni* ($\bar{D} = 0.563$) or between other closely related species. Comparable estimates of genetic differentiation between man and chimpanzee are obtained when methods other than electrophoresis are used, such as immunological assay of proteins, amino acid sequencing of proteins, and reannealing of nucleic acids.

A summary of the data discussed in this section is given in Table XIII. The average similarities between pairs of subspecies (0.769 in *Mus musculus* and 0.793 in *Peromyscus pectoralis* of the *boylii* group) are comparable but lower than those found in other groups of organisms. As pointed out above, *M. m. musculus* and *M. m. domesticus* are in the second stage of the speciation process, at least in the zone of hybridization from which the samples were taken.

The average similarities between pairs of species vary greatly from group to group. In *Thomomys, Geomys,* and *Sigmodon,* the similarity coefficients are like those often found among subspecies. The species studied in each of these three genera are mostly (*Thomomys* and *Geomys*) or completely (*Sigmodon*) allopatric; hybridization occurs where the two *Thomomys* species come into geographic contact. Indeed, Selander and Johnson (1973) consider the two species pairs of *Thomomys* and *Sigmodon* as two pairs of semispecies. The average interspecific similarities in *Dipodomys* and *Peromyscus* are of comparable magnitude to those found in other groups. It need be pointed out, however, that some of the species studied in these two genera are very closely related and may be considered siblings, while others are quite differentiated and belong to different groups of subgenera. The average similarity between the genera *Dipodomys* and *Perognathus* is about the same as that observed between Hawaiian species of *Drosophila* belonging to different groups.

"SALTATIONAL" SPECIATION IN RODENTS

The genus *Spalax* (mole rats) consists of three described species distributed in southern Europe, the Middle East, and North Africa. Chromosomal variation throughout the genus is extensive ($2n$ from 48 to 62). Four chromosomal types ($2n$ = 52, 54, 58, and 60) of *S. ehrenbergi* exist in Israel and vicinity, with largely allopatric distributions, although narrow zones of contact exist between contiguous forms. The distribution of the chromosomal forms follows an approximate north–south gradient, forms with lower chromosome numbers being in the north.

Reproductive isolation exists among the four chromosomal types of *S. ehrenbergi*. Indeed, speciation may be nearly complete. In some contact zones, hybridization between chromosomal types occurs when the types differ by only two chromosomes (52 with 54 and 58 with 60), but not when the contiguous forms differ by a larger number of chromosomes (such as 52 and 58). Furthermore, the hybrids are always rare and the hybrid zones are narrow. Chromosomal rearrangements including Robertsonian fusions and/or fissions, apparently provide effective postmating isolating mechanisms. There is evidence also of ethological isolation—in the laboratory there is a greater degree of aggression between individuals of different types than between individuals of the same type, and matings are assortative. The four chromosomal types of *S. ehrenbergi* may, then, be considered as sibling species emerged by rapid, or saltational, speciation. Presumably, chromosomal rearrangements provide the original isolating mechanism, which is later reinforced by ethological isolation and other differentiating traits.

Nevo and Shaw (1972) have studied allozyme variation at 17 loci in several populations of the four chromosomal types of *S. ehrenbergi*. Populations of these mole rats have low levels of genetic variation. The average frequency of heterozygous loci per individual is $\bar{H}$ = 0.037, with a range from 0.018 to 0.056. Ten loci are monomorphic and fixed for the same allele in all four chromosomal types. The other seven loci are polymorphic in some types, but no locus is polymorphic in all. In these polymorphic loci, the most common allele is always the same in the four chromosome types, with one single exception—at the 6PGD locus the allele most frequent in populations with 52, 54, and 58 chromosomes is the second most common in populations with 60 chromosomes.

The allelic frequencies given by Nevo and Shaw can be used to calculate Nei's index of genetic identity. The average value for all pairwise comparisons between different chromosomal types is $\bar{I}$ = 0.978 $\pm$ 0.007, with a range from 0.998 (between populations with 52 and 58 chromosomes) to 0.958 (between populations with 54 and 60 chromosomes). The average genetic distance between pairs of chromosomal types is $\bar{D} = 0.022$, i.e., about 2 electrophoretically detectable allelic substitutions per 100 loci are estimated to have occurred, on the average, in the separate evolutions of any two chromosomal types. This is about the same amount of genetic differentiation observed between local populations of the same species in organisms with fairly low geographic diversification. Speciation in *S. ehrenbergi* has been accompanied by little differentiation at the single locus level.

Nevo *et al.* (1974) have studied genetic differentiation in pocket gophers of the *Thomomys talpoides* complex, another group where speciation may have originated by a saltational process involving chromosomal rearrangements. The *T. talpoides* complex consists of more than eight distinct chromosomal types, with largely allopatric geographic distributions, which together extend over a large area in the north central and northwest United States and neighboring areas in southern Canada. Different chromosomal types often occur contiguously without interbreeding. The chromosomal types of *T. talpoides* may have reached species status, the basic reproductive isolation being due to the chromosomal rearrangements. Little or no morphological differentiation exists among most of the chromosomal types.

Allelic variation has been studied at 31 protein loci in ten populations of *T. talpoides,* belonging to six chromosome types ($2n = 40, 44, 46, 60$, and two morphologically distinguishable forms, one from Wisconsin and the other from New Mexico, with 48 chromosomes). Populations of *T. talpoides* have moderately low levels of genetic variation ($\bar{H} = 0.047$, with a range from 0.008 to 0.085). Populations of different chromosomal types are genetically very similar to each other; the average value of $\bar{I}$ is 0.925, with a range from 0.858 (between the form with 60 chromosomes and the form with 48 from New Mexico) to 0.996 (between the forms with 44 and 46 chromosomes). The average genetic distance between the chromosomal forms is $\bar{D} = 0.078$, larger than it is between the *S. ehrenbergi* chromosomal types, but very low for interspecific comparisons.

A summary of genetic differentiation in *S. ehrenbergi* and *T. talpoides* is given in Table XIV. If the chromosomal types of these two complexes are considered different species, speciation has been accompanied by little genetic differentiation.

GENETIC DIFFERENTIATION IN PLANTS

Fewer studies have been made to date in plants than in animals to measure genetically controlled protein variation by electrophoresis. Sufficiently large numbers of loci to establish average similarities between popu-

TABLE XIV. Genetic Identities and Genetic Distances (Calculated Using Nei's Method) between Different Chromosomal Forms of Two Rodent Groups

Organisms	Genetic identity		Genetic distance		Source
	Mean	Range	Mean	Range	
Spalax ehrenbergi	0.978	0.958–0.998	0.022	0.043–0.002	Calculated from Nevo and Shaw, 1972
Thomomys talpoides	0.925	0.858–0.996	0.078	0.153–0.004	Nevo et al., 1974

lations with some degree of confidence have been sampled in only a few annual plants. Clegg and Allard (1972) sampled five enzyme loci in 16 Californian populations and in nine Mediterranean regions of the slender wild oat, *Avena barbata*. The mean genetic identity (Nei's *I*) between all pairs of California populations is 0.715 ± 0.028; between the Mediterranean populations is 0.666 ± 0.031; and between the Californian and Mediterranean populations is 0.726 ± 0.029. The Californian and Mediterranian gene pools are not more heterogeneous, on the average, than those of different Californian populations. *A. barbata* is a self-fertilizing annual plant. Different local populations may be effectively identical genetically if they live in similar habitats, but extremely different if they grow in diverse habitats (*I* as low as 0.167 between populations less than 300 km apart).

Levin and Crepet (1973) have sampled 18 enzyme loci in 10 populations from Connecticut and 6 from New York of *Lycopodium lucidulum*, an evergreen herb that reproduces both sexually and asexually. The mean genetic identity between pairs of Connecticut populations is 0.983, between New York populations is 0.985, and between populations of different states is 0.969.

Babbel and Selander (1974) have studied allozyme variation in 8 to 14 local populations in each of four species of herbaceous plants, *Lupinus texensis*, *L. subcarnosus*, *Hymenopappus scabiosaeus*, and *H. artemisiaefolius*. Eight loci were sampled in *Lupinus*, seven in *Hymenopappus*. Local populations of a given species are genetically very similar to each other. The average genetic identities (Nei's *I*) between pairs of local populations are 0.957 for *L. texensis*, 0.975 for *L. subcarnosus*, 0.970 for *H. scabiosaeus*, and 0.942 for *H. artemisiaefolius*; the overall genetic identity between local populations is 0.961 ± 0.007. The two species of *Lupinus* are quite different from each other ($I = 0.353$), while those of *Hymenopappus* are quite similar ($I = 0.896$). *H. scabiosaeus* and *H. artemisiaefolius* are, nevertheless, considered different species because of some morphological and edaphic differences and their apparent failure to hybridize (except perhaps in southern Texas) despite extensive sympatry (Turner, 1956).

Gottlieb (1973*b*) has discovered a self-fertilizing species of the composite annual genus *Stephanomeria*. The new species, designated "Malheurensis," has apparently arisen *in situ* from a geographically peripheral population of *Stephanomeria exigua* ssp. *coronaria*, which is a widespread outcrossing species with self-incompatible individuals. Both species are diploid with $2n = 16$. "Malheurensis" probably originated by a rare mutation which permitted an individual of the progenitor species to self-pollinate. "Malheurensis" is found in a single locality in eastern Oregon and consists of a few hundred individuals growing interspersed with individuals of the parental species. The different breeding system, together with the acquisition of at least one translocation reducing the fertility of the hybrids, are the main mechanisms maintaining reproductive isolation between "Malheurensis" and *coronaria*. The two species are morphologically very similar in most features, but statistically significant differences have been established in some traits when the plants are grown in uniform garden conditions.

If Gottlieb's interpretation is correct, "Malheurensis" has arisen by saltational speciation, i.e., in a series of rapid events not requiring geographical separation and subsequent genetic differentiation. A comparison of the genetic constitutions of "Malheurensis" and *coronaria* at 13 gene loci coding for enzymes has been made by Gottlieb (1973b). "Malheurensis" possesses little genetic variation (average proportion of heterozygous loci per individual, $H = 0.027 \pm 0.019$) while $S.$ *exigua* ssp. *coronaria* is moderately polymorphic ($\bar{H} = 0.092 \pm 0.008$). More importantly, all alleles found in "Malheurensis" with the possible exception of a rare one, are also present in *coronaria,* and the most common allele at each locus is the same in both species. The average genetic identity between both species can be calculated approximately as 0.945 (an exact calculation is not possible since the precise allelic frequencies of *coronaria* at four loci are not given by Gottlieb). The great genetic similarity between "Malheurensis" and $S.$ *exigua* ssp. *coronaria* is consistent with the hypothesis of a saltational origin of "Malheurensis," and suggests that this species may have arisen fairly recently without enough time for its genetic differentiation from the progenitor species. A sample of 14 gene loci in 11 populations of the closely related $S.$ *exigua* ssp. *carotifera* gives an average genetic identity between local populations of 0.980 ± 0.002, with a range from 0.951 to 0.995 (Gottlieb, 1975). The genetic differentiation between "Malheurensis" and ssp. *coronaria* is only slightly greater than between populations of ssp. *carotifera*.

Lewis (1962, 1966) has proposed that several species of the annual plant genus *Clarkia* have arisen by rapid speciation events. One instance of saltational speciation involves the diploid species *Clarkia biloba* and *C. lingulata*. Both species are outcrossers, although capable of self-fertilization, and are native to California, although *C. lingulata* is known from only two sites in the Merced River Canyon at the southern geographical periphery of the distribution of *C. biloba*. The two species are very similar in external morphology (differences exist in the shape of the flower petals), but their karyotypes differ by a translocation, several paracentric inversions, and an extra chromosome in *C. lingulata* which is homologous to parts of two chromosomes of *C. biloba* (Lewis, 1962). The derivative species, *C. lingulata* is presumed to have arisen by a rapid series of events involving extensive chromosomal reorganization which maintains their reproductive isolation.

Gottlieb (1974) has studied the genetic composition of two populations of *C. lingulata* and three of *C. biloba* at eight enzyme loci. Five of the eight loci are polymorphic in both species, but *C. lingulata* has fewer alleles than *C. biloba*. The average heterozygosity over the eight loci is 0.08 and 0.15 in *C. lingulata* and *C. biloba*, respectively. This difference in heterozygosity suggests that inbreeding has occurred in *C. lingulata* either owing to its origin from one or a few individuals or because of its small population size during its history. The mean genetic identity between the two species is 0.88 ± 0.02, but one population of *C. lingulata* is intermediate between the other population of *C. lingulata* ($I = 0.909$) and the three populations of *C. biloba*. Only two new alleles, both at the same locus, *Est-1*, seem to have become established in *C. lingulata* since its origin; the locus *Est-1* is indeed diagnostic between the species. According to Gottlieb the evidence supports the hypothesis of a fairly recent origin of *C. lingulata* from *C. biloba* by rapid speciation.

Clarkia franciscana is another species thought to have emerged *in situ* by saltational speciation (Lewis and Raven, 1958). The presumed progenitor is *C. rubicunda* an ecologically diversified outcrosser, while *C. franciscana* is a self-pollinator consisting of a single population with 600–4000 individuals living in San Francisco, California, close to the northern boundary of the distribution of *C. rubicunda*. The two species are similar morphologically but their hybrids, possessing substantial karyotypic differences, have very low fertility caused by meiotic irregularities. Gottlieb (1973a) has assayed eight different enzyme systems, controlled by 12 to 15 gene loci, in

TABLE XV.   Genetic Identities (Nei's *I*) between Local Populations and between Species of Various Annual Plants

Organisms	Local populations	Species	Source
Avena barbata	0.714	—	Clegg and Allard, 1972
Lycopodium lucidulum	0.976	—	Levin, 1973
Lupinus	0.966	0.353	Babbel and Selander, 1974
Hymenopappus	0.956	0.896	Babbel and Selander, 1974
Stephanomeria	0.980	0.945[a]	Gottlieb, 1973b
Clarkia biloba and *C. lingulata*	0.909	0.88[a]	Gottlieb, 1974
Clarkia rubicunda and *C. franciscana*	—	0.280	Gottlieb, 1973a

[a] Comparisons between one species arisen by saltational speciation and its progenitor species.

four populations of *C. rubicunda* and in the single population of *C. francis-cana*. All but one locus are monomorphic in *C. franciscana*, while *C. rubicunda* is highly polymorphic. At most loci the allelic frequencies are very different in both species. Genetic identity between *C. rubicunda* and *C. fran-ciscana* based on eight loci for which allelic frequencies are given by Gottlieb is $I = 0.280$. Six of the eight loci are species diagnostic, since the two species have completely nonoverlapping allelic configurations. Gottlieb concludes that if *C. franciscana* arose by rapid speciation from *C. rubicunda*, the speciation events must have occurred long ago to allow for the enormous genetic differentiation between them. Alternatively, *C. franciscana* may be the last surviving population of an otherwise extinct species.

A summary of genetic identities between local populations and between species of annual plants is shown in Table XV.

RATES OF CLADOGENESIS AND RATES OF PHYLETIC EVOLUTION

The evolutionary process has two dimensions. One is anagenesis or phyletic evolution—changes occurring through time in a single line of descent. The other is cladogenesis or splitting—the branching of one lineage into two or more separate lines of descent. The basic cladogenetic process is speciation. Are cladogenesis and phyletic evolution directly correlated? That is, is it the case that lineages changing at a faster rate through time are also more cladogenetic than lines of descent evolving at a slower rate? With respect to genetic change the question can be formulated: do highly cladogenetic, or species-rich, lineages evolve genetically at a faster rate per unit time than species-poor lineages?

Note that even if such correlation exists, the cause–effect relationships may not be unambiguously established. The rate of cladogenesis in a lineage might conceivably be simply a result of the rate of anagenetic change—the faster a lineage would change through time, the more cladogenetic (specia-tion) events would occur. But the causal process could conceivably be the reverse. Once the process of speciation is started, the rate of genetic change might be accelerated by the speciation process itself. Natural selection favors the development of reproductive isolation between hybridizing popu-lations exhibiting postmating isolating barriers. If the development of re-productive isolating mechanisms would involve changes in a substantial fraction of the genome, speciation *per se* might significantly increase the rate of genetic change through time.

The question of the correlation between the rate of cladogenesis and the rate of genetic change has been raised by Avise and Ayala (1975), who have suggested empirical tests to ascertain whether the two rates are positively correlated or not. We will consider here two alternative models: (1) the amount of genetic change in a lineage is simply a function of time, and is uncorrelated to the number of cladogenetic events in the lineage; (2) the amount of genetic change in a lineage is proportional to the number of cladogenetic events. It should be noted that if model (2) is correct, it is still possible that the rate of genetic change may also be a function of time: both the time elapsed and the rate of speciation events could conceivably contribute to the rate of genetic change.

Model 1

The rate of genetic change per unit time is constant, and is not affected by the rate at which cladogenetic events occur or by the number of species generated at each cladogenetic event. We assume that genetic changes occur at a constant rate, a, per unit time, in such a way that the genetic distance, d_i, between any two contemporaneous species with a common ancestor t_i time units earlier is simply $d_i = at_i$. We are interested in calculating the average genetic distance among all extant species of a phylad.

Let us consider a phylad derived from a single ancestral species t_i time units ago; cladogenetic events occur at regular intervals, once every m_i time units from the beginning of the phylad, and a constant number of species, l_i, is generated at each cladogenetic event. For simplicity we will assume that no species becomes extinct, and we shall ignore parallel and convergent genetic changes in different lineages, since the frequency of such changes is likely to be small relative to the frequency of divergent genetic changes. The number of extant species, s_i, in the phylad is (dropping the subscripts for typographical simplicity): $s = l^{t/m}$. Given that genetic distance between any two species is proportional to the number of time units since the two species have a common ancestor, we can drop the proportionality constant a, and calculate distances simply as the number of time units. The average genetic distance $\bar{d}$ among all extant species is simply the sum of genetic distances for all pairwise comparisons between extant species, divided by the number of pairwise comparisons. Therefore,

$$\bar{d} = \frac{[m(l-1)l^k/2] \sum_{i=1}^{i=k} i \cdot l^{i-1}}{l^k(l^k-1)/2} = \frac{m(l-1) \sum_{i=1}^{i=k} i \cdot l^{i-1}}{l^k-1} \tag{4}$$

where $k = t/m$ (t is assumed to be an exact multiple of m). The solution of (4) can be shown to be

$$\bar{d} = m\left(k - \frac{1}{l-1} + \frac{k}{l^k - 1}\right) \qquad (5)$$

Assume, now, that we have two phylads, R and P, both started at the same time, t_i time units ago, and both generating l_i species at each cladogenetic event, but with cladogenetic events occurring at different rates. Let R be a species-rich or speciose phylad in which cladogenetic events occur every m_r time units, and P be a species-poor or depauperate phylad in which cladogenetic events occur every m_p time units ($m_r < m_p$). From equation (5), the average genetic distance among extant species is, for the R phylad:

$$\bar{d}_r = m_r\left(k_r - \frac{1}{l-1} + \frac{k_r}{l^{k_r} - 1}\right)$$

$$= m_r\left(\frac{t}{m_r}\right) - \frac{m_r}{l-1} + \frac{m_r(t/m_r)}{l^{t/m_r} - 1}$$

$$= t - \frac{m_r}{l-1} + \frac{t}{l^{t/m_r} - 1} \qquad (6)$$

Similarly, the average genetic distance among extant species of the P phylad is

$$\bar{d}_p = t - \frac{m_p}{l-1} + \frac{t}{l^{t/m_p} - 1} \qquad (7)$$

If the length of time, t, since the origin of the phylads is large relative to m and l, the average genetic distance is approximately t in either phylad; the ratio of d_r to d_p is approximately 1.

Model 2

The rate of genetic change in a phylad is directly proportional to the number of cladogenetic events in the phylad, i.e., phylads generating more species per unit time are also changing genetically at a faster rate than phylads generating fewer species. Assume that the rate of genetic change in a phylad is such that the genetic distance, d_i, between two contemporaneous species separated by c_i cladogenetic events is $d_i = bt_i$. We assume as before that cladogenetic events occur regularly every m_i time units, and that l_i

species are generated at each cladogenetic event. Ignoring the constant b, we can calculate the average genetic distance $\bar{d}$ among all extant species of a phylad as the average number of cladogenetic events separating any two species. Dropping subscripts as before for typographical simplicity,

$$\bar{d} = \frac{[(l-1)l^k/2]\sum_{i=1}^{i=k}(2i-1)\cdot l^{i-1}}{l^k(l^k-1)/2} = \frac{(l-1)\sum_{i=1}^{i=k}(2i-1)\cdot l^{i-1}}{l^k-1} \tag{8}$$

The solution of (8) can be shown to be

$$\bar{d} = 2k - \frac{l+1}{l-1} + \frac{2k}{l^k-1} \tag{9}$$

The average genetic distances in a speciose phylad, R, and in a depauperate phylad, P, defined as in model 1, are

$$\bar{d}_r = \frac{2t}{m_r} - \frac{l+1}{l-1} + \frac{2t/m_r}{l^{t/m_r}-1} \tag{10}$$

and

$$\bar{d}_p = \frac{2t}{m_p} - \frac{l+1}{l-1} + \frac{2t/m_p}{l^{t/m_p}-1} \tag{11}$$

If t is large relative to m, the average genetic distance in a phylad approximates $2t/m_i - (l+1)/(l-1)$, and the ratio of $\bar{d}_r$ to $\bar{d}_p$ will not be very different from m_p/m_r. The average genetic distance in a speciose phylad which speciates twice as often as a depauperate phylad will be approximately twice as large as in the depauperate phylad; if the speciose phylad speciates three times as often as the depauperate phylad, the average genetic distance among extant species will be three times larger in the speciose than in the depauperate phylad, and so on.

A variant of models 1 and 2 could have been considered, namely, a situation where the rate of cladogenetic events is the same in speciose as in depauperate phylads, but at each cladogenetic event more species are generated in the former than in the latter. The results are qualitatively similar to those presented. For model 2, if we assume that the amount of genetic change is directly proportional to the number of species produced, the average genetic distance would be greater in a speciose than in a depauperate group. Such is not the case for model 1, i.e., when the amount of genetic change is proportional to the time elapsed.

The two models proposed are, of course, oversimplifications; the amount of genetic differentiation between species is likely to be affected by both the time elapsed and the number of speciation events that have occurred since they had a common ancestor. But we are interested in the main effects. Given monophyletic groups of species of approximately equal evolutionary age, the average genetic distance among extant species should be greater in speciose than in depauperate groups if the amount of genetic change is correlated with the number of speciation events. If speciation events do not increase substantially the amount of genetic change, (or if the number of speciation events is not a function of the rate of genetic change), the average genetic distance among extant species should be nearly the same in speciose and in depauperate groups.

Avise and Ayala (1975) have compared the amount of genetic differentiation between species of two groups of fish, the genus *Lepomis* and the California minnows. The centrarchid genus *Lepomis* consists of 11 extant species, all endemic to North America. The genus *Lepomis* appears in the fossil record at the Miocene–Pliocene boundary, and has never consisted of many species (Miller, 1965). The North American minnows are a very species-rich group consisting of 250 species, most of which belong to the Leuciscinae, a subfamily which first appears in the fossil record of North America in the Miocene. The California minnows include a species, *Notemigonus crysoleucus,* which may belong to the subfamily Abramidinae. *Notemigonous* may be about as distant phylogenetically from other California minnows as the most divergent North American minnows are from each other.

Relative to the highly speciose minnows, the genus *Lepomis* represents a species-depauperate group. Both groups are of about equal evolutionary age, tracing back to the mid-Miocene, although it is possible that the North American minnows may be somewhat older than *Lepomis*. Ten of the 11 extant species of *Lepomis* have been studied by electrophoresis (Avise and Smith, 1974a,b). The average genetic distance for all pairwise comparisons between the ten species is $\bar{D} = 0.627 \pm 0.029$. Avise and Ayala (1975) have studied electrophoretically 24 loci in nine species of California minnows, including *Notemigonus crysoleucus*; the average genetic distance among all nine species is $\bar{D} = 0.568 \pm 0.052$.

There is no evidence of greater average genetic differentiation in the highly speciose minnows than in the depauperate *Lepomis*. This result is consistent with model 1 but not with model 2, suggesting that the amount of genetic change is not positively correlated with the rate of speciation. Needless to say, this is a far from definitive test. Many difficulties remain, particularly concerning the evolutionary age of the two groups, although as pointed out the minnows are likely to be a somewhat older group, and therefore their average genetic distance even by the first model should be larger than in *Lepomis*; it is not. Another difficulty concerns the true number of cladogenetic events that have occurred in both groups; such a number is unknown but there is little doubt that it is greater in the minnows than in *Lepomis*. One more possible difficulty derives from the relative small number (nine) of minnow species sampled. It is possible that the minnow species sampled may include a disproportionately large number of closely related species. If comparisons among the four most closely related species of minnows are excluded, the average genetic distance for the minnows becomes $\bar{D} = 0.65$, still not significantly greater than the average genetic distance among *Lepomis* ($\bar{D} = 0.63$).

The Hawaiian *Drosophila* provide excellent materials that could be used to test the predictions of models 1 and 2. Most species are endemic to a single island; the upper limit to the evolutionary age of all monophyletic species in an island is given by the known geological history of the islands. Species-rich, species-depauperate, and intermediate groups exist in the Hawaiian archipelago. The *crassifemur* group consists of only six known species; the white-tipped scutellum group of about 10 times (60 species) and the picture-winged of about 20 times (120 species) as many species as the *crassifemur* group. For species of about equal evolutionary age the average genetic distance should be the same in all three groups according to model 1; but it should be greatest among picture-wings, intermediate among white-tips, and least among *crassifemur* species, if model 2 is more nearly correct.

Unfortunately, not enough species have been adequately studied. Thirty-one loci have been assayed in four picture-wing species and in three white-tip species, all collected in the same locality in the island of Maui (see Table VIII). The average genetic distance among three picture-wing species belonging to a single group (*adiostola*) is $\bar{D} = 0.573 \pm 0.091$, ostensibly less than the average distance among the three white-tip species, $\bar{D} = 0.715 \pm 0.195$; this result would seem to favor model 1. However if we include *D. planitibia* the average $\bar{D}$ among the speciose picture-wings becomes 0.821 ± 0.129, somewhat larger (although not significantly so) than the average distance among the less species-rich group of white-tips. Clearly, no conclusion can be drawn from the limited information available for Hawaiian species.

The approach suggested in this section provides, in any case, a possible way to test whether rates of speciation and rates of genetic change are positively correlated. Doubtless, there are in many groups of organisms, phylads of about equal evolutionary age but very different in their rate of cladogenesis.

A FINAL OVERVIEW

Biological evolution consists of changes in the genetic constitution of populations, i.e., of changes in the frequencies of genes and genotypes. Genetic changes are produced by the processes of mutation (including chromosomal changes), genetic drift, natural selection, and gene exchange (migration) between populations. Mutation is the ultimate source of all genetic variation, and in that sense it is the most fundamental evolutionary process. In outcrossing sexual organisms, a mutation first occurring in a single organism may be passed on through the sexual process to all members of the population some generations later. Through gene exchange, the mutation may be eventually extended to all individuals of the species. A genetic change, however, cannot be passed on from members of one species to members of another species. The living world consists of independently evolving gene pools—species; the genetic discontinuities between species are maintained by reproductive isolating mechanisms.

This article has investigated the degree of genetic differentiation concomitant to the development of reproductive isolating mechanisms, and therefore to the formation of species. Reproductive isolation and speciation may come about by a variety of processes. The geographic mode of speciation, perhaps the most general, involves two stages. First, populations geographically separated become genetically differentiated due to genetic drift and to natural selection promoting adaptation to the local environment. The second stage occurs when populations, genetically differentiated while allopatric, regain the opportunity to interbreed. If interpopulational hybrids have low fitness owing to the disruption of coadapted genetic systems, natural selection will favor further development of reproductive isolating mechanisms. Complete reproductive isolation, and with it speciation, may ensue.

The *willistoni* group of *Drosophila* includes populations at the two stages of the process of geographic speciation. Pairs of subspecies represent the first stage of speciation; they are groups of populations that have diverged to the point that hybrid males are sterile, either always or at least when the mothers belong to a given one of two subspecies. The second stage of speciation is present among the semispecies of the *D. paulistorum* complex. Reproductive isolation is virtually complete among semispecies wherever they are sympatric, and it is well advanced elsewhere. The average genetic distance between pairs of subspecies is $\bar{D} = 0.230$, i.e., about 23 allelic substitutions per 100 loci are estimated to have occurred in the evolution of two subspecies from a common ancestor. This is approximately

10 times more genetic differentiation than that which exists between geographic populations.

On the other hand, the average genetic distance between the *D. paulistorum* semispecies is no greater than the average genetic distance between subspecies. Apparently the completion of reproductive isolation through ethological (sexual) mechanisms does not require changes in a substantial proportion of the genome. An alternative explanation may be advanced, however. It is possible that the completion of reproductive isolation may require many genetic changes, but not in the class of genes studied by electrophoretic techniques. The genes studied by electrophoresis are structural genes coding for soluble proteins and enzymes that are for the most part involved in basic cell metabolism. The changes required to accomplish reproductive isolation might involve other types of structural genes and/or regulatory genes. This possibility will be further discussed at the end of this section, but for now I shall consider only the classes of genes that are studied by electrophoresis (the same also studied by protein sequencing and immunological techniques), since these are the only genes about which we have direct information relevant to the questions at hand.

In the *D. willistoni* group, a substantial proportion of genes are changed in the first stage of speciation but few are changed during the second stage. It follows that the development of reproductive isolation *per se* may not require many gene changes. This inference has important implications. Assume that as a result of chromosomal rearrangements or some other process, a situation arises where natural selection favors the development of reproductive isolation between two populations that are quite similar at the single gene level. The process of speciation might be completed with relatively few additional gene changes; the two emerged species might remain quite similar at the single gene level. This suggestion is supported by findings reported in previous sections of this paper.

The literature reviewed in this paper supports on the whole the results obtained in the *D. willistoni* group, although there is, of course, some heterogeneity among results obtained with diverse organisms. The heterogeneity must be due in part to sampling errors, but it may also reflect the great heterogeneity of the living world. What is remarkable is the degree of similarity among results obtained in very different groups of organisms.

In general, subspecies, representing populations well advanced into the first stage of speciation, show moderate but substantial degrees of genetic differentiation. Between 10 and 25 allelic substitutions per locus are estimated to have occurred in the separate evolution of pairs of subspecies in organisms as diverse as several *Drosophila* species, sunfishes, *Taricha* salamanders, iguanid lizards, and rodents.

Subspecies and semispecies have not both been studied in any single group of closely related species other than *D. willistoni*. Nevertheless, as in the *willistoni* group, the degree of genetic differentiation between populations presumed to be in the second stage of speciation is about the same as the amount of differentiation observed between subspecies. This observation is consistent, as pointed out above, with the hypothesis that the development of reproductive isolation *per se* does not involve changes in a large proportion of the genes. Further support for this hypothesis derives from two sources. First, the amount of genetic differentiation among species is no greater in a species-rich group (the California minnows) than in a depauperate group (the sunfish genus *Lepomis*) of about equal evolutionary age. This suggests that genetic differentiation is a function of time, but is not substantially increased by the speciation events themselves. Second, in some rodents (*Spalax* and *Thomomys*, Table XIV) and in some annual plants (*Stephanomeria* and *Clarkia*, Table XV) speciation has been achieved, or is nearly completed, without many changes at the single gene level. In the instances of saltational speciation, reproductive isolation was initiated by chromosomal rearrangements and/or changes in mating systems; further development of reproductive isolating barriers has been accomplished without many allelic changes.

Evolution is a gradual process. Except for polyploidy and similar saltational events, no discontinuities exist in the evolutionary divergence of populations. Whether some groups of populations are classified as semispecies or as species may in certain intermediate situations be a matter of opinion or personal taste. By the criteria used to define species in some groups of fishes or of rodents reviewed in this paper, the semispecies of *D. paulistorum* might be classified as different species. It is, therefore, not surprising that the degree of genetic differentiation between populations classified as congeneric species ranges all the way from what is typical of semispecies (in the neighborhood of 20 allelic substitutions per 100 loci) to many times greater. Well-defined but closely related species often differ by about 50 allelic substitutions per 100 loci. Other congeneric species, however, have gone much farther along the path of evolutionary and genetic divergence. One or more allelic substitutions per locus are estimated to have occurred in the evolution of congeneric species belonging to different *Drosophila* groups, between the *Anolis* species in Bimini, between *Lupinus* species, and between *Clarkia rubicunda* and *C. franciscana*. Species belonging to different closely related genera are found to differ by about one or more allelic substitutions per locus in starfishes (*Pteraster* versus *Diplopteraster*), Scianidae fishes, and some rodents (*Dipodomys* versus *Perognathus*). At levels of evolutionary divergence beyond closely related genera, the techniques of electrophoresis provide no quantitatively precise information.

The process of evolution involves changes in structural genes as well as in gene regulation. Wallace (1963), Stebbins (1969), Britten and Davidson (1969) and others have proposed models endowing regulatory genes with a crucial evolutionary role. Most recently Wilson et al. (1974a,b; see also King and Wilson, 1975) have argued that reproductive isolation incompatibilities as well as morphological evolution may depend more on changes in gene regulation than on amino acid substitutions in proteins. Wilson and his colleagues point out that evolutionary divergence as measured by protein differentiation on one side and by morphological divergence and reproductive incompatibility (judged by the ability to produce viable hybrids) on the other do not always go hand in hand. Much morphological divergence and reproductive incompatibility with only moderate protein differentiation is observed among mammals. Moderate morphological differentiation and reproductive incompatibility with relatively great genetic differentiation is observed in other groups, like some amphibians and birds.

Discrepancies between morphological and genetic differentiation are apparent also in the literature reviewed in this paper. The sibling species of the D. willistoni group for example have average genetic distances of $\bar{D} = 0.581$, but they are morphologically nearly indistinguishable. On the other hand the California minnows, Hesperoleucus symmetricus and Lavinia exilicauda, differ in body size, overall morphology and in ecological preferences, but are genetically quite similar, $\bar{D} = 0.055$. Man and chimpanzee so notably different in morphology and behavior are genetically no more different ($D = 0.620$) than some D. willistoni siblings. Discrepancies between reproductive isolation and genetic differentiation have been pointed out above—genetic differentiation does not increase significantly during the second stage of speciation.

A possible interpretation of this discrepancy is the one suggested by Wilson and his colleagues, i.e., that morphological differentiation and reproductive incompatibilities are primarily effected by changes in gene regulation. But there are other possible explanations. One possible explanation is that conspicuous morphological changes and reproductive isolating mechanisms may come about by changes in only a small proportion of genes. Another possible explanation is that there are several classes of structural genes. Genes modifying morphology or sexual behavior might be structural genes different from the kinds of genes studied by electrophoresis that for the most part are involved in basic cell metabolism. At present, there seems to be no sufficient evidence to decide among these various explanations.

One final note: Not all amino acid substitutions in proteins are detectable by gel electrophoresis. Therefore, electrophoretic studies underestimate the degree of genetic differentiation between populations, although at present it is not possible to say by how much.

APPENDIX

Allelic Frequencies as 31 Gene Loci Coding for Enzymes in Seven Species of Drosophila from Waikamoi, Maui, Hawaii

Locus	Alleles	Picture-wings group			White-tips group			
		D. planitibia	*D. adiostola*	*D. truncipenna*	*D. nigra*	*D. nigella*	*D. dolichotarsis*	*D. crassifemur*
Acph	74	0.32	0.08	—	—	—	—	0.13
	75	—	—	—	—	—	—	—
	76	0.68	0.92	—	—	—	—	0.79
	78	—	—	—	—	—	—	—
	79	—	—	—	0.50	—	0.05	0.09
	80	—	—	—	0.50	—	0.95	—
	81	—	—	0.19	—	—	—	—
	83	—	—	0.81	—	—	—	—
	85	—	—	—	—	—	—	—
	108	—	—	—	—	0.95	—	—
	110	—	—	—	—	0.05	—	—
	Het[a]	0.434	0.140	0.305	0.500	0.087	0.087	0.359
Adh	104	0.98	—	—	—	—	—	0.98
	106	—	—	—	—	—	—	0.02
	108	—	—	—	—	—	1.00	—
	110	0.02	—	—	—	—	—	—
	113	—	—	—	1.00	1.00	—	—
	118	—	—	1.00	—	—	—	—
	120	—	0.02	—	—	—	—	—

[a] Expected frequency of heterozygotes assuming Hardy–Weinberg equilibrium. The symbols for the loci and the electrophoretic procedures are as in Ayala *et al.* (1974c) and references therein. The alleles are symbolized in reference to control alleles of *D. willistoni*; e.g., alleles 88 and 104 at the *Xdh* locus migrate, respectively 12 mm. less and 4mm. more than allele *100* in *D. willistoni*. Flies of every species have been run next to flies of every other species in the same gels in order to verify the electrophoretic mobilities of the allozymes.

[b] A locus is considered polymorphic: (1) when the frequency of the second most common allele is at least 0.01; (2) when the frequency of the most common allele is no greater than 0.95.

APPENDIX (Continued)

Locus	Alleles	Picture-wings group			White-tips group			
		D. planitibia	*D. adiostola*	*D. truncipenna*	*D. nigra*	*D. nigella*	*D. dolichotarsis*	*D. crassifemur*
	124	—	0.02	—	—	—	—	—
	127	—	0.90	—	—	—	—	—
	130	—	0.05	—	—	—	—	—
	133	—	0.02	—	—	—	—	—
	Het	0.044	0.193	0.000	0.000	0.000	0.000	0.035
Adk-1	91	—	0.02	—	—	—	—	0.24
	93	—	0.86	—	—	—	—	—
	94	0.07	0.12	—	—	—	—	0.39
	95	—	—	—	—	—	—	—
	96	0.16	—	—	—	—	—	0.35
	97	0.77	—	0.69	0.11	—	—	—
	99	—	—	—	—	—	1.00	—
	100	—	—	—	0.89	0.95	—	0.02
	101	—	—	—	—	0.05	—	—
	102	—	—	0.31	—	—	—	—
	Het	0.373	0.239	0.430	0.194	0.087	0.000	0.668
Adk-2	95	—	—	—	0.02	—	—	—
	98	—	—	—	0.02	—	—	—
	99	—	0.02	—	—	—	—	—
	100	0.98	—	—	0.96	0.50	1.00	0.98
	102	—	0.98	0.94	—	—	—	—
	103	—	—	—	—	—	—	—
	105	—	—	—	—	—	—	—
	106	—	—	—	—	0.50	—	0.02
	109	0.02	—	0.06	—	—	—	—
	Het	0.044	0.030	0.117	0.084	0.500	0.000	0.035

Locus	Allele	1	2	3	4	5	6	7
Ald-1	87	—	—	—	—	—	—	0.09
	89	0.73	0.85	0.75	0.80	0.70	0.09	0.91
	93	0.27	0.15	0.25	0.17	0.30	0.86	—
	96	—	—	—	0.02	—	0.05	—
	98	—	—	—	—	—	—	—
	Het	0.397	0.257	0.375	0.322	0.420	0.244	0.163
Ald-2	80	—	—	—	—	—	—	0.80
	84	0.02	—	—	—	—	—	0.20
	90	0.64	0.50	0.75	—	—	—	—
	92	—	—	—	—	—	—	—
	93	0.34	0.50	0.25	—	—	—	—
	94	—	—	—	—	—	—	—
	96	—	—	—	0.85	0.10	0.77	—
	97	—	—	—	0.15	0.90	0.23	—
	98	—	—	—	—	—	—	—
	Het	0.478	0.500	0.375	0.258	0.180	0.351	0.316
Ao-1	95	—	—	—	0.04	—	—	—
	97	0.02	—	—	—	—	—	—
	98	—	0.06	—	0.89	0.80	0.23	—
	99	0.36	0.02	0.56	—	0.05	—	—
	100	0.55	0.76	0.44	0.02	—	0.77	0.02
	101	—	—	—	—	0.15	—	—
	102	0.07	—	—	—	—	—	—
	103	—	—	—	—	—	—	—
	104	—	—	—	0.04	—	—	0.04
	105	—	0.17	—	—	—	—	0.74
	107	—	—	—	—	—	—	0.15
	110	—	—	—	—	—	—	0.06
	Het	0.565	0.394	0.492	0.201	0.335	0.351	0.425
Ao-2	92	0.02	—	—	—	—	0.05	—
	94	0.02	—	—	—	—	0.95	—
	96	0.36	—	—	—	—	—	—
	97	0.59	—	—	—	—	—	—

APPENDIX (Continued)

Locus	Alleles	Picture-wings group			White-tips group			
		D. planitibia	*D. adiostola*	*D. truncipenna*	*D. nigra*	*D. nigella*	*D. dolichotarsis*	*D. crassifemur*
	99	—	0.08	0.94	—	—	—	—
	100	—	—	—	0.88	—	—	—
	101	—	0.92	0.06	—	—	—	1.00
	103	—	—	—	0.12	—	—	—
	104	—	—	—	—	0.90	—	—
	106	—	—	—	—	0.10	—	—
	Het	0.518	0.140	0.117	0.204	0.180	0.087	0.000
Aph	77	—	—	—	—	0.14	—	—
	79	—	—	—	—	0.77	—	—
	81	—	—	—	—	0.09	—	—
	101	—	—	0.88	—	—	—	—
	102	—	—	0.12	—	—	—	—
	104	—	—	—	—	—	—	0.71
	106	—	—	—	—	—	—	0.29
	108	—	—	—	0.91	—	0.95	—
	109	—	—	—	0.09	—	0.05	—
	112	0.95	0.98	—	—	—	—	—
	114	0.05	0.02	—	—	—	—	—
	Het	0.087	0.031	0.219	0.159	0.376	0.087	0.408
Est-4	94	—	—	—	0.09	0.09	—	0.02
	96	—	—	0.13	—	—	—	—
	98	0.02	0.02	—	—	—	—	—
	99	0.91	0.95	0.81	—	—	0.09	0.02
	100	0.07	0.03	0.06	0.89	0.91	0.91	0.93

Locus	Allele	1	2	3	4	5	6	7
	101	0.04	—	—	—	—	—	—
	102	—	—	—	0.02	—	—	—
	Het	0.136	0.165	0.165	0.198	0.320	0.088	0.168
Est-5	87	—	—	—	—	—	—	—
	94	—	—	0.95	0.41	—	—	—
	104	—	0.05	0.05	0.57	—	—	—
	107	—	0.91	—	—	—	—	0.18
	109	—	—	—	0.02	—	—	—
	110	—	0.05	—	—	—	0.82	0.45
	112	0.43	—	—	—	—	0.17	0.36
	114	—	—	—	—	—	0.02	—
	116	0.57	—	—	—	0.25	—	—
	117	—	—	—	—	0.75	—	—
	119	—	—	—	—	—	—	—
	Het	0.490	0.169	0.087	0.509	0.375	0.303	0.628
Fum	94	1.00	—	1.00	0.94	—	—	—
	95	—	—	—	0.06	—	—	—
	97	—	—	—	—	1.00	1.00	1.00
	100	—	1.00	—	—	—	—	—
	102	—	—	—	—	—	—	—
	Het	0.000	0.000	0.000	0.122	0.000	0.000	0.000
Got	86	—	1.00	1.00	—	0.06	—	1.00
	90	0.05	—	—	—	0.94	0.94	—
	92	—	—	—	0.87	—	—	—
	93	0.91	—	—	—	—	—	—
	97	—	—	—	—	—	0.06	—
	99	0.02	—	—	—	—	—	—
	103	0.02	—	—	0.13	—	—	—
	123	—	—	—	—	—	—	—
	Het	0.167	0.000	0.000	0.227	0.117	0.114	0.000

APPENDIX (Continued)

Locus	Alleles	Picture-wings group			White-tips group			
		D. planitibia	*D. adiostola*	*D. truncipenna*	*D. nigra*	*D. nigella*	*D. dolichotarsis*	*D. crassifemur*
α-Gpd	97	—	—	—	—	—	0.05	—
	101	1.00	1.00	1.00	—	—	0.86	1.00
	105	—	—	—	1.00	—	—	—
	107	—	—	—	—	1.00	0.09	—
	108	—	—	—	—	—	—	—
	Het	0.000	0.000	0.000	0.000	0.000	0.244	0.000
G3pd	92	—	—	—	0.89	0.95	—	—
	94	—	—	—	0.09	0.05	—	—
	96	—	—	—	0.02	—	—	—
	97	0.20	—	—	—	—	—	0.95
	98	—	—	—	—	—	0.59	—
	99	0.75	—	—	—	—	—	0.05
	100	—	0.92	—	—	—	0.41	—
	102	0.05	0.08	—	—	—	—	—
	103	—	—	0.79	—	—	—	—
	105	—	—	0.21	—	—	—	—
	Het	0.394	0.140	0.337	0.198	0.095	0.483	0.101
G6pd	88	0.86	—	—	—	—	—	—
	90	0.07	—	—	—	—	—	—
	92	0.07	0.45	0.63	—	0.05	—	—
	94	—	0.55	0.31	—	—	—	—
	95	—	—	—	—	—	—	0.21

Locus	Allele	1	2	3	4	5	6	7
	97	0.79	0.23	0.30	0.43	0.06	—	—
	99	—	0.77	—	—	—	—	—
	100	—	—	0.65	0.54	—	—	—
	101	—	—	—	0.02	—	—	—
	103	—	—	—	—	—	—	—
	Het	0.337	0.351	0.485	0.515	0.508	0.496	0.245
Hbdh	98	0.09	—	—	—	—	—	—
	102	0.89	—	—	—	—	—	0.05
	108	—	0.18	—	0.02	—	0.98	0.95
	110	—	0.82	—	0.96	—	0.02	—
	112	0.02	—	1.00	0.02	1.00	—	—
	114	—	—	—	—	—	—	—
	122	—	—	—	—	—	—	—
	Het	0.195	0.298	0.000	0.084	0.000	0.030	0.087
Hk-1	98	0.04	—	0.05	—	—	—	—
	100	0.89	—	0.05	—	—	—	—
	102	0.05	—	0.65	—	0.12	—	—
	104	—	—	0.25	0.85	—	—	—
	105	0.02	—	—	—	0.88	0.02	—
	106	—	—	—	0.02	—	—	—
	107	—	—	—	—	—	0.08	—
	109	—	—	—	—	—	0.89	—
	111	—	1.00	—	0.13	—	0.02	—
	114	—	—	—	—	—	—	0.18
	115	—	—	—	—	—	—	0.75
	116	—	—	—	—	—	—	0.05
	118	—	—	—	—	—	—	0.02
	Het	0.198	0.000	0.510	0.264	0.219	0.195	0.402
Hk-2	94	—	—	—	—	0.06	—	—
	98	—	—	—	—	0.63	—	—

APPENDIX (Continued)

Locus	Alleles	Picture-wings group			White-tips group			
		D. planitibia	*D. adiostola*	*D. truncipenna*	*D. nigra*	*D. nigella*	*D. dolichotarsis*	*D. crassifemur*
	100	—	0.79	—	—	—	—	—
	102	—	—	0.31	—	—	—	—
	104	0.75	0.03	—	0.98	0.90	1.00	0.96
	108	0.25	0.18	—	—	0.05	—	—
	114	—	—	—	0.02	0.05	—	0.04
	Het	0.375	0.345	0.508	0.044	0.185	0.000	0.074
Hk-3	96	—	—	—	—	—	—	0.04
	101	—	—	0.06	0.02	0.05	—	—
	102	—	—	—	—	—	—	0.96
	103	0.02	—	0.88	—	—	—	—
	105	0.98	—	0.06	—	—	—	—
	107	—	0.98	—	0.98	0.95	1.00	—
	113	—	0.02	—	—	—	—	—
	Het	0.044	0.030	0.227	0.043	0.095	0.000	0.069
Idh	93	—	—	—	—	—	—	0.87
	95	—	0.03	—	—	—	—	0.13
	97	0.93	0.95	—	0.93	0.92	0.87	—
	98	—	—	—	—	—	—	—
	99	0.07	0.02	—	0.07	0.08	—	—
	100	—	—	0.38	—	—	0.13	—
	104	—	—	0.62	—	—	—	—
	106	—	—	—	—	—	—	—
	Het	0.127	0.088	0.469	0.124	0.153	0.219	0.224

Locus	Allele	1	2	3	4	5	6	7
Lap-5	94	—	—	—	—	—	—	0.11
	96	—	—	—	—	—	—	0.84
	98	0.16	—	—	—	—	—	0.04
	99	—	—	—	—	—	0.09	—
	100	0.05	—	—	—	—	—	—
	101	—	—	—	—	—	0.91	—
	102	0.02	—	—	—	—	—	—
	103	0.77	0.94	0.88	0.04	—	—	—
	105	—	—	0.12	0.96	1.00	—	—
	107	—	—	—	—	—	—	—
	109	—	0.06	—	—	—	—	—
	113	—	—	—	—	—	—	0.02
	116	—	—	—	—	—	—	—
	Het	0.375	0.114	0.219	0.083	0.000	0.165	0.283
Mdh	91	—	—	—	—	—	—	1.00
	92	1.00	1.00	—	1.00	1.00	1.00	—
	102	—	—	1.00	—	—	—	—
	Het	0.000	0.000	0.000	0.000	0.000	0.000	0.000
Me-1	102	—	0.02	—	—	—	—	—
	104	0.09	—	—	—	—	—	0.91
	105	—	—	—	—	—	—	0.09
	106	0.86	0.92	0.75	—	1.00	0.09	—
	108	0.05	0.03	0.25	1.00	—	0.23	—
	110	—	0.03	—	—	—	0.68	—
	Het	0.244	0.144	0.375	0.000	0.000	0.475	0.163
Odh	102	—	1.00	—	—	—	—	—
	106	0.98	—	0.04	—	—	—	0.02
	107	—	—	—	—	—	—	—
	109	—	—	—	—	—	—	0.02

APPENDIX (Continued)

Locus	Alleles	Picture-wings group			White-tips group			
		D. planitibia	*D. adiostola*	*D. truncipenna*	*D. nigra*	*D. nigella*	*D. dolichotarsis*	*D. crassifemur*
	110	0.02	—	—	1.00	0.96	1.00	—
	112	—	—	—	—	—	—	0.09
	114	—	—	—	—	—	—	0.86
	118	—	—	—	—	—	—	0.02
	119	—	—	1.00	—	—	—	—
	Het	0.044	0.000	0.000	0.000	0.087	0.000	0.256
6Pgd	88	—	1.00	—	—	—	—	—
	90	0.05	—	—	—	—	—	0.07
	92	—	—	—	—	—	—	—
	93	0.23	—	—	—	—	—	—
	96	0.02	—	—	—	—	1.00	0.86
	97	0.61	—	—	—	—	—	—
	98	—	—	1.00	1.00	1.00	—	0.07
	99	0.09	—	—	—	—	—	—
	Het	0.561	0.000	0.000	0.000	0.000	0.000	0.255
Pgi	101	—	—	—	—	—	—	0.98
	102	1.00	1.00	0.88	1.00	0.96	—	—
	105	—	—	0.12	—	—	—	—
	106	—	—	—	—	0.04	1.00	—
	108	—	—	—	—	—	—	—
	115	—	—	—	—	—	—	0.02
	Het	0.000	0.000	0.219	0.000	0.087	0.000	0.035

Locus	Allele							
Pgm	94	—	—	—	—	—	0.06	0.89
	96	—	—	—	—	0.81	0.94	0.11
	100	0.71	0.91	0.95	0.83	0.19	—	—
	103	0.27	—	—	—	—	—	—
	104	0.02	0.09	0.05	0.11	—	—	—
	107	—	—	—	—	—	—	—
	120	—	—	—	0.06	—	—	—
	Het	0.418	0.165	0.087	0.302	0.305	0.114	0.201
To	99	—	—	—	—	1.00	1.00	1.00
	108	—	0.05	0.05	0.98	—	—	—
	109	1.00	0.95	0.95	0.02	—	—	—
	111	—	—	—	—	—	—	—
	113	—	—	—	—	—	—	—
	Het	0.000	0.087	0.087	0.044	0.000	0.000	0.000
Tpi-2	94	1.00	—	—	—	—	—	—
	98	—	—	—	—	—	1.00	1.00
	102	—	1.00	1.00	1.00	1.00	—	—
	107	—	—	—	—	—	—	—
	Het	0.000	0.000	0.000	0.000	0.000	0.000	0.000
Xdh	84	—	—	—	—	—	0.87	—
	88	—	—	—	—	—	0.13	—
	92	—	—	—	—	0.06	—	0.09
	94	—	—	—	—	0.81	—	0.68
	95	—	—	—	—	0.13	—	0.20
	96	—	—	—	—	—	—	0.02
	100	—	—	—	0.04	—	—	—
	101	—	—	0.75	0.85	—	—	—
	102	0.04	—	—	—	—	—	—
	103	—	—	0.25	0.11	—	—	—
	104	0.64	0.09	—	—	—	—	—

APPENDIX (Continued)

Locus	Alleles	Picture-wings group			White-tips group			
		D. planitibia	*D. adiostola*	*D. truncipenna*	*D. nigra*	*D. nigella*	*D. dolichotarsis*	*D. crassifemur*
	106	—	—	—	—	—	0.86	0.32
	108	—	—	—	—	—	0.05	—
	Het	0.485	0.219	0.320	0.269	0.375	0.244	0.482
Frequency of heterozygous loci per individual[a]:								
	Mean	0.236	0.140	0.224	0.160	0.150	0.138	0.203
	s.e.	0.038	0.026	0.032	0.028	0.030	0.027	0.032
Proportion of polymorphic loci per population[b]:								
	(1)	0.774	0.742	0.710	0.742	0.677	0.581	0.806
	(2)	0.645	0.613	0.710	0.548	0.516	0.484	0.645
Number of genomes sampled per locus:								
	Mean	43.7	63.8	15.7	44.1	20.1	21.4	53.8
	s.e.	0.3	1.9	0.3	1.0	0.8	0.5	1.2

REFERENCES

Avise, J. C., and Ayala, F. J., 1975, Genetic differentiation in speciose versus depauperate phylads: evidence from the California minnows, *Evolution,* **29:** (in press).

Avise, J. C., and Smith, M. H., 1974a, Biochemical genetics of sunfish. I. Geographic variation and subspecific intergradation in the bluegill, *Lepomis macrochirus, Evolution* **28:**42-56.

Avise, J. C., and Smith, M. H., 1974b, Biochemical genetics of sunfish. II. Genetic similarity between hybridizing species, *Amer. Natur.* **108:**458-472.

Avise, J. C., Smith, M. H., and Selander, R. K., 1974a, Biochemical polymorphism and systematics in the genus *Peromyscus.* VI. The *boylii* species group, *J. Mammalogy* **55:**751-763.

Avise, J. C., Smith, M. H., Selander, R. K., Lawlor, T. E., and Ramsey, P. R., 1974b, Biochemical polymorphism and systematics in the genus *Peromyscus.* V. Insular and mainland species of the subgenus *Haplomylomys, Syst. Zool.* **23:**226-238.

Avise, J. C., Smith, J. J., and Ayala, F. J., 1975, Adaptive differentiation with little genetic change between two native California minnows, *Evolution* **29:**(in press).

Ayala, F. J., 1970, Competition, coexistence, and evolution, in: *Essays in Evolution and Genetics in Honor of Theodosius Dobzhansky* (M. K. Hecht and W. C. Steere, eds.), pp. 121-158, Appleton-Century-Crofts.

Ayala, F. J., and Dobzhansky, T., 1974, A new subspecies of *Drosophila pseudoobscura, Pan Pacific Entomologist* **50:**211-219.

Ayala, F. J., and Tracey, M. L., 1973, Enzyme variability in the *Drosophila willistoni* group. VIII. Genetic differentiation and reproductive isolation between two subspecies, *J. Heredity* **64:**120-124.

Ayala, F. J., and Valentine, J. W., 1974, Genetic variability in a cosmopolitan deep-water ophiuran, *Ophiomusium lymani, Marine Biology* **27:**51-57.

Ayala, F. J., Mourão, C. A., Pérez-Salas, S., Richmond, R., and Dobzhansky, T., 1970, Enzyme variability in the *Drosophila willistoni* group. I. Genetic differentiation among sibling species, *Proc. Natl. Acad. Sci.* **67:**225-232.

Ayala, F. J., Hedgecock, D., Zumwalt, G. S., and Valentine, J. W., 1973, Genetic variation in *Tridacna maxima,* an ecological analog of some unsuccessful evolutionary lineages, *Evolution* **27:**177-191.

Ayala, F. J., Tracey, M. L., Hedgecock, D., and Richmond, R. C., 1974a, Genetic differentiation during the speciation process in *Drosophila, Evolution* **28:**576-592.

Ayala, F. J., Tracey, M. L., Barr, L. G., and Ehrenfeld, J. G., 1974b, Genetic and reproductive differentiation of *Drosophila equinoxialis caribbensis, Evolution* **28:**24-41.

Ayala, F. J., Valentine, J. W., Barr, L. G., and Zumwalt, G. S., 1974c, Genetic variability in a temperate intertidal phoronid, *Phoronopsis viridis, Biochem. Genetics* **18:**413-427.

Ayala, F. J., Valentine, J. W., DeLaca, T., and Zumwalt, G. S., 1975a, Genetic variability of the Antarctic brachiopod *Liothyrella notorcadensis* and its bearing on mass extinction hypotheses, *J. Paleontology* **49:**1-9.

Ayala, F. J., Valentine, J. W., Hedgecock, D., and Barr, L. G., 1975b, Deep-sea asteroids: high genetic variability in a stable environment, *Evolution* **29:**203-212.

Ayala, F. J., Valentine, J. W., and Zumwalt, G. S., 1975c, An electrophoretic study of the Antarctic zooplankter *Euphausia superba, Limnology and Oceanography* (in press).

Babbel, G. R., and Selander, R. K., 1974, Genetic variability in edaphically restricted and widespread plant species, *Evolution* **28:**619-630.

Bishop, S. C., 1941, Notes on salamanders with descriptions of several new forms, *Occ. Papers Mus. Zool. Univ. Michigan* **451**:1–21.

Britten, R. J., and Davidson, E. H., 1969, Gene regulation for higher cells: a theory, *Science* **165**:349–357.

Campbell, C. A., Ayala, F. J., and Valentine, J. W., 1975, High genetic variability in a population of *Tridacna maxima* from the Great Barrier Reef, *Marine Biology* (in press).

Carson, H. L., 1970, Chromosome tracers of the origin of species, *Science* **168**:1414–1418.

Carson, H. L., 1971*a*, Speciation and the founder principle, *Stadler Symp. Vol.* **3**:51–70.

Carson, H. L., 1971*b*, Polytene chromosome relationships in Hawaiian species of *Drosophila*. V. Additions to the chromosomal phylogeny of the picture-winged species, in: *Studies in Genetics VI* (M. R. Wheeler, ed.), pp. 183–191.

Clegg, M. T., and Allard, R. W., 1972, Patterns of genetic differentiation in the slender wild oat species *Avena barbata, Proc. Natl. Acad. Sci.,* **69**:1820–1824.

Davis, W. C., and Twitty, V. C., 1964, Courtship behavior and reproductive isolation in the species of *Taricha* (Amphibia, Caudata), *Copeia* **1964**:601–610.

Dobzhansky, Th., 1956, What is an adaptive trait?, *Amer. Nat.* **90**:337–347.

Dobzhansky, Th., 1970, *Genetics of the Evolutionary Process,* Columbia Univ. Press, New York.

Dobzhansky, Th., 1972, Species of *Drosophila, Science* **177**:664–669.

Farris, J. S., 1972, Estimating phylogenetic trees from distance matrices, *Amer. Nat.* **106**:645–668.

Fowler, H. W., 1945, A study of the fishes of the southern piedmont and coastal plain, *Acad. Nat. Sci. Philadelphia Monogr.* **7**:1–144.

Gottlieb, L. D., 1973*a*, Enzyme differentiation and phylogeny in *Clarkia franciscana, C. rubicunda,* and *C. amoena, Evolution* **27**:205–214.

Gottlieb, L. D., 1973*b*, Genetic differentiation, sympatric speciation, and the origin of a diploid species of *Stephanomeria, Amer. J. Bot.* **60**:545–553.

Gottlieb, L. D., 1974, Genetic confirmation of the origin of *Clarkia lingulata, Evolution* **28**:244–250.

Gottlieb, L. D., 1975, Allelic diversity in the outcrossing annual plant *Stephanomeria exigua* ssp. *carotifera (compositae), Evolution* **29**:213–225.

Hall, W. P., and Selander, R. K., 1973, Hybridization of karyotypically differentiated populations in the *Sceloporus grammicus* complex (Iguanidae), *Evolution* **27**:226–242.

Hedgecock, D., 1974, Protein variation and evolution in the genus *Taricha* (Salamandridae), Ph.D. Thesis, Univ. of California, Davis.

Hedgecock, D., and Ayala, F. J., 1974, Evolutionary divergence in the genus *Taricha* (Salamandridae), *Copeia* **1974**:738–747.

Hubbs, C. L., 1955, Hybridization between fish species in nature, *Syst., Zool.* **4**:1–20.

Hubby, J. L., and Throckmorton, L. H., 1968, Protein differences in Drosophila. IV. A study of sibling species, *Amer. Natur.* **102**:193–205.

Hunt, W. G., and Selander, R. K., 1973, Biochemical genetics of hybridization in European house mice, *Heredity* **31**:11–33.

Johnson, W., 1973, Dissertation for M.S., University of Arizona.

Johnson, W. E., and Selander, R. K., 1971, Protein variation and systematics in Kangaroo rats (genus *Dipodomys*), *Syst. Zool.* **20**:377–405.

Johnson, W. E., Selander, R. K., Smith, M. H., and Kim, Y. J., 1972, Biochemical genetics of sibling species of the cotton rat (*Sigmodon*), in: *Studies in Genetics, VII,* (M. R. Wheeler, ed.), pp. 297–305, *Univ. Texas Publ.* **7213**.

Kaneshiro, K. Y., 1969, The *Drosophila crassifemur* group of species in a new subgenus, *Univ. of Texas Publ.* **6918**:79–84.

King, M. C., and Wilson, A. C., 1975, Evolution at two levels: molecular similarities and biological differences between humans and chimpanzees, *Science,* **188:**107–116.

Lakovaara, S., Saura, A., and Falk, C. T., 1972, Genetic distance and evolutionary relationships in the *Drosophila obscura* group, *Evolution* **26:**177–184.

Levin, D. A., and Crepet, W. L., 1973, Genetic variation in *Lycopodium lucidulum:* a phylogenetic relic, *Evolution* **27:**622–632.

Lewis, H., 1962, Catastrophic selection as a factor in speciation, *Evolution* **16:**257–271.

Lewis, H., 1966, Speciation in flowering plants, *Science* **152:**167–172.

Lewis, H., and Raven, P. H., 1958, Rapid speciation in *Clarkia, Evolution* **12:**319–336.

Lewis, H., and Roberts, M. R., 1956, The origin of *Clarkia lingulata, Evolution* **10:**126–138.

Lewontin, R. C., and Hubby, J. L., 1966, A molecular approach to the study of genic heterozygosity in natural populations. II. Amount of variation and degree of heterozygosity in natural populations of *Drosophila pseudoobscura, Genetics* **54:**595–609.

Mayr, E., 1963, *Animal Species and Evolution,* Harvard Univ. Press, Cambridge.

McKinney, C. O., Selander, R. K., Johnson, W. E., and Yang, S. Y., 1972, Genetic variation in the side-blotched lizard (*Uta stansburiana*), in: *Studies in Genetics VII* (M. R. Wheeler, ed.), pp. 307–318, *Univ. Texas Publ.* **7213.**

Mettler, L. E., and Nagle, J. J., 1966, Corroboratory evidence for the concept of the sympatric origin of isolating mechanisms, *Drosophila Information Service* **41:**76.

Morgan, T. H., 1919, *The Physical Basis of Heredity,* Philadelphia, Lippincott.

Myers, G. S., 1942, Notes on Pacific Coast *Triturus, Copeia* **1942:**77–82.

Nair, P. S., Brncic, D., and Kojima, K., 1971, Isozyme variations and evolutionary relationships in the *mesophragmatica* species group of *Drosophila,* in: *Studies in Genetics VI* (M. R. Wheeler, ed.), pp. 17–28.

Nei, M., 1972, Genetic distance between populations, *Amer. Natur.* **106:**283–291.

Nevo, E., and Shaw, C. R., 1972, Genetic variation in a subterranean mammal, *Spalax ehrenbergi, Biochem. Genetics* **7:**235–241.

Nevo, E., Kim, Y. J., Shaw, C. R., and Thaeler, C. S., 1974, Genetics variation, selection and speciation in *Thomomys talpoides* pocket gophers, *Evolution* **28:**1–23.

Patton, J. L., Selander, R. K., and Smith, H. H., 1972, Genic variation in hybridizing populations of gophers (genus *Thomomys*), *Syst. Zool.* **21:**263–275.

Reeves, J. D., and Moore, G. A., 1951, *Lepomis marginatus* (Holbrook) in Oklahoma, *Proc. Oklahoma Acad. Sci.* **29:**41–42.

Rockwood, E. S., Kanapi, C. G., Wheeler, R. R., and Stone, W. S., 1971, Allozyme changes during the evolution of Hawaiian Drosophila, in: *Studies in Genetics VI* (M. R. Wheeler, ed.), *Univ. Texas Publ.* **7103:** 193–212.

Rogers, J. S., 1972, Measures of genetic similarity and genetic distance, in: *Studies in Genetics VII* (M. R. Wheeler, ed.), *Univ. Texas Publ.* **7213:**145–153.

Schopf, T. J. M., and Murphy, L. S., 1973, Protein polymorphism of the hybridizing seastars *Asterias forbesi* and *Asterias vulgaris* and implications for their evolution, *Biol. Bull.* **145:**589–597.

Selander, R. K., and Johnson, W. E., 1973, Genetic variation among vertebrate species, *Ann. Rev. Ecol. Systematics* **4:**75–91.

Selander, R. K., and Kaufman, D. W., 1973, Genic variability and strategies of adaptation in animals, *Proc. Natl. Acad. Sci.* **70:**1875–1877.

Selander, R. K., Hunt, W. G., Yang, S. Y., 1969, Protein polymorphism and genic heterozygosity in two European subspecies of the house mouse, *Evolution* **23:**379–390.

Selander, R. K., Yang, W. Y., Lewontin, R. C., and Johnson, W. E., 1970, Genetic variation in the horseshoe crab (*Limulus polyphemus*), a phylogenetic "relic," *Evolution* **24:**402–414.

Selander, R. K., Kaufman, D. W., Baker, R. J., and Williams, S. L., 1974, Genic and chromosomal differentiation in pocket gophers of the *Geomys bursarius* group, *Evolution* **28**:557–564.

Shaw, C. R., 1970, How many genes evolve?, *Biochem. Genet.* **4**:275–283.

Spassky, B., Richmond, R. V., Pérez-Salas, S., O. Pavlovksy, Mourão, C. A., Hunter, A. S., Hoenigsberg, H., Dobzhansky, T., and Ayala, F. J., 1971, Geography of the sibling species relates to *Drosophila willistoni,* and of the semispecies of the *Drosophila paulistorum* complex, *Evolution* **25**:129–143.

Stebbins, G. L., 1969, *The Basis of Progressive Evolution,* Univ. of North Carolina Press.

Turner, B. J., 1974, Genetic divergence of Death Valley pupfish species: Biochemical versus morphological evidence, *Evolution* **28**:281–294.

Turner, B. L., 1956, A cytotaxonomic study of the genus *Hymenopappus* (Compositae), *Rhodora* **58**:164–308.

Twitty, V. C., 1961, Second generation hybrids of the species of *Taricha, Proc. Natl. Acad. Sci.* **47**:1461–1468.

Twitty, V. C., 1964, Fertility of *Taricha* species-hybrids and viability of their offspring, *Proc. Natl. Acad. Sci. U.S.A.* **51**:156–161.

Utter, F. M., Allendorf, F. W., and Hodgins, H. O., 1973, Genetic variability and relationships in Pacific salmon and related trout based on protein variations, *Syst. Zool.* **22**:257–270.

Valentine, J. W., and Ayala, F. J., 1975, Genetic variation in *Frieleia halli,* a deep-sea brachiopod, *Deep-Sea Research* **22**:37–44.

Wallace, B., 1963, Genetic diversity, genetic uniformity and heterosis, *Can. J. of Genetics* and *Cytology* **5**:239–253.

Webster, T. P., Selander, R. K., and Yang, S. Y., 1973, Genetic variability and similarity in the *Anolis* lizards of Bimini, *Evolution* **26**:523–535.

Wilson, A. C., Maxson, L. R., and Sarich, V. M., 1974a, Two types of molecular evolution. Evidence from studies of interspecific hybridization, *Proc. Natl. Acad. Sci.* **71**:2843–2847.

Wilson, A. C., Sarich, V. M., and Maxson, L. R., 1974b, The importance of gene rearrangement in evolution: evidence from studies on rates of chromosomal, protein and anatomical evolution, *Proc. Natl. Acad. Sci.* **71**:3028–3030.

White, M. J. D., 1973, *Animal Cytology and Evolution,* 3rd ed., Cambridge Univ. Press, London and New York.

Yang, S. Y., Wheeler, L. L., and Bock, I. R., 1972, Isozyme variations and phylogenetic relationships in the *Drosophila bipectinata* species complex, in: *Studies in Genetics* (M. R. Wheeler, ed.), pp. 213–227, *Univ. Texas Publ.* **7213**.

Yang, S. Y., Soulé, M., and Gorman, G. C., 1974, *Anolis* lizards of the eastern Caribbean: a case study in evolution. I. Genetic relationships, phylogeny, and colonization sequence of the *roquet* group, *Syst. Zool.* **23**:387–399.

Zouros, E., 1973, Genic differentiation associated with the early stages of speciation in the *mulleri* subgroup of *Drosophila, Evolution* **27**:601–621.

2

Protein Variation in Natural Populations of Animals*

JEFFREY R. POWELL

Department of Biology
Yale University
New Haven, Connecticut 06520

INTRODUCTION

For about ten years evolutionary geneticists have been widely applying the technique of gel electrophoresis to studies of genetic variation in natural populations. To be sure, this technique, as well as other biochemical techniques, has long been used in a limited manner to study genetics of populations (e.g., blood groups). However, it is only since 1966 with the seminal publications of Harris, Johnson *et al.,* and especially Lewontin and Hubby that electrophoresis has become a common method of analyzing the genetic structure of populations. The purpose of this paper is to review briefly the results of electrophoretic studies on natural populations of animals, excluding man.

In a rapidly moving large field such as that being reviewed here, two things are inevitable. First, pertinent literature will be overlooked. I have included in this compilation all studies of which I was aware which are extensive (i.e., several loci studied in several populations) or are of particular evolutionary interest. I apologize for oversights. Second, with the lag of publication time, it is inevitable that many important studies will have appeared since this compilation was made. The major literature search here ended in November 1974.

* Research supported by NSF Grant GB-42220.

A NOTE ON THE TECHNIQUE

The technique of electrophoresis separates proteins on the basis of net electrical charge and to a lesser degree by shape and size. Since proteins are the primary gene product, variation in a given protein usually represents variation in the DNA coding for the protein. Thus, if two individuals of the same species have homologous proteins which migrate different distances during electrophoresis, we may conclude they have different alleles at the locus coding for the protein. To be certain of this conclusion, it is desirable to do the basic genetic crosses to show that the differences are owing to a single Mendelian factor which maps to the location of the structural gene for the protein. Obviously this cannot be done with most organisms. However, when the genetics have been done (primarily with *Drosophila* and mice), with very few exceptions, the assumption that electrophoretic variation is the result of allelic substitutions at structural loci is borne out. Thus, with the experience from genetically well-studied species it is possible to infer, with considerable confidence, the genetic basis of variation of electrophoretic patterns of species which do not lend themselves easily to genetic analysis.

Many different kinds of proteins can be analyzed by electrophoresis. Most commonly, specific enzyme stains are used to visualize the position of the enzyme (proteins) in the gel. In addition, general protein stains have been used to visualize the position of proteins of unknown function as well as proteins such as hemoglobins, albumins, and transferrins.

To use studies on several protein loci as an estimate of the total genetic variation in a genome or in a population, one should bear in mind certain qualifications. First, one must assume the loci studied represent a random sample of all the structural genes in the genome. This is probably a fairly good assumption as the only selective criterion is whether procedures are known for staining the protein in a gel. (However, a possible bias is discussed below.) This means either a specific enzyme stain is available or the protein is present in high enough concentration to be detected by a general protein stain. Second, electrophoresis detects only a fraction of the variation in a fraction of the genome. Only those DNA base changes in structural genes which result in an amino acid substitution which changes the charge of a protein are detected (with some possible exceptions). Thus, contrary to some authors, finding no electrophoretic variation in 20 or so proteins does not indicate absence of genetic variation in a population or species. As will be briefly discussed below, new techniques are beginning to reveal genetic variation undetected by electrophoresis.

The results of electrophoretic studies can be used for three purposes:

(1) to estimate quantitatively the genetic variability in a population or a species; (2) for studies of geographic or temporal patterns of genetic variation within a species; and (3) to determine degrees of genetic divergence among different taxonomic levels. This review is concerned primarily with the first use.

PROTEINS AND SPECIES STUDIED

Table I lists 30 enzymes and nonenzymatic proteins which have been assayed in studies of animal populations. These represent a variety of protein types. In the first category, nonenzymatic proteins, are included hemoglobins, albumins, transferrins, and proteins of unknown function. Among these 30 enzymes there may be a bias in the type of enzymes studied; 17 of them are either dehydrogenases or oxidases. This bias may be due to technical reasons: the tetrazolium system of staining for these types of enzymes is quite simple and sensitive. If this category of enzymes is more or less variable than most proteins, this could bias heterozygosity estimates. However, there is no evidence to suggest whether this is, or is not, the case.

Table II is a compilation of species which have been studied electrophoretically. Heterozygosity estimates are given only for species for which ten or more loci have been studied. Different authors calculate this mean heterozygosity in different ways. For purposes of consistency, I have recalculated this statistic in the same way for all studies. Mean heterozygosities per locus studied were calculated for each population. The "Het" figures in Table II are unweighted means of these means over all populations; that is, no weight was given to unequal sample sizes from different populations. Whenever possible, I have used observed heterozygosities rather than expectations from Hardy–Weinberg equilibrium. The average sample size (in number of genomes) is the average over all loci studied. Since it is often difficult to put standard errors on heterozygosity estimates, these average sample sizes at least convey an idea of the accuracy of these estimates and of the extensiveness of the studies. References marked with asterisks came to my attention too late to be included in the calculations in the remainder of this paper.

Heterozygosities range from 0 for several species of vertebrates to 24.2% for *Drosophila bifasciata*. The overall mean heterozygosity is 9.3%. Since this is an underestimate of total heterozygosity by a factor of about 3 (estimates range from about 1.5 to 5) it may be concluded that, on the average, animals are heterozygous at about 28% of their structural loci. This is a considerable amount of variation, more than that predicted by

TABLE I. Number Code, Abbreviations, and Enzyme Commission Numbers (E.C. No.) of Proteins and Enzymes Used in This Paper

Code Number	Name	Abbreviation	E.C. No.
1	Nonenzymatic proteins	Pt	
2	Esterases	EST	3.1.1.2
3	Peptidases, leucine amino peptidase	PEP/LAP	3.4.1.1
4	Acid phosphatase	ACPH	3.1.3.2
5	Akaline phosphatase	APH	3.1.3.1
6	Malate dehydrogenase	MDH	1.1.1.37
7	α-Glycerophosphate dehydrogenase	α-GPDH	1.1.1.8
8	Lactate dehydrogenase	LDH	1.1.1.28
9	Alcohol dehydrogenase	ADH	1.1.1.1
10	Isocitrate dehydrogenase	IDH	1.1.1.42
11	Octanol dehydrogenase	ODH	1.1.1.1
12	6-Phosphogluconate dehydrogenase	6-PGDH	1.1.1.43
13	Glucose-6-phosphate dehydrogenase	G-6-PDH	1.1.1.49
14	Xanthine dehydrogenase	XDH	1.2.3.2
15	Malic enzyme	ME	1.1.1.40
16	Glutamate dehydrogenase	GDH	1.4.1.2
17	Tetrazolium oxidase/indophenol oxidase/ superoxide dismutase	TO	—
18	Phosphoglucoisomerase	PGI	5.3.1.9
19	Hexokinase	HK	2.7.1.1
20	Aldolase	ALDO	4.1.2.13
21	Acetaldehyde oxidase	AO	1.2.1.3
22	Phosphoglucomutase	PGM	2.7.5.1
23	Glutamate-oxaloacetate aminotransferase	GOT	2.6.1.1
24	Adenylate kinase	ADKIN	2.7.4.3
25	Fumerase	FUM	4.2.1.2
26	Galactose-6-phosphate dehydrogenase	GAL-6-PDH	—
27	Amylase	AMY	3.2.1.1
28	Glyceraldehyde-3-phosphate dehydrogenase	GLY-3-PDH	1.2.2.12
29	Triosephosphate isomerase	TPI	5.3.1.1
30	Sorbital dehydrogenase	SDH	1.1.1.14
31	Hydroxybutyrate dehydrogenase	HBDH	1.1.1.30

some models of the genetic structure of populations. Lewontin (1974) has discussed in detail the implications of these findings for evolutionary genetics.

It is particularly interesting to note that parthenogenetic animals maintain considerable genetic variation. The best examples are the curculionid beetles studied by Suomalainen and Saura (1973). It is misleading to calculate heterozygosities for these species as they would not really be comparable to heterozygosities for sexually out-crossing species. Nevertheless, the results show that within any parthenogenetic species or population much genetic variation exists. Sexual reproduction is not a prerequisite to maintain extensive enzyme polymorphism.

GENERAL PATTERNS AMONG TAXA

To compare amounts of electrophoretic genetic variation in different groups of animals, I have calculated mean heterozygosities for all species studied within different groups. Only species for which ten or more loci have been studied are included. Table III shows the results of these calculations. N includes all species, subspecies and semispecies studied in each group. If both island and continental populations of a species were studied, the average of the two heterozygosities is given in Table III. The standard errors associated with thse means should be viewed with caution for two reasons. First, they were calculated assuming no errors associated with the heterozygosity estimates for each species. It is often impossible to calculate such errors from information presented in publications. Second, these errors assume normal distributions of heterozygosities; there is no reason to accept (or reject) this assumption.

Despite the uncertainties in the statistical interpretation of these figures, it is almost certainly safe to conclude that invertebrates are more variable than vertebrates (Student's t-test, $p < 0.001$). Selander and Kaufman (1973b) have pointed out this difference and explained it in terms of Levins' (1968) models of evolution in changing environments. They concluded that vertebrates tend to experience more "fine-grained" environments and invertebrates experience more "coarse-grained" environments. This difference in the "grain" of their environment plus the fact that some vertebrates are homeothermic would tend to make less genetic variability in vertebrates and more in invertebrates good "evolutionary strategies."

Interestingly, amphibians tend to have levels of genetic variability intermediate between non-amphibian vertebrates and invertebrates. (Student's t-tests indicate amphibians have greater heterozygosities than any other group of vertebrates, $p < 0.01$. For reasons mentioned above these tests should be viewed with reservations.) This observation could fit into Levin's theory as interpreted by Selander and Kaufman. Amphibians are poikilotherms which experience two radically different environments as juveniles and adults. Almost all the insects listed in Table II have radically different environments in the adult and larval stages.

The following alternative, but not mutually exclusive, theory could

TABLE II. Compilation of Electrophoretic Population Studies of Animal Species[a]

	Het	Number of populations	Number of loci	Proteins or enzymes	Average sample	Reference
PROTOZOA						
Ciliophora						
Hymenostomatida						
Paramecidae						
Paramecium aurelia	—	—	—	2	—	Allen *et al.*, 1971 Allen and Gibson, 1971
Tetrahymenidae						
Tetrahymena pyriformis	—	—	—	2, 4	—	Allen and Weremink, 1971
MOLLUSCA						
Gastropoda						
Caenogastropoda						
Littorinidae						
Littorina saxatilis	—	22	4	2, 3, 18	710	Snyder and Gooch, 1973 Berger, 1973
L. obtusata	—	15	3	2	596	''
L. littorea	—	2	2	2	300	Vuilleumier and Matteo, 1972 Berger, 1973
Subulinidae						
Rumina decollata	0	25	25	2, 3, 6, 12, 17, 18, 22, 23	—	Selander and Kaufman, 1973a Selander and Hudson, MS
Styrommatophorida						
Helicidae						
Arianta arbustorum	—	1	8	2, 3, 6, 12, 13, 17, 18, 22	34	Levan and Fredga, 1972
Helix pomatia	—	3	2	2, 17	—	Wahren and Tegelstrom, 1973
H. aspersa	—	43	5	3, 6, 22, 23	4436	Selander and Kaufman, MS

Species	Het		Loci	Code numbers	N	Reference
Cepaea nemoralis	—	3	9	2, 3, 6, 10, 12, 13, 17, 18, 22	70	Levan and Fredga, 1972; Manwell and Baker, 1968
C. hortensis	0.198	2	10	2, 3, 6, 10, 12, 13, 17, 18, 22	75	Levan and Fredga, 1972; Manwell and Baker, 1968
Arionidae						
Arion ater ater	—	20	2	2, 17	1168	Burnet, 1972
A. ater rufus	—	20	2	2, 17	2866	"
Endodontidae						
Mesodon roemeri	0.142	—	17	—	—	Selander and Johnson, 1972
Bivalvia						
Veneroida						
Veneridae						
Saxidomus giganteus	—	5	2	1, 17	877	A. Johnson and Utter, 1973
Mercenaria mercenaria	—	4	4	2, 6, 8, 17	400	Pesch, 1974
M. campechiensis	—	2	4	2, 6, 8, 17	200	"
Tridacna maxima	0.202	1	30	1, 2, 3, 6, 8, 10, 11, 15, 16, 17, 22, 24, 28, 29	179	Ayala *et al.*, 1973a
Nuculoida						
Nuculanidae						
Nuculana pantonia	0.230	1	12	1, 2, 3, 6, 7, 18, 22	22	Gooch and Schopf, 1972
Malletiidae						
Malletia sp.	0.118	1	10	"	16	"

[a] "Het" is mean heterozygosity per locus studied (equivalent to commonly used mean heterozygosity per individual); only in studies in which 10 or more loci were studied is this statistic given. (These heterozygosity figures may be different from those given by the original authors because of different methods of calculation; see text for details). Heterozygosity figures are not given for parthenogenetic or polyploid species. "Proteins or enzymes" indicates what systems have been studied; numbers refer to code in Table I. Average sample is average over all loci studied. Sample size is expressed in terms of number of genomes sampled, equal to twice the number of individuals studied (for diploids). Dashes indicate the information is not available in the published works. Species marked with asterisks are not included in calculations in rest of paper as they were found late.

TABLE II. (Continued)

	Het	Number of populations	Number of loci	Proteins or enzymes	Average sample	Reference
Mytiloida						
Mytilidae						
Mytilus edulis	—	—	4	3, 6	>1000	Milkman and Beaty, 1970 Mitton and Koehn, 1973 Koehn and Mitton, 1972 Koehn, 1975
M. californianus	—	1	3	3, 6, 18	≃2000	Tracey *et al.,* MS Koehn and Mitton, 1972
Modiolus demissus	—	—	3	3, 6, 17	>2000	Koehn *et al.,* 1973
Pterioida						
Ostreidae						
Ostrea edulis	—	4	3	2, 18	670	Wilkins and Mathers, 1973
Crassostrea gigas	—	2	2	2, 18	191	Mathers *et al.,* 1974
C. angulata	—	1	2	2, 18	100	"
ARTHROPODA						
Merostomata						
Xiphosurida						
Limulidae						
Limulus polyphemus	0.064	4	24	1, 3, 6, 7, 8, 10, 12, 17, 18, 20, 22, 23	128	Selander *et al.,* 1970
Crustacea						
Decapoda						
Galatheidae						
Munidopsis diomedeae	0.123	—	12	1, 2, 3, 6, 7, 18, 22	12	Gooch and Schopf, 1972
Pandalopsis ampla	0.072	—	15	"		"

Taxon					Reference
Homaridae					
Homarus americana	0.038	6	43	1, 2, 3, 4, 6, 10, 12, 13, 15, 17, 18, 19, 22, 23, 25, 28	252 Tracey *et al.*, MS
Ocypodidae					
Uca pugilator	—	8	35	1, 2, 18, 22, 23	— Selander *et al.*, 1971
U. pugnax	—	8	35	"	— "
Cladocera					
Daphniidae					
Daphnia magna	—	30	3	2, 6	— Hebert, 1974*a* and *b*
Insecta					
Orthoptera					
Gryllidae					
Gryllus integer	0.145	—	20	—	— Selander and Johnson, 1972
Homoptera					
Cicadidae					
Magicicada tredecassini	0.191	2	15	1, 2, 6, 7, 13, 17, 18, 22, 23	— Krepp and Smith, 1974
M. tredecula	0.169	2	15	"	— "
Cercopidae					
Philaenus spumarius	0.076	6	23	2, 5, 6, 7, 10, 12, 14, 15, 17, 19, 20, 21, 22, 24, 28, 29	405 Saura *et al.*, 1973
Coleoptera					
Curculionidae					
Otiorrhynchus scaber (parthenogenetic)	—	9	26	3, 4, 6, 7, 9, 10, 12, 15, 19, 22, 27, 29	— Suomalainen and Saura, 1973
O. scaber (sexual)	0.309	1	24	"	85 "
O. singularis (parthenogenetic)	—	4	24	"	— "
Strophosomus melanogrammus (parthenogenetic)	—	3	21	"	— "
S. capitatus	0.157	3	19	"	111
Carabidae					
Rhadine subterranea	—	2	3	6, 18, 22	— Avise and Selander, 1972

TABLE II. (Continued)

	Het	Number of populations	Number of loci	Proteins or enzymes	Average sample	Reference
Lepidoptera						
Lycaenidae						
Hemiargus isola	—	4	1	2	—	Burns and Johnson, 1971
Pieridae						
Colias alexandra	0.146	2	13	6, 7, 13, 15, 16, 19, 22, 24, 25, 29	132	G. B. Johnson, pers. com.
*C. meadii**	0.160	1	12	6, 7, 13, 15, 19, 22, 24, 25, 29	80	G. B. Johnson, pers. com.
*C. philodice**	0.200	1	12	"	80	"
Satyridae						
Maniola jurtina	—	20	2	2	2072	Handford, 1973a and b
Psychidae						
Solenobia triquetrella (sexual)	0.182	2	16	2, 6, 7, 10, 19, 24, 25, 29	98	Lokki et al., 1975
S. triquetrella (parthenogenetic)	—	5	16	"	118	"
Diptera						
Tephritidae						
Dacus tryoni	—	—	4	2, 9, 11	—	McKechnie, 1974
D. neohumeralis	—	—	4	2, 9, 11	—	"
D. oleae	—	—	1	2	—	Krimbas and Tsakas, 1971
Drosophilidae						
Drosophila busckii	0.044	18	30	1, 2, 3, 6, 7, 9, 11, 14, 17, 19, 21, 24, 25, 27	—	Prakash, 1973

Willistoni Group

D. willistoni willistoni						
Continental	0.200	81	36	2, 3, 4, 5, 6, 7, 9, 10, 11, 13, 14, 15, 17, 19, 20, 21, 22, 23, 24, 25, 28, 29	>3000	Ayala et al., 1972a; Ayala and Powell, 1972a; Ayala et al., 1974; Ayala et al., 1971
Island	0.183	12	28	2, 3, 4, 5, 6, 7, 9, 10, 11, 14, 15, 17, 19, 20, 21, 22, 24, 28, 29	2441	Ayala et al., 1971; Ayala and Tracey, 1974
D. w. quechua	0.171	1	25	2, 3, 4, 5, 6, 7, 9, 10, 11, 14, 15, 17, 19, 22, 24, 28, 29	77	Ayala and Tracey, 1973
D. equinoxialis equinoxialis	0.220	85	36	2, 3, 4, 5, 6, 7, 9, 10, 11, 13, 14, 15, 17, 19, 20, 21, 22, 23, 24, 25, 28, 29	943	Ayala and Powell, 1972a; Ayala et al., 1972b; Ayala et al., 1974
D. e. caribbensis						
Continental	0.222	1	23	2, 3, 4, 5, 6, 7, 9, 10, 11, 15, 17, 19, 22, 24	333	Ayala et al., 1974
Island	0.181	5	27	2, 3, 4, 5, 6, 7, 9, 10, 11, 14, 15, 17, 19, 20, 21, 22, 24, 28, 29	>1000	Ayala et al., 1974
D. tropicalis						
Continental	0.198	49	34	2, 3, 4, 5, 6, 7, 9, 10, 11, 13, 14, 15, 17, 19, 20, 21, 22, 23, 24, 28, 29, 31	>1000	Ayala and Powell, 1972a; Ayala et al., 1974

TABLE II. (Continued)

	Het	Number of popu- lations	Number of loci	Proteins or enzymes	Average sample	Reference
D. tropicalis (cont.) Island	0.167	5	26	2, 3, 4, 5, 6, 7, 9, 10, 11, 14, 15, 17, 19, 20, 21, 22, 24, 28, 29	370	Ayala and Tracey, 1974
D. paulistorum Orinocan	0.205	15	33	2, 3, 4, 5, 7, 9, 10, 11, 13, 14, 15, 17, 19, 20, 21, 22, 23, 24, 29, 31	150	Richmond, 1972 Ayala *et al.*, 1974
Andean	0.189	32	15	2, 3, 4, 5, 7, 9, 11, 15, 17, 29	483	Richmond, 1972
Interior	0.165	11	14	2, 3, 4, 6, 7, 9, 15, 17, 29	518	''
Amazonian	0.205	32	33	2, 3, 4, 5, 7, 9, 10, 11, 13, 14, 15, 17, 19, 20, 21, 22, 23, 24, 29, 31	563	Ayala *et al.*, 1974 Richmond, 1972
Centro-American	0.134	6	13	2, 3, 4, 6, 7, 9, 15, 17	27	Richmond, 1972
Transitional	0.164	7	15	2, 3, 4, 6, 7, 9, 11, 15, 17, 29	231	''
D. nebulosa Continental	0.188	2	30	2, 3, 4, 6, 7, 9, 10, 11, 13, 14, 15, 17, 19, 20, 21, 22, 23, 24, 25, 29	97	Ayala *et al.*, 1974

Island	0.170	5	2, 3, 4, 5, 6, 7, 9, 10, 11, 14, 15, 17, 19, 20, 22, 24, 29	25	203	Ayala and Tracey, 1974
Melanogaster Group						
D. melanogaster	0.154	25	2, 3, 4, 5, 6, 7, 9, 10, 11, 12, 13, 14, 15, 17, 19, 20, 21, 22, 25	21	—	O'Brien and MacIntyre, 1969; Berger, 1970; Kojima *et al.*, 1970; Johnson and Schaffer, 1973; Mukai *et al.*, 1974; Triantaphyllidis and Christodoulou, 1974
D. simulans	0.107	3	2, 3, 4, 5, 6, 7, 9, 10, 11, 12, 13, 14, 15, 17, 19, 20, 21, 22, 25	20	—	Berger, 1970; O'Brien and MacIntyre, 1969; Kojima *et al.*, 1970
D. bipectinata	0.241	19	2, 3, 4, 5, 6, 7, 9, 10, 11, 12, 13, 15, 18, 19, 21, 22, 25	19	392	Yang *et al.*, 1972
D. parabipectinata	0.134	12	''	19	440	''
D. pseudoananassae pseudoananassae	0.197	2	''	19	16	''
D. p. nigrens	0.209	2	''	19	72	''
D. malerkotliana malerkotliana	0.186	3	''	19	120	''
D. m. pallens	0.235	7	''	19	112	''
D. ananassae	0.135	11	2, 3, 4, 5, 6, 7, 9, 11, 13, 15, 19, 20, 25	13	1705	Gillespie and Kojima, 1968; F. Johnson, 1971
Obscura Group						
D. obscura	0.109	57	2, 3, 5, 6, 7, 9, 10, 11, 13, 14, 15, 17, 20	30	898	Lakovaara and Saura, 1971a

TABLE II. (Continued)

	Het	Number of populations	Number of loci	Proteins or enzymes	Average sample	Reference
D. subobscura	0.105	30	34	2, 3, 5, 6, 7, 9, 10, 11, 13, 14, 15, 17, 19, 21	865	Lakovaara and Saura, 1971b; Saura et al., 1973b
D. bifasciata	0.242	23	21	2, 3, 5, 6, 7, 9, 10, 11, 14, 15, 17, 19, 21, 22, 29	710	Saura, 1974
D. pseudoobscura pseudoobscura	0.125	12	27	1, 2, 3, 4, 6, 7, 11, 13, 14, 15, 17, 21, 22, 27, 19	—	Prakash et al., 1969; Prakash et al., in Lewontin, 1974; Dobzhansky and Ayala, 1973; Ayala and Powell, 1972b
D. pseudoobscura bogotana	0.051	1	24	1, 2, 3, 4, 6, 7, 11, 13, 14, 17, 21, 27	80	Prakash et al., 1969
D. persimilis	0.106	3	27	1, 2, 3, 4, 6, 7, 11, 13, 14, 15, 17, 21, 22, 27, 19	≃100	Prakash, 1969; Dobzhansky and Ayala, 1973; Ayala and Powell, 1972b
D. athabasca	0.146	1	15	2, 5, 6, 7, 9, 11, 14, 15, 20, 22, 25	≃100	Kojima et al., 1970
D. affinis	0.238	1	14	2, 4, 5, 6, 7, 9, 10, 11, 14, 15, 17, 21, 25	≃80	″
Quinaria Group						
D. falleni	—	1	5	—	—	Sabath, 1974
D. palustris	—	1	6	—	—	″

Virilis Group						
D. montana	—	3	4	2	474	Roberts and Baker, 1973
Testacea Group						
D. putrida	—	1	6	—	—	Sabath, 1974
Tripunctata Group						
D. tripunctata	—	—	6	—	—	"
Repleta Group						
D. pachea	—	11	4	2, 4, 6	1525	Rockwood-Sluss et al., 1973
D. mojavensis B1	0.068	2	16	1, 2, 3, 6, 7, 9, 11, 20, 25	242	Zouros, 1973
D. mojavensis A	0.054	1	17	"	7	"
D. mojavensis B11	0.082	6	17	"	473	"
D. arizonensis	0.128	2	17	"	227	"
D. mulleri	0.113	1	16	"	157	"
D. aldrichi	0.115	1	16	"	162	"
Robusta Group						
D. robusta	0.110	8	40	1, 2, 3, 4, 6, 7, 9, 11, 13, 14, 15, 17, 19, 20, 21, 25, 27, 24	527	Prakash, 1973
Immigrans Group						
D. nasuta	—	6	3	2, 4, 11	2410	Stone et al., 1968
D. immigrans	—	1	6	—	—	Sabath, 1974
Mesophragmatica Group						
D. pavani	0.192	14	16	2, 3, 4, 5, 6, 7, 10, 12, 18, 19, 22, 23	≅1000	Nair and Brncic, 1971
D. gaucha	—	—	24	2, 5, 6, 7, 9, 10, 11, 12, 13, 15, 19, 20, 21, 22, 25	—	Nair et al., 1971
D. viracochi	—	—	24	"		"
D. mesophragmatica	—	—	24	"		"
D. gasici	—	—	24	"		"
D. brncici	—	—	24	"		"

TABLE II. (Continued)

	Het	Number of populations	Number of loci	Proteins or enzymes	Average sample	Reference
Hawaiian Species						
37 species	—	1–3	4–8	2, 3, 4, 5, 6, 7, 9, 11	1420	Rockwood et al., 1971
D. mimica	—	2	5	4, 5, 6, 7, 11	≃2000	Rockwood, 1969
Antopocercus aduncus	—	2	5	2, 3, 6, 7, 9	60	Rockwood et al., 1971
A. tanythrix	—	2	5	"	60	"
Exalloscaptomyza throckmortoni	—	1	8	2, 3, 4, 6, 7, 9, 11	70	"
Other Drosophilidae						
Scaptomyza adjusta	—	1	6	—	—	Sabath, 1974
Leucophenga varia	—	1	6	—	—	"
Chymomyza amoena	—	1	6	—	—	Sabath, 1974
Hymenoptera						
Formicidae						
Aphaenogaster rudis	—	3	1	6	47	Crozier, 1973, 1974
Pogonomyrmex barbatus	—	31	3	2, 6	5400	Johnson et al., 1969
Apidae						
Apis mellifera	—	4	2	1, 2	1144	Mestrines and Contel, 1972
BRYOZOA						
Gymnolaemata						
Cheilostomata						
Bugulidae						
Bugula stolonifera	0.094	1	11	2, 3, 6	262	Gooch and Schopf, 1970
Schizoporellidae						
Schizoporella unicornis	—	1	8	2, 3, 6	252	"
Schizoporella errata	0.071	15	11	2, 3, 5, 6, 17, 23	>1000	Gooch and Schopf, 1971 Schopf, 1974

ECHINODERMATA
Asteroidea
  Forcipulatida
   Asteriidae

VERTEBRATA
Osteichthyes
  Anguilliformes
   Anguillidae
  Salmoniformes
   Salmonidae
  Cypriniformes
   Cyprinidae

Taxon						Reference
Asterius vulgaris	0.024	1	26	1, 3, 6, 17, 18, 30, and others	≃50	Schopf and Murphy, 1973
A. forbesi	0.041	1	27	"	≃55	"
Ophiuroidea Ophiurae Ophiuridae *Ophiomusium lymani**	0.170	1	15	2, 3, 5, 6, 14, 15, 17, 18, 19, 22, 29	476	Ayala and Valentine, 1974
Anguilla rostrata	—	5	5	2, 6, 9, 18, 30	1427	Williams *et al.*, 1973
Oncorhynchus gorbucha	0.020–	21	5	6, 7	≃5000	Aspinwall, 1974
*O. keta**	0.032	1	>28	1, 2, 6, 7, 8, 13, 22	204	Altukov *et al.*, 1972
Hesperoleucus symmetricus	0.068	14	24	1, 2, 6, 7, 8, 9, 10, 12, 17, 18, 22, 23, 29	≃500	Avise *et al.*, MS
Lavinia exilicanda	0.053	11	24	"	≃600	"
Pimephales promelus	—	24	1	8	3266	Merritt, 1972
Notropis stramineus	—	14	2	2, 8	2500	Koehn *et al.*, 1971

TABLE II. (Continued)

	Het	Number of populations	Number of loci	Proteins or enzymes	Average sample	Reference
Characidae						
Astyanax mexicanus surface populations	0.128	6	17	1, 2, 3, 6, 7, 8, 10, 12, 18, 22, 23	—	Avise and Selander, 1972
cave populations	0.044	3	17	"	—	"
Lophiiformes						
Macrouridae						
Coryphaenoides acrolepis	—	1	6	—	36	Somero and Soulé, 1974
Gadiformes						
Zoarcidae						
Zoarces viviparus	0.090	2	32	1, 2, 4, 5, 6, 8, 10, 12, 14, 15, 16, 17, 18, 19, 22, 23, 24, 25, 28	758	Frydenberg and Simonsen, 1973; Christiansen and Frydenberg, 1974
Antheriniformes						
Cyprinodontidae						
Cyprinodon 5 species	—	—	31	1, 2, 3, 4, 5, 6, 7, 8, 9, 10, 12, 13, 14, 15, 17, 22, 24, 25, 30	65	Turner, 1974
Fundulus heteroclitus	0.180	10	25	1, 2, 3, 4, 6, 7, 8, 12, 14, 17, 19, 28	≃600	Mitton and Koehn, 1975
Atherinidae						
Leuresthes tenuis	0.036	1	33	1, 2, 3, 4, 6, 7, 8, 9, 10, 12, 13, 14	40	Somero and Soulé, 1974

*Menidia menidia**	0.054	5	24	2, 3, 6, 8, 10, 13, 17, 18, 22, 23, 24, 30	—	M. S. Johnson, 1973, 1974, 1975
*M. peninsulae**	0.055	5	24	"	—	"
*M. beryllina**	0.042	8	24	"	—	"
*M. audens**	0.070	3	24	"	—	"
*M. extensa**	0.033	1	24	"	—	"
Scorpaeniformes						
Scorpaenidae						
Sebastes elongatus	0.025	2	23	1, 3, 6, 7, 8, 10, 13, 15, 18, 22, 26	345	A. G. Johnson et al., 1973
S. alutus	0.038	3	23	"	≥2000	"
S. caurinus	0.014	1	23	"	480	"
Perciformes						
Centrarchidae						
Lepomis humilis	0.049	1	14	1, 2, 3, 6, 8, 10, 12, 17, 18, 22, 23	70	Avise and Smith, 1974a
L. microlophus	0.037	3	14	"	140	"
L. auritus	0.071	2	14	"	90	"
L. gulosus	0.030	2	14	"	78	"
L. megalotis	0.114	1	14	"	60	"
L. cyanellus	0.074	1	14	"	56	"
L. punctatus	0.113	1	14	"	58	"
L. gibbosus	0.067	1	14	"	58	"
L. marginatus	0.069	1	14	"	68	"
L. macrochirus	0.041	47	15	1, 2, 6, 8, 10, 13, 17, 18, 22, 23	4830	Avise and Smith, 1974b
Nototheniidae						
Trematomus borchgrevinki	0.005	1	21	1, 2, 3, 4, 6, 7, 8, 9, 10, 12, 13, 14	18	Somero and Soulé, 1974
T. hansoni	0.025	1	26	"	52	"
T. bernacchii	0.033	1	26	1, 2, 3, 4, 6, 7, 8, 9, 10, 12, 13, 14	60	Somero and Soulé, 1974

TABLE II. (Continued)

	Het	Number of populations	Number of loci	Proteins or enzymes	Average sample	Reference
Pomacentridae						
Dascyllus reticulatus	0.107	1	29	"	20	"
Amphiprion clarkii	0.091	1	27	"	22	"
Abudejduf troschelli	0.050	1	20	"	32	"
Labridae						
Halichores sp.	0.057	1	28	"	20	"
Clinidae						
Gibbonsia metzi	0.043	1	28	"	56	"
Gobiidae						
Bathygobius ramosus	0.005	1	23	"	32	"
Gillichthys mirabilis	0.046	1	29	"	60	"
Mugilidae						
Mugil cephalus	0.071	1	30	"	40	"
Amphibia						
Anura						
Hylidae						
Acris crepitans	—	27	21	1, 2, 6, 7, 8, 15, 16, 23	—	Dessauer and Nevo, 1969
Bufonidae						
Bufo cognatus	0.135	5	10	1, 2, 6, 8, 10, 23	254	Rogers, 1973
B. speciosus	0.097	6	10	"	222	"
B. viridis						
Continental	0.134	11	26	1, 2, 3, 4, 6, 7, 8, 10, 12, 16, 17	1014	Dessauer et al., MS
Island	0.029	1	26	"	20	"

	H		No. loci	Loci	N	Reference
Urodela						
Salamandridae						
*Taricha rivularis**	0.109	3	18	1, 2, 6, 7, 8, 10, 11, 15, 16, 17	151	Hedgecock and Ayala, 1975
*T. granulosa**	0.081	2	18	"	80	"
*T. torosa**	0.053	3	18	"	281	"
*T. t. sierrae**	0.094	1	18	"	131	"
Reptilia						
Squamata						
Colubridae						
Lampropeltis getulus	0.052(?)	17	10(?)	1, 4, 6, 8, 12, 13	≃100	Dessauer and Pough, 1975
Iguanidae						
Uta stansburiana						
Continental	0.051	17	18	1, 2, 6, 7, 17, 18, 22, 23	824	McKinney *et al.*, 1972
Island	0.055	14	20	1, 2, 6, 7, 8, 10, 12, 17, 22, 23	916	Soulé and Yang, 1973
Anolis carolinensis						
Continental	0.049	3	29	1, 2, 3, 6, 7, 8, 9, 10, 12, 14, 15, 18, 22, 23	268	Webster *et al.*, 1972
Island	0.064	1	28	"	118	"
A. distichus	0.043	1	27	1, 2, 3, 6, 7, 8, 9, 10, 12, 14, 15, 18, 22, 23	110	Webster *et al.*, 1972
A. angusticeps	0	1	25	2, 6, 7, 8, 10, 12	76	"
A. sagrei	0.020	2	26	"	396	"
A. brevirostris	—	12	6	"	955	Webster and Burns, 1973
Sceloporus graciosus	0.070	5	22	1, 2, 3, 6, 7, 8, 10, 12, 17, 18, 22, 23	402	Tinkle and Selander, 1973; Hall and Selander, 1973

TABLE II. (Continued)

	Het	Number of populations	Number of loci	Proteins or enzymes	Average sample	Reference
Aves						
Galliformes						
Phasianidae						
Coturnix coturnix	—	1	9	1, 2, 6, 7, 8, 12, 13, 20	124	Manwell and Baker, 1969
Scolopacidae						
Philomachus pugnax	—	1	—	1, 2, 3, 5, 6, 8, 27	—	Segre et al., 1970
Passeriformes						
Sturnidae						
Aplonis metallica	0.047	15	18	1, 2, 3, 5, 6, 8, 10, 14	≃700	Corbin et al., 1974
A. cantoroides	0.012	4	18	''	≃200	''
Fringillidae						
Zonotrichia capensis	0.035	4	24	1, 3, 6, 7, 8, 9, 10, 12, 18, 22, 23	154	Nottebohm and Selander, 1972
Mammalia						
Carnivora						
Phocidae						
Mirounga angustirostris	0	5	24	1, 2, 3, 6, 8, 12, 13, 17, 18, 20, 23, 24, 30	224	Bonnell and Selander, 1974
M. leonina	0.030	1	20	1, 2, 3, 4, 5, 6, 8, 12, 18, 22, 24, 27	84	McDermid et al., 1972

Rodentia						
Muridae						
Mus musculus musculus						
Continental	0.087	3	39	1, 2, 5, 6, 7, 8, 9, 10, 12, 13, 14, 15, 17, 18, 19, 20, 22, 25	100	Selander et al., 1969
Island	0.122	3	36	''	38	Berry and Murphy, 1970 Selander et al., 1969
M. m. domesticus						
Continental	0.069	2	37	''	60	Selander et al., 1969
Island	—	6	—	1, 2, 6, 9, 10, 8, 12, 13, 18, 22	≃1000	Wheeler and Selander, 1972
M. m. brevirostris	0.102	1	40	1, 2, 5, 6, 7, 8, 9, 10, 12, 13, 14, 15, 17, 18, 19, 20, 22, 25	274	Selander and Yang, 1969
M. musculus, lab strains	0.100	6	17	2, 6, 7, 8, 10, 12, 13	168	Ruddle et al., 1969
*Rattus norvegicus**	—	—	21	1, 2, 4, 5, 8, 12, 13, 17, 23	—	Jimenez-Marin and Dessauer, 1973
Geomyidae						
Geomys bursarius	0.038	12	23	1, 2, 6, 7, 8, 9, 10, 12, 17, 22, 23, 30	—	Selander et al., in press
G. personatus	0.044	5	23	''	—	''
G. arenarius	0.050	2	25	''	74	''
G. tropicalis	0	1	34	''	60	''
Critcetidae						
Peromyscus gossypinus	0.051	—	—	—	—	Smith et al., in Selander et al., in press
P. leucopus	0.070	—	—	—	—	''
P. floridanus	0.055	4	41	1, 2, 3, 6, 7, 8, 9, 10, 12, 13, 14, 15, 17, 18, 22, 23, 30	142	Smith et al., 1973

TABLE II. (Continued)

	Het	Number of populations	Number of loci	Proteins or enzymes	Average sample	Reference
P. polionotus Continental	0.059	26	32	1, 2, 3, 6, 7, 8, 10, 12, 13, 17, 18, 22, 23	≃800	Selander *et al.*, 1971
Island	0.051	4	32	"	≃200	"
Sigmodon arizonae	0.028	1	23	1, 6, 7, 8, 9, 10, 12, 13, 17, 18, 22, 23, 30	100	W. Johnson *et al.*, 1972
S. hispidus	0.018	5	23	"	404	W. Johnson *et al.*, 1972
Heteromyidae						
Dipodomys merriami	0.0511	7	18	1, 6, 7, 8, 10, 12, 17, 22, 23	484	W. Johnson and Selander, 1971
D. ordii	0.018	9	17	"	798	"
D. panamintinus	0	2	17	"	20	"
D. elator	0.002	1	17	"	46	"
D. microps	0.007	7	17	"	206	"
D. spectabilis	0.008	2	17	"	96	"
D. deserti	0.010	4	17	"	44	"
D. compactus	0.023	6	17	1, 6, 7, 8, 10, 12, 17, 22, 23	68	W. Johnson and Selander, 1971
D. agilis	0.040	1	17	"	28	"
D. nitratoides	0.040	1	17	"	14	"
D. hermanni	0.042	3	17	"	98	"

	Het		Loci	Code		Reference
Geomyidae						
Thomomys bottae	0.070	5	27	"	148	Patton *et al.*, 1972
T. umbrinus	0.033	1	27	"	60	"
T. talpoides	0.055	10	31	1, 2, 3, 4, 6, 7, 8, 9, 10, 12, 17, 18, 20, 22, 23, 24, 29, 20	532	Nevo *et al.*, 1974
Spalacidae						
Spalax ehrenbergi	0.039	47	17	3, 4, 5, 6, 7, 8, 10, 12, 13, 17, 20, 22	670	Nevo and Shaw, 1972

[a] "Het" is mean heterozygosity per locus studied (equivalent to commonly used mean heterozygosity per individual); only in studies in which 10 or more loci were studied is this statistic given. (These heterozygosity figures may be different from those given by the original authors because of different methods of calculation; see text for details.) Heterozygosity figures are not given for parthenogenetic or polyploid species. "Proteins or enzymes" indicates what systems have been studied; numbers refer to code in Table I. Average sample is average over all loci studied. Sample size is expressed in terms of number of genomes sampled, equal to twice the number of individuals studied (for diploids). Dashes indicate the information is not available in the published works. Species marked with asterisks are not included in calculations in rest of paper as they were found late.

TABLE III. Heterozygosity Estimates[a] (per Locus or per Individual) Over All Species, Subspecies, and Semispecies Listed in Table II

Grouping	N	Mean heterozygosity $\pm$ S.E.
Drosophila	38	0.157 ± 0.009
Other Insects	7	0.170 ± 0.027
Noninsect Invertebrates	13	0.102 ± 0.021
Fishes	31	0.058 ± 0.006
Amphibians	3	0.105 ± 0.016
Reptiles	9	0.043 ± 0.008
Birds	3	0.031 ± 0.010
Mammals	25	0.039 ± 0.006
Invertebrates	58	0.146 ± 0.009
Vertebrates	71	0.050 ± 0.004

[a] Only studies in which at least ten loci were analyzed.

explain the differences in levels of enzyme variability between invertebrates and vertebrates. There are two ways in which the metabolic flexibility and diversity of enzyme systems can increase as a result of evolutionary processes. One is to maintain multiple alleles at enzyme structural loci, i.e., maintain polymorphism. Alternatively, duplications of genetic loci may be selected for, so a single genome has several loci producing a product with similar functions, i.e., multiple-loci isozymes. These different loci could then become specialized to function optimally in different tissues or developmental stages. These two ways of diversifying metabolic systems should not be viewed as mutually exclusive; different evolutionary lines may exhibit various degrees of both strategies. In general, however, it appears that invertebrates have diversified their metabolic systems more often by polymorphism rather than duplication. Many more tissue specific, multiple-locus isozymes are known in vertebrates than in invertebrates (see, for example, Markert *et al.* eds., 1975). However, one should be cautious that the apparent relative lack of multiple-loci enzymes in invertebrates is not an artifact of the technical difficulties which may be associated with their detection. Nevertheless it appears that vertebrates have tended to use the second mechanism more often than the first. In other words, this theory predicts a negative correlation between the number of loci coding for an enzyme type and the degree of polymorphism at each locus.

G. B. Johnson (1974) has predicted just the opposite, i.e., multiple-loci enzymes should have more variability. Fortunately these different conclusions are testable.

LIFE ZONE PATTERNS

Table IV presents a summary of variability in what I will call different life zones. The mean heterozygosities in this table were calculated similarly to those in Table III, and thus the standard errors should be viewed with caution. Using Student's t-test, the only statistically significant differences are (1) fresh water vertebrates are more heterozygous than terrestrial ($p <$ 0.01) and (2) continental invertebrates are more heterozygous than island ($0.05 > p > 0.01$). Again considering the statistical ambiguities, these differences should not be taken too seriously. In general, there are no differences among the different life zone categories used here. It is interesting to note that the greater variability of invertebrates as compared to vertebrates is borne out in every "life zone" in Table IV.

TABLE IV. Heterozygosities per Locus or per Individual for Different Life Zones[a]

	Invertebrates		Vertebrates	
	N	Mean Het ± S.E.	N	Mean Het ± S.E.
Tropical zone	12	0.109 ± 0.009	12	0.047 ± 0.010
Temperate zone	35	0.132 ± 0.012	50	0.049 ± 0.005
Terrestrial	48	0.157 ± 0.009	30	0.039 ± 0.005
Marine	10	0.098 ± 0.022	17	0.050 ± 0.008
Fresh water	—	—	14	0.068 ± 0.008
Continental	4	0.202 ± 0.007	5	0.076 ± 0.019
Island	4	0.175 ± 0.004	5	0.064 ± 0.018

[a] Only studies with 10 or more loci studied are included. In cases of ambiguity (e.g., are amphibians aquatic or terrestrial?) the studies are not included. Continental/island comparisons include only species for which *both* continental and island populations were studied. The same species may be included in more than one category.

METABOLIC FUNCTIONS

Different enzymes tend to have different levels of genetic variation. Table V lists 31 different protein types and the mean heterozygosities in each group of organisms as well as the total over all species. These means and standard errors were calculated in the same manner as in the two previous tables. As has been noted by others, esterases tend to be more variable than other enzymes. However, the *most* polymorphic enzyme in this list is A0. This enzyme has only been studied in insects (22 *Drosophila* species, one coleopteran, and one homopteran). It would be of interest to study this enzyme in other taxa.

Since the relative degree of variability of any particular enzyme tends to cut across taxonomic lines (e.g., EST is very polymorphic in all taxa, IDH is very monomorphic in most taxa, etc.), it seems reasonable to speculate that the degree of variation of a given enzyme is related to its function. Gillespie and Kojima (1968) were the first to point this out and proffer an explanation for these consistencies. They divided enzymes into two classes, those involved in glucose metabolism and those not involved in glucose metabolism. Glucose-metabolizing enzymes tend to have less variability than non-glucose-metabolizing enzymes. This observation has been confirmed in a variety of organisms and is clearly supported by the extensive data summarized in Table V. Kojima *et al.* (1970) reasoned that most enzymes in known metabolic pathways (e.g., glucose metabolism) have a single substrate which is in relatively constant concentration. Other enzymes (e.g., esterases, phosphatases) act on a variety of substrates which may vary greatly in concentrations; many of the substrates of these enzymes originate outside the organism and thus reflect variation in the environment. The conclusion of this reasoning is that the latter type of enzymes should be more genetically variable than the former. Data support this conclusion.

More recently, G. B. Johnson (1974) extended these ideas and has divided enzymes into three classes: variable-substrate, regulatory, and nonregulatory. For reasons mentioned above, variable-substrate enzymes should have high levels of genetic variability. Johnson further reasoned that within a given metabolic pathway, those enzymes which were regulating the flow of metabolites would be the most important in determining the fitness of the individual. Thus, these regulatory enzymes catalyze more sensitive steps in pathways on which natural selection can operate as compared to steps where enzymes do not regulate flow. Johnson presents data from some *Drosophila* species, small vertebrates, and man which show that the most polymorphic group of enzymes are the variable substrate enzymes, while regulatory enzymes are less variable and nonregulatory enzymes are least.

Table VI presents an analysis of the data from the studies listed in Table II along the lines of Johnson's scheme. The variable-substrate enzymes are EST, PEP, ACPH, APH, ODH, and TO. The regulatory enzymes are ADH, XDH, ME, PGI, HK, AO, PGM, ADKIN, and GLY-3-PDH. Nonregulatory enzymes are MDH, LDH, 6-PGDH, G-6-PDH, ALDO, GOT, FUM, AMY, TPI, and SDH; α-GPDH is considered nonregulatory in insects and regulatory in other groups. These designations are in accord with Johnson (1974) and the rationale behind these categorizations is presented in his paper. Although the results in Table VI are not as clear-cut as Johnson's, the same general pattern is obtained. These calculations are based on more extensive data than Johnson's and differ from his in other ways. Johnson did not include TO data from *Drosophila* and used only polymorphic species for his vertebrate calculations. Furthermore, he lumped all esterase loci into one heterozygosity estimate rather than using the heterozygosity at each locus separately as I have done. One result different from Johnson's is that in most groups regulatory enzymes are as polymorphic as, or even more polymorphic than, variable-substrate enzymes. The overall total, however, shows the same pattern as obtained by Johnson, though less pronounced. Birds would seem to have an opposite pattern of variability. However, this exception should not be given too much weight as it is based on a very limited number of studies.

It seems safe to conclude that the levels of polymorphism of enzymes are related to their metabolic functions. This is a first step in understanding more precisely the nature of selective forces acting on enzyme polymorphisms.

ALTERNATIVE TECHNIQUES

Electrophoresis has been used very extensively to study amounts and patterns of genetic variation in natural populations. The popularity of this technique is due to several factors: (1) ease of use; (2) ability to analyze large samples relatively quickly; (3) relative low cost; and (4) generally unambiguous results. However, as mentioned above, the technique has limitations. The two main limitations are that only structural loci (i.e., loci coding for proteins) can be studied and only a fraction of the variation at these loci is detected. Base substitutions which do not change the electrophoretic mobilities of proteins remain undetected. In order to overcome this latter difficulty, other, possibly more sensitive, techniques have been used to study genetic variation in natural populations.

Isoelectric focusing is a technique which separates proteins on the basis

TABLE V.   Mean Heterozygosities

Code number	Name	Drosophila	Other insects	Noninsect invertebrates	Fish
1	PT	0.071 ± 0.031	0.139 ± 0.106	0.122 ± 0.078	0.016 ± 0.0
2	EST	0.341 ± 0.024	0.277 ± 0.050	0.278 ± 0.044	0.187 ± 0.0(
3	LAP/PEP	0.235 ± 0.027	0.588 ± 0.054	0.234 ± 0.054	0.096 ± 0.0;
4	ACPH	0.259 ± 0.039	0.265 ± 0.265	0.062	0
5	APH	0.217 ± 0.036	0.188	0	0
6	MDH	0.044 ± 0.044	0.118 ± 0.054	0.159 ± 0.051	0.010 ± 0.0(
7	α-GPDH	0.0128 ± 0.005	0.082 ± 0.047	0.037	0.127 ± 0.0.
8	LDH	—	—	0.391 ± 0.202	0.080 ± 0.0;
9	ADH	0.152 ± 0.032	0.375	—	0.536
10	IDH	0.127 ± 0.040	0.081 ± 0.041	0.208 ± 0.067	0.003 ± 0.0(
11	ODH	0.105 ± 0.024	—	0.327	—
12	6-PGDH	0.037 ± 0.028	0.059 ± 0.058	0.120 ± 0.080	0.001 ± 0.0(
13	G-6-PDH	0.168 ± 0.036	0.448 ± 0.044	0.002 ± 0.002	0
14	XDH	0.364 ± 0.058	0.211	—	0
15	ME	0.117 ± 0.029	0.394 ± 0.217	0.093 ± 0.034	0.121 ± 0.1;
16	GDH	—	0.200	0.136	—
17	TO	0.070 ± 0.018	0.129 ± 0.129	0.112 ± 0.050	0.033 ± 0.0;
18	PGI	0.353 ± 0.049	0.362 ± 0.041	0.126 ± 0.039	0.071 ± 0.0'
19	HK	0.077 ± 0.018	0.059 ± 0.040	0.008	0.003
20	ALDO	0.095 ± 0.036	0.238	0	—
21	AO	0.321 ± 0.050	0.349 ± 0.276	—	—
22	PGM	0.199 ± 0.032	0.283 ± 0.092	0.102 ± 0.062	0.235 ± 0.0(
23	GOT	0.075 ± 0.047	0	0.121 ± 0.086	0.193 ± 0.1;
24	ADKIN	.0.215 ± 0.119	0.064 ± 0.045	0.128	0.006
25	FUM	0.050 ± 0.019	0.060	0.002	0.006
26	GAL-6-PDH	—	—	—	0
27	AMY	0.321 ± 0.124	0.065 ± 0.065	—	—
28	GLY-3-PDH	0.131 ± 0.022	0.267	0.239	0
29	TPI	0.012 ± 0.006	0.130 ± 0.081	0.272	—
30	SDH	—	—	0	0.106
31	HBDH	0.026 ± 0.007	—	—	—

ᵃ Numbers before proteins and abbreviations are same as in Table I. Dashes indicate the tandard Errors for Different Proteins[a]

Amphibians	Reptiles	Birds	Mammals	Total
0.100 ± 0.063	0.0281 ± 0.009	—	0.060 ± 0.018	0.066 ± 0.014
0.308 ± 0.040	0.191 ± 0.054	0.265 ± 0.046	0.129 ± 0.017	0.277 ± 0.016
0.111	0.002 ± 0.001	0	0.011 ± 0.011	0.192 ± 0.022
0.039	0	—	0.038 ± 0.038	0.224 ± 0.052
—	—	0	0.013 ± 0.013	0.160 ± 0.034
0.087 ± 0.046	0.045 ± 0.043	0	0.055 ± 0.029	0.064 ± 0.012
0	0.063 ± 0.053	0.050 ± 0.049	0.035 ± 0.017	0.039 ± 0.010
0.092 ± 0.032	0.028 ± 0.015	0.277 ± 0.157	0.050 ± 0.026	0.102 ± 0.028
—	0.018 ± 0.016	—	0.083 ± 0.052	0.140 ± 0.026
0.091 ± 0.091	0.011 ± 0.010	0	0.055 ± 0.021	0.082 ± 0.017
—	—	—	—	0.118 ± 0.024
0.034	0.051 ± 0.014	0.327 ± 0.076	0.105 ± 0.032	0.083 ± 0.014
—	0	0.290	0.033 ± 0.030	0.121 ± 0.025
—	0	0	0	0.208 ± 0.046
—	0	—	0.140 ± 0.111	0.131 ± 0.030
0	—	—	—	0.084 ± 0.050
0.012	0.125 ± 0.122	—	0.049 ± 0.041	0.080 ± 0.019
—	0.045 ± 0.025	0.072	0.097 ± 0.024	0.134 ± 0.022
—	—	—	—	0.087 ± 0.015
—	—	—	0	0.086 ± 0.028
—	—	—	—	0.323 ± 0.049
0.453	0.146 ± 0.047	0.069	0.079 ± 0.021	0.170 ± 0.020
0.035 ± 0.014	0.031 ± 0.014	0	0.027 ± 0.017	0.057 ± 0.018
—	—	—	0.017 ± 0.017	0.136 ± 0.032
—	—	—	0	0.041 ± 0.015
—	—	—	—	0
—	—	—	0	0.156 ± 0.090
—	—	—	—	0.159 ± 0.030
—	—	—	0.045	0.054 ± 0.035
—	—	—	0.003 ± 0.003	0.018 ± 0.015
—	—	—	—	0.026 ± 0.007

rotein was not studied in that group.

TABLE VI. Average Heterozygosities at Enzyme Loci whose Products Have Different Metabolic Functions[a]

Enzyme category	Drosophila	Other insects	Noninsect invertebrates	Fishes	Amphibians	Reptiles	Birds	Mammals	Total
Variable substrate	0.205	0.289	0.169	0.063	0.118	0.079	0.088	0.048	0.175
Regulatory enzymes	0.210	0.281	0.100	0.110	0.227	0.039	0.096	0.056	0.161
Nonregulatory enzymes	0.086	0.094	0.122	0.066	0.062	0.039	0.151	0.032	0.073

[a] Categorization of enzymes follows scheme of G.B. Johnson (1974). No standard errors are given because of statistical ambiguities and lack of necessary data.

of isoelectric points. A pH gradient is set up in a gel. Proteins then migrate to their isoelectric point, i.e., to the area of the gel with a pH at which the protein has no charge. This is probably a more sensitive technique than gel electrophoresis, i.e., more amino acid substitutions can be detected by isoelectric focusing. For example, in a population of *Drosophila willistoni* which appeared monomorphic for ADH by electrophoretic analysis, isoelectric focusing detected a total of three alleles (Powell, unpublished data). Unfortunately, to date, this technique does not have advantages 1–3 listed above and its use in population studies remains limited.

Some workers have applied immunological techniques to problems of genetic differentiation. Irwin's (1953) studies on pigeons (Columbidae) is a classic example; Rasmussen (1969) cites work from the 1930s utilizing the technique. Basically, the technique involves obtaining antibodies to specific antigens, usually proteins. The degree of cross-reactivity of two antigens to the same antibody is taken as a measure of the relatedness of the antigens. Although this technique may be more sensitive than electrophoresis in discriminating protein differences, it has several drawbacks for widespread use in population studies. First it is technically more difficult than electrophoresis. Second, and more importantly, results may be more difficult to interpret than when electrophoresis is used. Results depend on the nature of the immune response, a not very well understood process. For example, two laboratories that have obtained antibodies against a certain antigen may not be dealing with the same antibody. Different strains or even different individuals of the same strain may produce different antibodies against the same antigen. The discriminating ability of different antibodies may be different.

Recently Bernstein *et al.* (1973) and Singh *et al.* (1974 and unpublished manuscript) have exploited the fact that proteins with different amino acid sequences, but identical electrophoretic mobilities, may be differentially sensitive to heat denaturation. Approximately two to three times as many alleles could be detected by using electrophoresis coupled with heat denaturation than with electrophoresis alone. To date this technique has only been applied to two enzymes in *Drosophila,* XDH and ODH. If it is applicable to a variety of enzymes, it could be very useful for population surveys. However, there are still some ambiguities with the technique and it is, at present, more difficult than electrophoresis. Lewontin (personal communication) has found that relative stability in high urea concentrations may be a more reliable method than temperature to detect differences in protein stabilities.

Thus, despite some limitations, electrophoresis is still the most reliable and practical technique for routinely studying genetic variation in natural populations. The above-mentioned techniques, or others, may eventually be applicable on a wide scale.

CONCLUSIONS AND SUMMARY

Since 1966, a remarkable number of studies have been published which use the technique of electrophoresis to detect genetic variation in natural populations of animals (Table II). Some generalizations can be drawn from these studies.

1. Most natural populations harbor a wealth of genetic variability. This appears to support the "balance view" of the genetic structure of populations (see Dobzhansky, 1970 and Lewontin, 1974). There are, however, interesting exceptions, e.g., the elephant seal, *Mirounga* (Bonnell and Selander, 1974). Such cases should be carefully evaluated to determine why these particular species appear to have little genetic variation. This may help to understand why most species have high levels of variation.

2. Contrary to some earlier predictions, parthenogenetic species can be as high in genetic variability as sexual species (e.g., Suomalainen and Saura, 1973). Thus this mode of reproduction is not necessarily an evolutionary "deadend."

3. On average, invertebrates are more genetically variable than vertebrates (Table III). Amphibians tend to be intermediate in variability. The level of genetic variability may reflect the different ecological niches of the groups and/or may be related to the evolutionary strategy followed by different phylogenetic lines.

4. There appears to be no strong relationship between amount of genetic variability and the "life zone" of different species (Table IV). For example, temperate and tropical species seem not to differ in degree of genetic variability.

5. The degree of variation of a particular enzyme is correlated among different groups (Table V) and is apparently related to the metabolic function of the enzyme (Table VI). Enzymes with several substrates and enzymes which control the flow of metabolites through a defined pathway have more variation, on average, than enzymes which are not exerting any control over a metabolic pathway.

The major unsolved problem precipitated by these studies is whether the enzyme polymorphisms are adaptive. If many or most are adaptive, what are the selective forces which make enzyme variation adaptive? This problem will be solved only by much more work and by new, creative approaches to the problem.

ACKNOWLEDGMENTS

The following people have provided me with unpublished data: J. C. Avise, H. Dessauer, G. B. Johnson, M. S. Johnson, R. Koehn, A. Saura, R. K. Selander, R. Singh, P. Spieth, and M. L. Tracey. The following have contributed to this work by their encouragement, discussions, advice, and criticisms: Th. Dobzhansky, W. Hartman, J. Darling, K. Scott, M. Watson, F. Vuilleumier, A. Saura, G. B. Johnson. To all of the above I offer my thanks.

REFERENCES

Allen, S. L., and Gibson, I., 1971, Intersyngenic variations in esterases of axenic stocks of *Paramecium aurelia, Biochem. Genet.* **5**:161–172.

Allen, S. L., and Weremuik, S. L., 1971, Intersyngenic variations in the esterases and acid phosphatases of *Tetrahymena pyriformis, Biochem. Genet.* **15**:119–133.

Allen, S. L., Byrue, B. C., and Cronkite, D. L., 1971, Intersyngenic variations in the esterases of bacterized *Paramecium aurelia, Biochem. Genet.* **5**:135–149.

Altukhov, Yu, P., Salmenkova, E. A., Omelchenko, V. T., Satchko, G. D., and Slynko, V. T., 1972, The number of monomorphic and polymorphic loci in populations of the salmon, *Oncorhynchus keta*—one of the tetraploid species, *Genetika* **8**:67–75.

Aspinwall, N., 1974, Genetic analysis of North American populations of the pink salmon, *Oncorhynchus gorbuscha,* possible evidence for the neutral mutation—random drift hypothesis, *Evolution* **28**:295–305.

Avise, J. C., and Selander, R. K., 1972, Evolutionary genetics of cave-dwelling fishes of the genus *Astyanax, Evolution* **26**:1–19.

Avise, J. C., and Smith, M. H., 1974a, Biochemical genetics of sunfish. II. Genic similarity between hybridizing species, *Am. Nat.* **108**:458–472.

Avise, J. C., and Smith, M. H., 1974b, Biochemical genetics of sunfish. I. Geographic variation and subspecific intergradation in the bluegill, *Lepomis macrochirus, Evolution* **28**:42–56.

Avise, J. C., Smith, J. J., and Ayala, F. J., In press, Adaptive differentiation with little genic change between two native California minnows, *Evolution,* submitted.

Ayala, F. J., and Powell, J. R., 1972a, Enzyme variability in the *Drosophila willistoni* group. VI. Levels of polymorphism and the physiological function of enzymes, *Biochem. Genet.* **7**:331–345.

Ayala, F. J., and Powell, J. R., 1972b, Allozymes as diagnostic characters of sibling species of *Drosophila, Proc. Natl. Acad. Sci.* **69**:1094–1096.

Ayala, F. J., and Tracey, M. L., 1973, Enzyme variability in the *Drosophila willistoni* group. VIII. Genetic differentiation and reproductive isolation between two subspecies, *J. Heredity* **64**:120–124.

Ayala, F. J., and Tracey, M. L., 1974, Genetic differentiation within and between species of the *Drosophila willistoni* group, *Proc. Natl. Acad. Sci.* **71**:999–1003.

Ayala, F. J., and Valentine, J. W., 1974, Genetic variability in the cosmopolitan deep-water Ophiuran *Ophiomusium lymani, Marine Biology.* **27**:51–57.

Ayala, F. J., Powell, J. R., and Dobzhansky, Th., 1971, Polymorphisms in continental and island populations of *Drosophila willistoni, Proc. Natl. Acad. Sci.* **68**:2480–2483.

Ayala, F. J., Powell, J. R., Tracey, M. L., Mourão, C. A., and Pérez-Salas, S., 1972*a*, Enzyme variability in the *Drosophila willistoni* group. IV. Genic variation in natural populations of *Drosophila willistoni, Genetics* **70**:113–139.

Ayala, F. J., Powell, J. R., and Tracey, M. L., 1972*b*, Enzyme variability in the *Drosophila willistoni* group. V. Genic variation in natural populations of *Drosophila equinoxialis, Genet. Res. Camb.* **20**:19–42.

Ayala, F. J., Hedgecock, D., Zumwalt, G. S., and Valentine, J. W., 1973, Genetic variation in *Tridacna maxima,* an ecological analog of some unsuccessful evolutionary lineages, *Evolution* **27**:177–191.

Ayala, F. J., Tracey, M. L., Barr, L. G., and Ehrenfeld, J. G., 1974*a*, Genetic and reproductive differentiation of the subspecies, *Drosophila equinoxialis caribbensis, Evolution* **28**:24–41.

Ayala, F. J., Tracey, M. L., Barr, L. G., McDonald, J. F., and Pérez-Salas, S., 1974*b*, Genetic variation in natural populations of five Drosophila species and the hypothesis of the selective neutrality of protein polymorphisms, *Genetics* **77**:343–384.

Berger, E., 1970, A comparison of gene-enzyme variation between *Drosophila melanogaster* and *D. simulans, Genetics* **66**:677–683.

Berger, E. M., 1973, Gene-enzyme variation in three sympatric species of *Littorina, Biol. Bull.* **145**:83–90.

Bernstein, S., Throckmorton, L. H., and Hubby, J. L., 1973, Still more genetic variability in natural populations, *Proc. Natl. Acad. Sci.* **70**:3928–3931.

Berry, R. J., and Murphy, H. M., 1970, The biochemical genetics of an island population of the house mouse, *Proc. Roy. Soc. Lond. B.* **176**:87–103.

Bonnell, M. L., and Selander, R. K., 1974, Elephant seals: Genetic variation and near extinction, *Science* **184**:908–909.

Bowers, J. H., Baker, R. J., and Smith, M. H., 1973, Chromosomal, electrophoretic, and breeding studies of selected populations of deer mice (*Peromyscus maniculatus*) and black-eared mice (*P. melanotis*), *Evolution* **27**:378–386.

Burnet, B., 1972, Enzyme protein polymorphism in the slug *Arion ater, Genet. Res. Camb.* **20**:161–173.

Burns, J. M. and Johnson, F. M., 1971, Esterase polymorphism in the butterfly *Hemiargus isola*: stability in a variable environment, *Proc. Natl. Acad. Sci.* **68**:34–37.

Christiansen, F. B., and Frydenberg, O., 1973, Geographical patterns of four polymorphisms in *Zoarces viviparus* as evidence of selection, *Genetics* **77**:765–770.

Corbin, K. W., Sibley, C. G., Ferguson, A., Wilson, A. C., Brush, A. H., and Ahlquist, J. E., 1974, Genetic polymorphism in the New Guinea starlings of the genus *Aplonis, Condor* **76**:307–318.

Crozier, R. H., 1973, Apparent differential selection at an isozyme locus between queens and workers of the ant *Aphaenogaster rudis, Genetics* **73**:313–318.

Crozier, R. H., 1974, Allozyme analysis or reproductive strategy in the ant *Aphaenogaster rudis, Isozyme Bull.* **7**:18.

Dessauer, H. C., and Nevo, E., 1969, Geographic variation of blood and liver proteins in cricket frogs, *Biochem. Genet.* **3**:171–188.

Dessauer, H. C., and Pough, F. H., 1975, Geographic variation of blood proteins and the systematics of kingsnakes (*Lampropeltis getulus*), *Comp. Biochem. Physiol.* **50B**:9–12.

Dessauer, H. C., Nevo, E., and Chuang, K., High genetic variability in an ecologically variable vertebrate, *Bufo viridis, Genetics,* submitted.

Dobzhansky, Th., 1970, *Genetics of the evolutionary process,* Columbia University Press, New York.

Dobzhansky, Th., and Ayala, F. J., 1973, Temporal frequency changes of enzyme and chromosomal polymorphisms in natural populations of *Drosophila, Proc. Nat. Acad. Sci.* **70**:680–683.

Frydenberg, O., and Simonsen, V., 1973, Genetics of *Zoarces* populations. V. Amount of protein polymorphism and degree of genic heterozygosity, *Hereditas* **75**:221–232.

Gillespie, J. H., and Kojima, K., 1968, The degree of polymorphism in enzymes involved in energy production compared to that in nonspecific enzymes in two *Drosophila ananassae* populations, *Proc. Natl. Acad. Sci.* **61**:582–585.

Gooch, J. L., and Schopf, T. J. M., 1970, Population genetics of marine species of the phylum Ectoprocta, *Biol. Bull.* **138**:138–156.

Gooch, J. L., and Schopf, T. J. M., 1971, Genetic variation in the marine Ectoprot *Schizoporella errata, Biol. Bull.* **141**:235–246.

Gooch, J. L., and Schopf, T., 1972, Genetic variability in the deep sea: relation to environmental variability, *Evolution* **26**:545–552.

Hall, W. P., and Selander, R. K., 1973, Hybridization of karyotypically differentiated populations in the *Sceloporus grammicus* complex (Iguanidae), *Evolution* **27**:226–242.

Handford, P. T., 1973a, Patterns of variation in a number of genetic systems in *Maniola jurtina:* the Isles of Scilly, *Proc. Roy. Soc. Lond. B.* **183**:285–300.

Handford, P. T., 1973b, Patterns of variation in a number of genetic systems in *Maniola jurtina*: the boundary region, *Proc. Roy. Soc. Lond. B.* **183**:265–284.

Harris, H., 1966, Enzyme polymorphism in man, *Proc. Roy. Soc. Lond. B.* **164**:298–310.

Hebert, P. D., 1974a, Enzyme variability in natural populations of *Daphnia magna*. II. Genotypic frequencies in permanent populations, *Genetics* **77**:323–334.

Hebert, P. D., 1974b, Enzyme variability in natural populations of *Daphnia magna*. III. Genotypic frequencies in intermittent populations, *Genetics* **77**:335–341.

Hedgecock, D., and Ayala, F. J., 1974, Evolutionary divergence in the genus *Taricha* (Salamandridae), *Copeia* **1974**:738–747.

Hubby, J. L., and Lewontin, R. C., 1966, A molecular approach to the study of genic heterozygosity in natural populations. I. The number of alleles at different loci in *Drosophila pseudoobscura, Genetics* **54**:577–594.

Irwin, M. R., 1953, Evolutionary patterns of antigenic substances of the blood corpuscles in Columbidae, *Evolution* **7**:31–50.

Jimenez-Marin, D., and Dessauer, H. C., 1973, Protein phenotype variation in laboratory populations of *Rattus norvegicus, Comp. Biochem. Physiol.* **46B**:487–492.

Johnson, A. G., and Utter, F. M., 1973, Electrophoretic variation of adductor muscle protein and tetrazolium oxidase in the smooth Washington clam, *Saxidomus giganteus* (Deshayes 1839), *Anim. Blood. Biochem. Genet.* **4**:147–152.

Johnson, A. G., Utter, F. M., and Hodgins, H. O., 1973, Estimates of genetic polymorphism and heterozygosity in three species of rockfish (Genus *Sebastes*), *Comp. Biochem. Physiol.* **44B**:397–406.

Johnson, F. M., 1971, Isozyme polymorphisms in *Drosophila ananassae*: genetic diversity among island populations in the south Pacific, *Genetics* **68**:77–95.

Johnson, F. M., Kanapi, C., Richardson, R. H., Wheeler, M. R., and Stone, W. S., 1966, An analysis of polymorphisms among isozyme loci in dark and light *Drosophila ananassae* strains from American and Western Samoa, *Proc. Natl. Acad. Sci.* **56**:119–125.

Johnson, F. M., and Schaffer, H. E., 1973, Isozyme variability in species of the genus *Drosophila*. VII. Genotype-environment relationships in populations of *D. melanogaster* from the eastern United States, *Biochem. Genet.* **10**:149–163.

Johnson, F. M., Schaffer, H. E., Gillaspy, J. E., and Rockwood, E. S., 1969, Isozyme genotype-environment relationships in natural populations of the harvester ant, *Pogonomyrmex barbatus,* from Texas, *Biochem. Genet.* **3**:429–450.

Johnson, G. B., 1974, Enzyme polymorphism and metabolism, *Science* **184**:28–37.

Johnson, M. S., 1973, An electrophoretic study of enzyme variation in fishes of the genus *Menidia* (Teleostei, Atherinidae), Ph.D. thesis, Yale University.

Johnson, M. S., 1974, Comparative geographic variation in *Menidia, Evolution* **28**:607–618.

Johnson, M. S., 1975, Biochemical systematics of the atherinid genus *Menidia, Copeia,* in press.

Johnson, W. E., and Selander, R. K., 1971, Protein variation and systematics in kangaroo rats (Genus *Dipodomys*), *Syst. Zool.* **20**:377–405.

Johnson, W. E., Selander, R. K., Smith, M. H., and Kim, Y. J., 1972, Biochemical genetics of sibling species of the cotton rat (*Sigmodon*), *Univ. Texas Publ.* **7213**:298–305.

Koehn, R. K., 1975, Migration and population structure in the pelagically dispersing marine invertebrate, *Mytilus edulis, Proc. 3rd Int. Conf. Isozymes,* C. Markert *et al.* (eds).

Koehn, R. K., Perez, J. E., and Merritt, R. B., 1971, Esterase enzyme function and genetical structure of populations of the freshwater fish, *Notropis stramineus, Amer. Nat.* **105**:51–69.

Koehn, R. K., Turano, F. J., and Mitton, J. B., 1973, Population genetics of marine pelecypods. II. Genetic differences in microhabitats of *Modiolus demissus, Evolution* **27**:100–105.

Kojima, K., Gillespie, J., and Tobari, Y. N., 1970, A profile of *Drosophila* species enzymes assayed by electrophoresis. I. Number of alleles, heterozygosities, and linkage disequilibrium in glucose-metabolizing systems and some other enzymes, *Biochem. Genet.* **4**:627–637.

Kojima, K., Smouse, P., Yang, S., Nair, P. S., and Brncic, D., 1972, Isozyme frequency patterns in *Drosophila pavani* associated with geographical and seasonal variables, *Genetics* **72**:721–731.

Krepp, S. R., and Smith, M. H., 1974, Genic heterozygosity in the 13-year cicada, *Magicicada, Evolution* **28**:396–401.

Krimbas, C. B., and Tsakas, S., 1971, The genetics of *Dacus oleae.* V. Changes of esterase polymorphism in a natural population following insecticide control—selection or drift? *Evolution* **25**:454–460.

Lakovaara, S., and Saura, A., 1971*a*, Genetic variation in natural populations of *Drosophila obscura, Genetics* **69**:377–384.

Lakovaara, S., and Saura, A., 1971*b*, Genic variation in marginal populations of *Drosophila subobscura, Hereditas* **69**:77–82.

Levan, G., and Fredga, K., 1972, Isozyme polymorphism in three species of land snails, *Hereditas* **71**:245–252.

Levins, R., 1968, *Evolution in changing environments,* Princeton University Press, Princeton.

Lewontin, R. C., 1974, *The Genetic Basis of Evolutionary Change,* Columbia University Press.

Lewontin, R. C., and Hubby, J. L., 1966, A molecular approach to the study of genic heterozygosity in natural populations. II. Amount of variability and degree of heterozygosity in natural populations of *Drosophila pseudoobscura, Genetics* **54**:595–609.

Lokki, J., Suomalainen, E., Saura, A., and Lankinen, P., 1975, Genetic polymorphism and evolution in parthenogenetic animals. II. Diploid and polyploid *Solenobia triquetrella* (Lepidoptera: Psychidae), *Genetics* **79**:513–525.

Manwell, C., and Baker, C. M. A., 1968, Genetic variation of isocitrate, malate and 6-phosphogluconate dehydrogenases in snails of the genus *Cepaea*—introgressive hybridization, polymorphism and pollution? *Comp. Biochem. Physiol.* **26**:195–209.

Manwell, C., and Baker, C. M. A., 1969, Hybrid proteins, heterosis and the origin of species—I. Unusual variation of polychaete *Hyalinoecia* "nothing dehydrogenase" and of quail *Coturnix* erythrocyte enzymes, *Comp. Biochem. Physiol.* **28**:1007–1028.

Markert, C. L. (ed), 1975, *Proceedings of the IIIrd International Conference on Isozymes.* Academic Press, New York.

Mathers, N. F., Wilkins, N. P., and Walne, P. R., 1974, Phosphoglucose isomerase and esterase phenotypes in *Crassostrea angulata* and *C. gigas, Biochem. Sys. Ecol.* **2**:93–96.

McDermid, E. M., Ananthakrishnan, R., and Agar, N. S., 1972, Electrophoretic investigation of plasma and red cell proteins and enzymes of Macquarie Island elephant seals. Animal Blood Groups, *Biochem. Genet.* **3**:85–94.

McKechnie, S. W., 1974, Allozyme variation in the fruit flies *Dacus tryoni* and *Dacus neohumeralis* (Tephritidae), *Biochem. Genet.* **11**:337–346.

Mckinney, C. O., Selander, R. K., Johnson, W. E., and Yang, S. Y., 1972, Genetic variation in the side-blotched lizard (*Uta stansburiana*), *Univ. Texas Publ.* **7213**:307–318.

Merritt, R. B., 1972, Geographic distribution and enzymatic properties of lactate dehydrogenase allozymes in the fathead minnow, *Am. Nat.* **106**:173–184.

Mestriner, M. A., and Contel, E. P. B., 1972, The *P-3* and *EST* loci in the honeybee *Apis mellifera, Genetics* **72**:733–738.

Milkman, R., and Beaty, L. D., 1970, Large-scale electrophoretic studies of allelic variation in *Mytilus edulis, Biol. Bull.* **139**:430.

Mitton, J. B., and Koehn, R. K., 1973, Population genetics of marine pelecypods. III. Epistasis between functionally related isoenzymes of *Mytilus edulis, Genetics* **73**:487–496.

Mitton, J. B., and Koehn, R. K., 1975, Genetic organization and adaptive response of allozymes to ecological variables in *Fundulus heteroclitus, Genetics, 79*:97–111.

Mukai, T., Watanabe, T. K., and Yamaguchi, O., 1974, The genetic structure of natural populations of *Drosophila melanogaster.* XII. Linkage disequilibrium in a large local population, *Genetics* **77**:771–793.

Nair, P. S., and Brncic, D., 1971, Allelic variations within identical chromosomal inversions, *Am. Nat.* **105**:291–294.

Nair, P. S., Brncic, D., Kojima, K., 1971, Isozyme variations and evolutionary relationships in the *mesophragmatica* species group of *Drosophila, Univ. Texas Publ.* **7103**:17–28.

Nevo, E., and Shaw, C. R., 1972, Genetic variation in a subterranean mammal, *Spalax ehrenbergi, Biochem. Genet.* **7**:235–241.

Nevo, E., Kim, Y. J., Shaw, C. R., and Thaeler, C. S., 1974, Genetic variation, selection, and speciation in *Thomomys talpoides* pocket gophers, *Evolution* **28**:1–23.

Nottebohm, F., and Selander, R. K., 1972, Vocal dialects and gene frequencies in the chingolo sparrow (*Zonotrichia capensis*), *Condor* **74**:137–143.

O'Brien, S. J., and MacIntyre, R. J., 1969, An analysis of gene–enzyme variability in natural populations of *Drosophila melanogaster* and *D. simulans, Am. Nat.* **103**:97–113.

Patton, J. L., Selander, R. K., and Smith, M. H., 1972, Genic variation in hybridizing populations of gophers (Genus *Thomomys*), *Syst. Zool.* **21**:263–270.

Pesch, G., 1974, Protein polymorphisms in the hard clams *Mercenaria mercenaria* and *Mercenaria campechiensis, Biol. Bull.* **146**:393–403.

Prakash, S., 1969, Genic variation in a natural population of *Drosophila persimilis, Proc. Natl. Acad. Sci.* **62**:778–784.

Prakash, S., 1973, Patterns of gene variation in central and marginal populations of *Drosophila robusta, Genetics* **75**:347–369.

Prakash, S., Lewontin, R. C., and Hubby, J. L., 1969, A molecular approach to the study of genic heterozygosity in natural populations. IV. Patterns of genic variation in central, marginal and isolated populations of *Drosophila pseudoobscura, Genetics* **61**:841–858.

Rasmussen, D. I., 1969, Molecular taxonomy and typology, *Bioscience* **19**:418–420.

Richmond, R. C., 1972, Enzyme variability in the *Drosophila willistoni* group. III. Amounts of variability in the superspecies, *D. paulistorum, Genetics* **70**:87–112.

Roberts, R. M., and Baker, W. K., 1973, Frequency distribution and linkage disequilibrium of active and null esterase isozymes in natural populations of *Drosophila montana, Am. Nat.* **107**:709–726.

Rockwood, E. S., 1969, Enzyme variation in natural populations of *Drosophila mimica, Univ. Texas Publ.* **6918**:111–125.

Rockwood, E. S., Kanapi, C. G., Wheeler, M. R., and Stone, W. S., 1971, Allozyme changes during the evolution of Hawaiian *Drosophila, Univ. Texas Publ.* **7103**:193–212.

Rockwood-Sluss, E. S., Johnston, J. S., and Heed, W. B., 1973, Allozyme genotype-environment relationship. I. Variation in natural populations of *Drosophila pachea, Genetics* **73**:135–146.

Rogers, J. S., 1973, Protein polymorphism, genic heterozygosity and divergence in the toads *Bufo cognatus* and *B. speciosus, Copeia,* **1973**:322–330.

Ruddle, F. H., Roderick, T. H., Shows, T. B., Weigl, P. G., Chipman, R. K., and Anderson, P. K., 1969, Measurement of genetic heterogeneity by means of enzyme polymorphisms, *J. Heredity* **60**:321–322.

Sabath, M. D., 1974, Niche breadth and genetic variability in sympatric natural populations of *Drosophila* flies, *Am. Nat.* **108**:533–540.

Saura, A., 1974, Genic variation in Scandinavian populations of *Drosophila bifasciata, Hereditas* **76**:161–172.

Saura, A., Halkka, O., and Lokki, J., 1973a, Enzyme gene heterozygosity in small island populations of *Philaenus spumarius* (L). (Homoptera), *Genetica* **44**:459–473.

Saura, A., Lakovaara, S., Lokki, J., and Lankinen, P., 1973b, Genic variation in central and marginal populations of *Drosophila subobscura, Hereditas* **75**:33–46.

Schopf, T. J., 1974, Survey of genetic differentiation in a coastal zone invertebrate the Ectoprot *Schizoporella errata, Biol. Bull.* **146**:78–87.

Schopf, T. J. M., and Murphy, L. S., 1973, Protein polymorphism of the hybridizing seastars *Asterias forbesi* and *Asterias vulgaris* and implications for their evolution, *Biol. Bull.* **145**:589–597.

Segre, A., Richmond, R. C., and Wiley, R. H., 1970, Isozyme polymorphism in the ruff (Aves, *Philomachus pugnax*) a species with polymorphic plumage, *Comp. Biochem. Physiol.* **36**:589–595.

Selander, R. K., and Hudson, R. O., 1975, Animal population structure under close inbreeding: The land snail *Rumina* in southern France. MS.

Selander, R. K., and Johnson, W. E., 1972, Genetic variation among vertebrate species, *Proc. XVII. Int. Congr. Zool. Monaco.*

Selander, R. K., and Kaufman, D. W., 1973a, Self-fertilization and genetic population structure in a colonizing land snail, *Proc. Natl. Acad. Sci.* **70**:1186–1190.

Selander, R. K., and Kaufman, D. W., 1973b, Genic variability and strategies of adaptation in animals, *Proc. Natl. Acad. Sci.* **70**:1875–1877.

Selander, R. K., and Kaufman, D. W., MS., Genetic population structure in the brown snail (*Helix aspersa*). I. Microgeographic variation.

Selander, R. K., and Yang, S. Y., 1969, Protein polymorphism and genetic heterozygosity in a wild population of the house mouse (*Mus musculus*), *Genetics* **63**:653–667.

Selander, R. K., Hunt, W. G., and Yang, S. Y., 1969, Protein polymorphism and genic heterozygosity in two European subspecies of the house mouse, *Evolution* **23**:379–390.

Selander, R. K., Yang, S. Y., Lewontin, R. C., and Johnson, W. E., 1970, Genetic variation in the horseshoe crab (*Limulus polyphemus*), a phylogenetic "relic," *Evolution* **24**:402–414.

Selander, R. K., Johnson, W. E., and Avise, J. C., 1971, Biochemical population genetics of fiddler crabs (*Uca*), *Biol. Bull.* **141**:402.

Selander, R. K., Smith, M. H., Yang, S. Y., Johnson, W. E., and Gentry, J. B., 1971, Biochemical polymorphism and systematics in the genus *Peromyscus*. I. Variation in the old-field mouse (*Peromyscus polionotus*), *Univ. Texas Publ.* **7103**:49–90.

Selander, R. K., Kaufman, D. W., Baker, R. J., and Williams, S. L., 1974, Genic and chromosomal differentiation in pocket gophers of the *Geomys bursarius* group, *Evolution* **28**:557–564.

Singh, R. S., Hubby, J. L., and Lewontin, R. C., 1974, Molecular heterosis for heat-sensitive enzyme alleles, *Proc. Natl. Acad. Sci.* 71:1808–1810.

Singh, R. S., Hubby, J. L., and Throckmorton, L. H., MS., The study of genic variation by electrophoretic and heat denaturation techniques at the *octanol dehydrogenase* locus in members of the *Drosophila virilis* group.

Smith, M. H., Selander, R. K., and Johnson, W. E., 1973, Biochemical polymorphism and systematics in the genus *Peromyscus*. III. Variation in the Florida deer mouse (*Peromyscus floridanus*) a Pleistocene relic, *J. Mammalogy* 54:1–13.

Snyder, T. P., and Gooch, J. L., 1973, Genetic differentiation in *Littorina saxatilis* (Gastropoda), *Marine Biol.* 22:177–182.

Somero, G. N., and Soulé, M., 1974, Genetic variation in marine fishes as a test of the niche-variation hypothesis, *Nature* 249:670–672.

Soulé, M., and Yang, S. Y., 1973, Genetic variation in side-blotched lizards on islands in the Gulf of California, *Evolution* 27:593–600.

Suomalainen, E., and Saura, A., 1973, Genetic polymorphism and evolution in parthenogenetic animals. I. Polyploid Curculionidae, *Genetics* 74:489–508.

Stone, W. S., Wheeler, M. R., Johnson, F. M., and Kojima, K., 1968, Genetic variation in natural island populations of members of the *Drosophila nasuta* and *Drosophila ananassae* Subgroups, *Proc. Natl. Acad. Sci.* 59:102–109.

Tinkle, D. W., and Selander, R. K., 1973, Age-dependent allozymic variation in a natural population of lizards, *Biochem. Genet.* 8:231–237.

Tracey, M. L., Bellet, N. F., and Graven, C. D., 1975, Genetic variation and population structure in the mussel *Mytilus californianus*. MS.

Tracey, M. L., Nelson, K., Hedgcock, D., Shlesser, R. A., and Pressick, M. L., 1975, Biochemical genetics of lobsters (*Homarus*). I. Genetic variation and the structure of American lobster populations, MS. Submitted to *J. Fish. Res. Bd.*

Triantaphyllidis, C. D., Christodoulou, C., and Beckman, L., 1973, Polymorphism at two esterase loci in *Drosophila melanogaster* from northern Greece, *Hereditas* 74:25–29.

Turner, B. J., 1974, Genetic divergence of Death Valley pupfish species: biochemical versus morphological evidence, *Evolution* 28:281–294.

Vuilleumier, F., and Matteo, M. B., 1972, Esterase polymorphisms in American and European populations of the periwinkle, *Littorina littorea* (Gastropoda), *Experientia* 28:1241–1242.

Wahren, H., and Tegelstrom, H., 1973, Polymorphism of esterases and tetrazolium oxidases in the Roman snail, *Helix pomati*: A study of populations from Sweden and Germany, *Biochem. Genet.* 9:169–174.

Webster, T. P., and Burns, J. M., 1973, Dewlap color variation and electrophoretically detected sibling species in a Haitian lizard, *Anolis brevirostris, Evolution* 27:368–377.

Webster, T. P., Selander, R. K., and Yang, S. Y., 1972, Genetic variability and similarity in the *Anolis* lizards of Bimini, *Evolution* 26:523–535.

Wheeler, L. L., and Selander, R. K., 1972, Genetic variation in populations of the house mouse, *Mus musculus*, in the Hawaiian Islands, *Univ. Texas Publ.* 7213:269–296.

Wilkins, N. P., and Mathers, N. F., 1973, Enzyme polymorphisms in the European oyster, *Ostrea edulis* L. *Anim. Blood Grps. Biochem. Genet.* 4:41–477.

Williams, G. C., Koehn, R. K., and Mitton, J. B., 1973, Genetic differentiation without isolation in the American eel, *Anguilla rostrata, Evolution* 27:192–204.

Yang, S. Y., Wheeler, L. L., and Bock, I. R., 1972, Isozyme variations and phylogenetic relationships in the *Drosophila bipectinata* species complex, *Univ. Texas Publ.* 7213:213–227.

Zouros, E., 1973, Genic differentiation associated with the early stages of speciation in the *Mulleri* subgroup of *Drosophila, Evolution* 27:601–621.

3

Chemical Basis of Mutation

V. N. SOYFER

All-Union Institute of Applied
Molecular Biology and Genetics
Moscow, U.S.S.R.

INTRODUCTION

Mutation theory has become the most thoroughly explored branch of genetics. The induction of mutations by radiation and chemical agents has been established; mutagenesis has been studied in molecular terms; and the types of changes in the genetic code causing mutations have been revealed.

A new chapter begins now in the development of mutation theory. It is based on studies of enzymatic activity involved in the process of mutation production. Several reviews about the origin of mutations and their molecular nature have been published (Krieg, 1963*b*; Freese, 1964; Rieger, 1965; Orgel, 1965; Loveless, 1966; Auerbach, 1967 (ed.), 1969; Bartoshewich, 1966; Drake, 1969; Singer and Fraenkel-Conrat, 1969; Putrament and Baranowska, 1973; Sobels, 1973 and others).

The chemical nature of mutations was scrutinized in the monographs of Soyfer, *The Molecular Mechanisms of Mutagenesis* (1969) and Drake, *The Molecular Basis of Mutation* (1970).

THE FORERUNNERS OF MUTATION STUDIES

Mutation theory originated at the time that Mendel's laws were rediscovered. In 1899–1900, S. Korzhinsky and H. de Vries published the first information about the existence of genetically changed individuals. During the following quarter century, geneticists studied various natural mutants very intensively. Though there were attempts to produce genetically changed forms in the laboratory under the influence of various agents (in particular by T. H. Morgan), none succeeded.

The pioneer in the investigation of induced mutation was the outstanding American geneticist H. J. Muller (1927). However, a study of the literature (Soyfer, 1971a) showed that Muller's work was not the first in this field. The first case of radiation-induced mutation was described in 1925 by G. A. Nadson and G. S. Phillipov. In their experiments fresh fungal cells (18–20 hours after plating) of *Mucor genevensis* and *Zygorynchus molleri* were irradiated by X-rays. The authors found two types of mutants, both of which were morphological changes. The fact that they found only mosaic mutants was and is very interesting. Nadson and Phillipov studied one mutation through 8 generations and another through 13.

I now turn to the history of the investigation of chemical mutagenesis. C. Auerbach and I. A. Rapoport are considered to be the first investigators in this field. The forerunners of Auerbach and Rapoport—E. Baur (1916), V. Saccharov (1932), S. Gershenson (1934), M. Lobashov (1937)—are mentioned much less frequently. So far as I know, the first to cite these forerunners was H. Stubbe in 1937 and A. Gustafsson in 1959. Gustafsson wrote:

> Stubbe paid much attention to the results of Soviet-Russian geneticists (Sacharov and co-workers), who induced mutations in *Drosophila* by means of iodine treatment. These pioneer experiments are of interest in plant genetics, too, since iodine (and chlorine) act as mutagens also in barley and . . . *Oenothera* (Gustafsson, 1960, p. 14).

However, I have found that the first to conduct such investigations was a pupil of G. A. Nadson—Maxim N. Meissel. In 1928, Meissel published the results of 4-year experiments on the induction of morphological mutants of yeast, using chloroform as a mutagenic agent. He produced several types of mutants and managed to increase the frequency of mutations above the level of spontaneous mutation. He studied the mutant clones through 36 generations (Meissel, 1932, 1933) and showed genetical stability of these mutants. He also discussed the nature of induced mutagenesis and came to the conclusion that microbial cells have an hereditary structure. He wrote:

> The formation of these new races which we must regard as mutants is evidently the result of the chloroform action on internal factors, influencing hereditary characters. It is highly probable that this is a case of an effect on nuclear substances of the cell. Previous investigation showed that rather profound changes of the cytoplasm of cells under the influence of drugs are reversible and do not have lasting effects. The smallest change of the nuclear substance, we must believe, produces much more permanent changes in the cells which are transmitted by heredity (Meissel, 1928, p. 258).

This sounds as if it were written in our day. Ten years later, after the publication of Meissel's work, there was a discussion about the role of adaptations and mutations in the world of microbes, about the roles of cytoplasm and nucleus, etc. This discussion ended in complete victory for those who recognized the possibility of inducing mutations in microbes, the role of nucleotides in microbial cells, and the role of DNA molecules in hereditary transmission in microbial organisms. The point of view stated by Meissel (and later by Dobzhansky in *Genetics and the Origin of Species,* 1941, p. 189) as early as 1928 turned out to be true.

MOLECULAR PRINCIPLES OF HEREDITARY CHANGES

Classification of Mutations

The Types of Mutations

After Watson and Crick (1953) and Gamov (1954) propounded their concept of the genetic code, interpreting it as a sequence of bases in the DNA (and RNA) molecules, the study of the molecular nature of mutation became possible. Change in the sequence of nucleotides in the molecules of nucleic acids brings about mutations. The following kinds of changes can be distinguished (Fig. 1):

1. Substitution of one or several bases
2. Nucleotide deletions
3. Insertion of one or several nucleotides followed by restoration of sugar-phosphate bond in DNA
4. Removal of a segment of the polynucleotide chain (extended deletion)
5. A segment may join the DNA molecule in a different site or another molecule (translocation)
6. A segment may join at the site of removal but rotate by 180° (inversion)*

It became possible to give a **rigorous** description of the molecular nature of mutations within the **framework** of this phenomenological theory. Freese (1961) studied initially **only those** mutation types which resulted

---

* The processes of translocations and inversions are usually treated in the genetic literature as mere acts of joining DNA fragments (or chromosomes) to other regions of DNA with a rotation by 180° or preserving the same direction of a sugar-phosphate chain. But spontaneous joining is no doubt impossible, the acts of translocation and inversions should be mediated by special enzymatic systems. In the process of inversion, for example, enzymatic change of terminal nucleotides should occur providing joining of sugar-phosphate bonds for a new site.

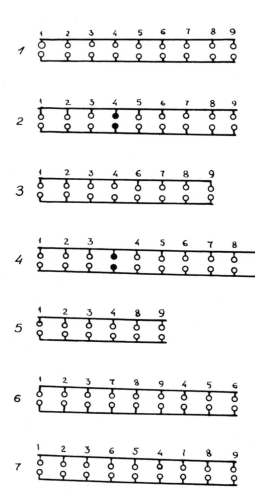

FIG. 1. Structural changes in DNA in molecular terms. Circles signify nucleotides, their order is signified by figures. 1. normal DNA structure; 2. substitution of bases in DNA (base 4 is substituted by base 4 ′, which leads to the substitution of a pair of bases; see black circles); 3. deletion (the fifth pair of nucleotides lost); 4. insertion (an extra pair inserted between the third and fourth pairs of nucleotides); 5. extended deletion (pairs 5, 6, 7 deleted); 6. translocation (regions 4, 5, 6 moved to a new place); 7. inversion (regions 4, 5, 6 rotated by 180°) (from Soyfer, 1970a).

from the substitution of one nucleotide for another; only two types of replacements may occur: a purine (or pyrimidine) nucleotide may be replaced by another purine nucleotide (or pyrimidine nucleotide, respectively); or a purine nucleotide may be replaced by a pyrimidine nucleotide, and pyrimidine nucleotide may be replaced by a purine nucleotide. Freese introduced the term *transition* for the substitution of a purine or pyrimidine nucleotide by a related nucleotide, and *transversion* for the replacement of purine nucleotide by a pyrimidine nucleotide and vice versa.

Later Freese (1963) showed that many chemical mutagens caused replacement in the code of a transversion type. He obtained experimental evidence for some mutagens (e.g., alkylating agents), which he interpreted as supporting the existence of transversions.

Freese's model lacks a class of mutations due to deletions or insertions of individual nucleotides. The existence of such mutations was assumed and then proved by Crick *et al.* (1961), in experiments on the general nature of the genetic code. It was shown that, because of the continuity of the code reading, insertions and deletions of nucleotides will lead to the shift of reading. These mutations were called *frame-shift mutations* or *sign mutations.*

Mutations in Terms of Changes in the Genetic Code

The experiments for the study of the exact structure of all 64 codons (Nirenberg, Matthei, Leder, Ochoa, Khorana, and others) made it possible to treat the phenomenology of mutagenesis in broader terms, distinguishing among three possible types of changes in the code. No matter whether they are caused by transitions, transversions or frame shifts, mutations would fall into:

(a) Missense-mutations, i.e., mutations affecting the codon sense, and finally resulting in substitution of a new amino acid in the coded protein. These mutations may be caused by substitutions of nucleotides bringing about any of 61 sense triplets. As a matter of fact, for the lack of the related transfer RNA in an organism, some codons may remain unidentified during translation turning out to be senseless in the organism.

(b) Nonsense-mutations, when a nonsense codon* arises unrelated to any amino acid and switching the code reading.

(c) The substitution of bases in codons producing *no* changes in the codon sense (within code degeneracy).

This approach to the phenomenology of mutagenesis provides a reasonable account of the consequences of mutational substitutions in the genetic code. However, it is not concerned with the causal aspects of mutagenesis, that is, with molecular mechanisms of origin of changes in the code.

Molecular Basis of Mutational Changes

The theoretical premises adopted by Freese 1959*a*) show that two principal ways of nucleotide substitution must exist. A purine base can be replaced by an alternative purine, and a pyrimidine by another pyrimidine.

---

* As a rule, these are UAG, UAA, and UGA-codons, though as was just mentioned, some other triplets may perform as nonsense codons in some organisms.

The second way is substitution of purine by pyrimidine or a pyrimidine by a purine. As was mentioned above, Freese called the first type of mutation transitions, and the second type transversions.

Transitions are mutations in which a pair GC is replaced by AT or AT by GC:

$$GC \rightarrow AT$$
$$AT \rightarrow GC$$

while transversions are

$$GC \rightarrow CG$$
$$GC \rightarrow TA$$
$$AT \rightarrow TA$$
$$AT \rightarrow CG$$

For some years geneticists supposed that molecular changes in DNA are limited to transitions and transversions only. However, Brenner *et al.* (1961) and Lerman (1961) found another type of molecular change in DNA. One base can be inserted or deleted. This type of mutation they called the frame-shift mutation. Point mutations can be classified according to their ability to change the genetic code. Transitions and transversions, as well as frame-shift mutations, can lead to substitutions of amino acid in proteins (missense mutations) or to formation of nonsense codons (nonsense mutations). In the latter case, the synthesis of a protein is stopped.

Ways of Substitution of Bases in DNA

There are at least two ways of introducing wrong bases into DNA. The mutagen can affect nonreplicating DNA. In this case, the wrong base will be inserted at the time of replication opposite to modified base. This process was recognized as a mistake in the replication (Fig. 2).

Another possibility of point mutation is the insertion of base analogues at the time of the replication of DNA (mistake in pairing or mistake in in-

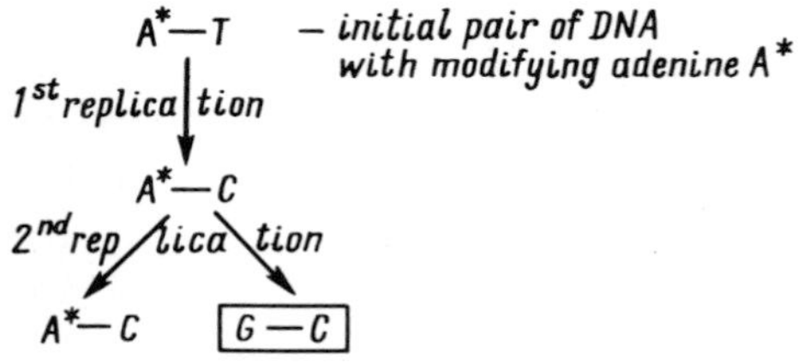

$A^{*}\!\!-\!T$ — initial pair of DNA with modifying adenine A^{*}

1st replication

$A^{*}\!\!-\!C$

2nd replication

$A^{*}\!\!-\!C$ $\boxed{G-C}$

FIG. 2. Induction of transition mutations as a result of mistakes in replication (from Soyfer, 1969).

corporation). In this case, the base analogue involved into the replicating chain of DNA due to wrong pairing of analogue with normal base in parent DNA chain. The synthesized molecule has now an analogue of base in one strand of DNA and a normal base in other chain. At the next replication a wrong base complementary to the analogue will be inserted.

Stabilization of the change will occur in the third act of replication, when a complementary partner will pair with the wrong base. In such a manner, the exchange of one pair of bases for another pair will be completed (Fig. 3).

Transitions

The orientation of purine-pyrimidine is conserved in transitions. We shall consider an example of bromouracil substitution. Bromouracil can pair with adenine and guanine (Lawley and Brookes, 1962):

Br

H—N—H————O

N

N—————H—N   N

N   N   O

Adenine   .   5-Bromouracil

(normal keto-form)

Such pairing with 5-bromouracil in the keto-form does not lead to the mutations. However, bromouracil may be converted sometimes into the enol form, and then it can pair with guanine but not with adenine:

Br

O————H—O

N

N—H————N   N

N   N   N—H————O

H

Guanine   5-Bromouracil

(rare enol-form)

The replacement of one pair by another can happen at the time of replication of DNA, or as a mistake of pairing (Freese, 1959*b*; Rudner, 1961). Transitions will occur in both cases (Fig. 4). In the first case, the AT pair will be replaced by a GC pair, in the second case there will take place a

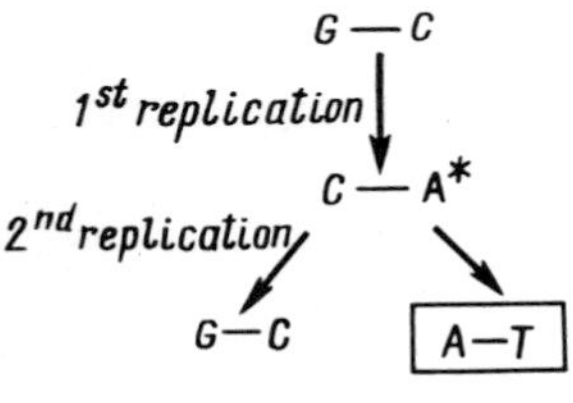

FIG. 3.   Induction of transitions as a result of mistakes of pairing (from Soyfer, 1969).

transition GC → AT. Transition mutations may be induced by treatment with alkylating compounds. According to Krieg (1963a), 7-alkyl-guanine may be paired with thymine:

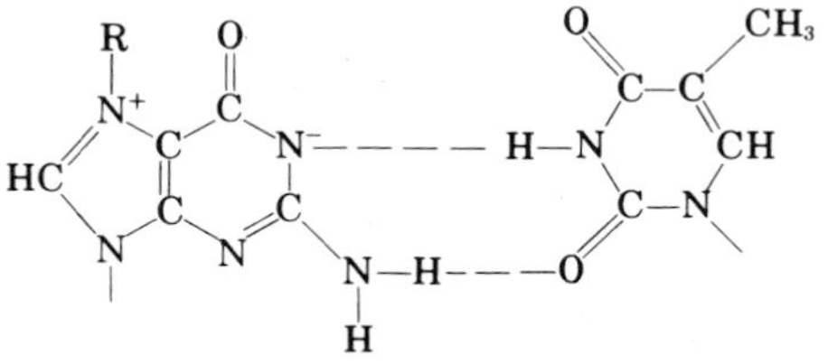

The ionized 7-alkyl-guanine   Thymine

This was confirmed by Lawley and Brookes (Lawley and Brookes, 1961; Brookes and Lawley, 1964). The abnormal tautomeric transition was considered by Nagata et al. (1963). The authors made an attempt to calculate the distribution of π-electrons in guanine and cytidylic acid molecules, and came to the conclusion that alkylguanine may pair with thymine after tautomerization, and alkylated cytosine with adenine:

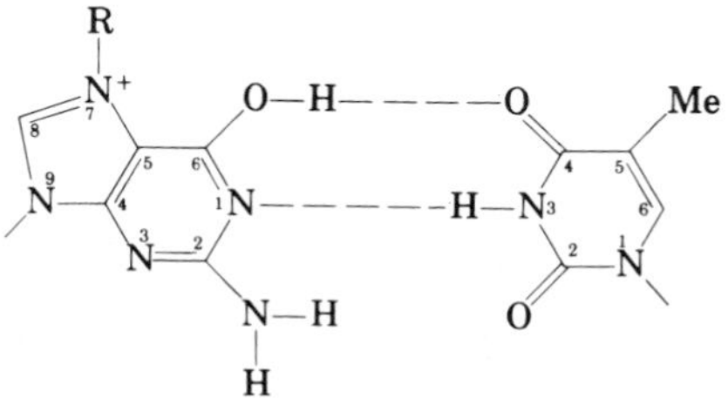

7-alkylguanine   Thymine
(rare tautomeric form)

Transversions

The evidence for transversions was less conclusive than for transitions at the first time. Bautz and Freese (1960) proposed a model for the origin of

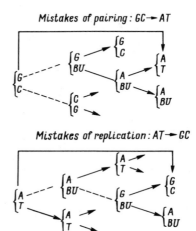

FIG. 4. Induction of transitions in DNA by 5-bromouracil incorporation (from Freese, 1962).

transversions as a result of gaps which can arise, for example, after depurinizations of DNA induced by alkylation or heating. At the next replication pyrimidine as well as purine bases can fill the gap. Incorporation of pyrimidine instead of purine will lead to transversion mutation. According to Freese, a low pH could also lead to transversions; he believed that a majority of spontaneous mutations were caused by transition substitutions.

The attempt to obtain evidence for transversions was made by Bautz and Freese (1960) and Freese (1961). In both cases revertants induced by different mutagens were studied. Only those mutants which were not transitions were taken into consideration. The authors supposed that if a transition appeared, it could not be reversed by mutagens which only induced transitions. When Freese (1961) obtained r-mutants of phage T4 by treatment with a low pH and ethylmethanesulfonate, he tried to revert them by nitrous acid (making transitions) but without success. He concluded that the mutants obtained were transversions.

Freese's conclusions were criticized by Brenner et al. (1961) and also by Krieg (1963a,b). First of all, these authors questioned the validity of tests used by Freese as evidence of transversions; they formulated an hypothesis about origin of point mutations by deletions and insertions of individual bases. Although the process of deletion–insertion undoubtedly takes place (particularly induced by proflavine and its analogues), the results could not be easily distinguished from transversions.

A direct proof of transversion substitutions was obtained in the experiments of the Yanofsky group (Yanofsky et al., 1966). For many years Yanofsky and collaborators were investigating a single enzyme—tryptophane synthetase of E. coli—and moreover the only one polypeptide chain of this enzyme—the A protein. But this limited experimental model gave very significant results. Substitution of amino acids in the A-protein of tryptophane synthetase by point mutations was studied. Yanofsky and co-workers isolated a set of mutants producing abnormal enzymes, in which the separate amino acids in the A-protein were substituted. Knowledge of the genetic code for these amino acids permitted the identification of the molecular changes caused by the mutations. It was found that a majority of spontaneous mutants and mutants induced by ethylmethanesulfonate were transversions, while 2-aminopurine induced the substitutions of a AT pair by a GC pair (transitions).

Mutations Induced by Deletion and Insertion of Bases

The logical foundation for Crick's theory of "deletion–insertion" was a supposition about the nature of genetic code and its disturbances. Crick proceeded from the following postulates:

1. The reading of genetic templates is accomplished by triplets of bases.
2. This reading is realized continuously.
3. The reading begins with a fixed point (zero-point) and proceeds in only one direction.

Thus, if we represent the genetic information in DNA as some sequence of the bases

ABC   ABC   ABC   ABC   ABC   ABC   ABC,

it will be divided into seven sense elements.

How to change this information without changing the bases? Evidently it is possible to accomplish change by inserting one or more bases at any place in the sequence of bases in DNA:

ABC   ABB   CAB   CAB   CAB   CAB   CAB   C
          ↑
         insertion
         of base B

or by deletion of one base:

ABC   ACA   BCA   BCA   BCA   BCA   BCA
         ↓
         deletion
         of base B

This formal scheme explains very well the nature of frame-shift mutations. However, it needed an experimental proof of its validity. This evidence for hypothesis about "deletion–insertion" of bases was obtained in the excellent work of Crick et al. (1961). The authors developed a clever but at the same time very simple method for testing the correctness of their idea. Crick and co-workers already pointed out (Crick et al., 1961) that proflavine may add or remove one or two bases. Now Crick et al. (1961), using the proflavine, obtained the first mutation in rII-region of bacteriophage T4. This mutant was treated repeatedly with proflavine. One phage particle from the progeny returned to initial wild type. The reversion could have been induced by one of two causes: either, according to Freese's theory, by substitution of a normal base by a wrong base or a substitution of a wrong base by a normal base was obtained after the second treatment. In accordance with Crick's theory, the first treatment might have led to insertion or deletion of bases, and the second treatment could have induced the same frame-shift mutation but opposite in sign (suppressor mutation) near the changed site.

Let us examine the second possibility in detail. It is necessary to return to the notation of bases:

ABC ABC ABC ABC ABC ABC ABC

Supposing one base in the template is removed as a result of the first treatment by proflavine ("−" mutation):

ABC ACA BCA BCA BCA BCA BC

(note, that the result will not be changed if we begin with the insertion of base but not with deletion of it).

The reversion to the wild phenotype is possible in this case if after the second treatment near the first mutation there occurs a suppressor mutation of opposite sign (in our example, the insertion of a base, + mutation):

ABC ACA BBC ABC ABC ABC ABC
            ↑
      insertion

As we see, the correct sequence of bases was restored nearly in the whole series except the short region ACA–BBC. If this short region does not drastically change the activity of the protein coded by this gene, then after the second mutation the wild (or more exactly, pseudowild) phenotype will be restored.

The restoration of pseudowild phenotype was found in experiments. It was necessary to know what reason (Freese's or Crick's mechanism) formed the basis of this event.

To answer this question Crick and co-workers isolated many mutants which were either addition (+) or deletion (−) mutants. If they really are + mutants, pseudowild phenotype restoration will be observed after crossing of three + mutants, and combining them in one phage genome (if the code is a triplet code).* This can be illustrated thus:

ABC	ABC	ABC	ABC	ABC	ABC	ABC	The normal sequence
ABC	ABC	ABC	AB*B*	CAB	CAB	CAB	The first + mutation
ABC	AB*A*	CAB	CAB	CAB	CAB	CAB	The second + mutation
ABC	ABC	ABC	*C*AB	CAB	CAB	CAB	The third + mutation
ABC	AB*A*	CAB	CCA	B*B*C	ABC	ABC	The pseudowild phage with three + mutations

Crick and co-workers confirmed the validity of their theory by restoration of the "wild" phenotype after the cross of three + mutants, and also after the cross of three (−) mutants. Thus, they proved that there exists a third type of molecular rearrangement producing point mutations, namely, insertions (additions) and deletions.

At present, many examples of frame-shift mutations are available from the experiments with different organisms. Very interesting results were obtained using the model of his D3052 mutation of the histidinol dehydrogenase gene of *Salmonella*. In the DNA region corresponding to this mutation repetite sequences of $\frac{GC}{CG}$ or $\frac{CC}{GG}$ pairs were found. This observation provided the possibility of proving that various mutagens (ICR-compounds, 2-nitrofluorene, indazole derivatives-hycanthone, lucanthone and others, X-ray irradiation) induced frame-shift mutations (Hartman *et al.*, 1973; Ino and Yourno, 1974; Isono and Yourno, 1974, and others).

THE CHEMICAL PRINCIPLES OF ACTION OF DIFFERENT MUTAGENS

CHEMICAL MUTAGENESIS

The Action of Mutagens on Resting DNA and RNA

Nitrous Acid

Deamination of Bases. Nitrous acid mainly deaminates bases in DNA and RNA leading to mutation production and inactivation (Gierer

---

* By the way, this experiment has proved the triplet nature of the genetic code.

and Mundry, 1958; Granoff, 1961; Freese and Freese, 1961a). Three nucleotides having NH groups may be deaminated (Schuster et al., 1960).

The basic regularities of the reaction of deamination were found by
Schuster and co-workers. They also observed that bases in RNA and DNA
react with nitrous acid as free bases (Schuster, 1960; Vielmetter and
Schuster, 1960). Adenine transforms to hypoxanthine, guanine to xanthine,
cytosine to uracil. Xanthine, as a guanine, pairs with cytosine and therefore
the deamination of guanine does not lead to mutations but has mainly an
inactivating effect (Freese, 1959a; Vielmetter and Schuster, 1960).

Hypoxanthine and uracil pair with cytosine and adenine correspondingly. These products of reaction lead to mutation of the transition type:

The rate of base deamination in DNA decreases in the following order: guanine, cytosine, adenine. Bases in RNA are deaminated at pH 4.2 with approximately the same rate. When the pH changes from 5.0 to 4.2 there is observed a 90-fold increase of the mutation frequency, but the lethal effect increased only 30 times. This result may be considered as one more proof of lethality of guanine deamination, because after the change of pH from 5.0 to 4.2 the rate of deamination of cytosine and adenine was increased in 90 times, but only in 30 times for guanine. However, we cannot exclude deamination of guanine in some cases which will finish with mutation (see, e.g., Tessman *et al.*, 1964). Herriott (1966) postulates that xanthine may sometimes pair with thymine and it will stimulate the appearance of mutation. Besides, guanine can react with nitrous acid differently (Schuster and Wilhelm, 1963). As it was shown by Shapiro (1966) nitrous acid can react with nitrogen in guanine ring converting it into 2-nitroinosine.

In principle, the same data were obtained after treatment with nitrous acid (*in vitro*) of transforming DNA of *Bacillus subtilis* (Anagnostopolous and Crawford, 1961), pneumococcus (Litman and Ephrussi-Taylor, 1959) and after the treatment of T4 phage DNA (Strack *et al.*, 1964) and *Neurospora* (Jha, 1970). The denaturation of transforming DNA of *H. influenzae* increases the mutagenic action of nitrous acid (Strack *et al.*, 1964; Horn and Herriott, 1962).

Baylor and Mahler (1962), comparing the data on mutagenesis of extracellular and intracellular phage T2, came to the conclusion that nitrous acid acted in the same manner *in vitro* as *in vivo*, because the rates of mutation coincided in both cases. The reaction of nitrous acid with DNA subunits obeys first order kinetics.

The conclusion about the character of reaction of nitrous acid with adenine, guanine, and cytosine, leading to the substitution of a GC pair by an AT pair, and an AT pair by a GC pair was completely confirmed in experiments in which the amino acid substitutions in coat protein of tobacco mosaic virus (TMV) (Tsugita, 1962; Wittman, 1962; Wittman and Wittman-Liebold, 1966) were studied. The data obtained by Wittman and Wittman-Liebold (1966) were especially significant. These authors analyzed nearly 200 mutants of TMV. In 55% of all studied mutants the deamination of cytosine and adenine led to substitution of amino acids in coat protein;

25% of all substitutions were lethal; and in 20%, the changed codons coded the same amino acids as in wild type proteins (because of the degeneracy of the genetic code). All changes except one (glutamic acid → aspartic acid) confirmed the accuracy of the assumption that nitrous acid deaminates bases, and as a result, adenine is changed to guanine, and cytosine to uracil.

The evidence of deamination of cytosine to uracil under the action of nitrous acid was obtained in experiments with t-RNA$_{gly}$. The anticodon of glycine t-RNA contains a UCC triplet. If cytosine is deaminated, it is possible to get two anticodons UUC and UCU. The first of these pairs with GAA codon for glutamic acid and the second with AGA codon for arginine. But as joining of amino acid with the deaminated t-RNA will be accomplished as before (amino acid center of t-RNA will be unchanged), then such molecules will insert glycine instead of arginine and glutamic acid. This supposition was checked by Carbon and Curry (1968). The prediction about exchange of UCC anticodon for UCU and UUC codons after deamination of cytosine was confirmed and the authors showed that such t-RNA began to insert glycine instead of arginine and glutamic acid.

The Crosslinking of DNA Chains. Apart from deamination, nitrous acid induces crosslinks in DNA (Geiduschek, 1961). The kinetics crosslinkage of chains of DNA after the nitrous acid treatment was defined by Alberts (1968). The observed kinetics exhibit a single-hit mechanism both in *B. subtilis*, and in the phage λ.

According to the calculations of Geiduschek (Becker *et al.*, 1964), one crosslink takes place for each four acts of deamination. Apparently the reaction of crosslinking of opposite chains of DNA takes place with the help of intermediate products of interaction of bases and nitrous acid; for example, ions of the diazonium $R - N = N^+$ type (Orgel, 1965), arising in one base and reacting with the amino group of the nearest base. Since the efficiency of this reaction depends upon the presence of an amino group of a neighboring base near the diazonium ion, the frequency of crosslinks in the DNA of various organisms will be different (Luzatti, 1962).

Orgel assumed that the appearance of extended deletions in DNA of T4 phage after nitrous acid treatment may be explained by existing crosslinks in DNA which lead to a subsequent inactivation in DNA replication. Extended deletions in DNA after nitrous acid treatment were described by Tessman (1962).

Alberts (1968) and Mulder and Doty (1968) criticized the unqualified statement that the appearance of crosslinks between chains in DNA leads to inactivation. They observed that 6% of transforming molecules of DNA of *B. subtilis* and *H. influenzae* had crosslinks but nevertheless possessed transforming activity. The number of normal λ phages with crosslinks in its DNA in natural conditions has 200-fold fewer crosslinks than transforming DNA, but the frequency and the number of appearing crosslinks in both

DNAs coincides under the treatment with nitrous acid. In addition, if each of the DNA molecules carries even up to six crosslinks, they still have up to 30% of biological activity. Consequently, a complete inactivation of DNA will not be observed in the case of crosslinks.

R may be H or a pentose (deoxyribose or ribose) residue
R' is H or CH_3
X is a nucleophilic radical of amino acid residue

Hydroxylamine and O-methylhydroxylamine

Hydroxylamine (NH_2OH) is a mutagen which is well studied and very often used in practice. O-methylhydroxylamine, or methoxyamine (an oxymethylated derivative of hydroxylamine—NH_2OCH_3), particularly intensively studied in the laboratories of Kochetkov and Budowsky, has been widely used in recent years.

Mutagenic and inactivating action of hydroxylamine and methoxyamine was studied in practically all representatives of the living world—in DNA and RNA—containing viruses and phages (Freese *et al.*, 1961; Schuster, 1961; Tichonenko *et al.*, 1970; Levisohn, 1970; Herrington and Takahachi, 1973; Budowsky *et al.*, 1974), bacteria (Isabaeva *et al.*, 1967), fungi (Malling, 1966a; Hartley, 1970; Malling and DeSerres, 1971; Putrament *et al.*, 1973), tissue culture cells of mammals and man (Somers and Hsu, 1962; Engel *et al.*, 1967; Reznik and Shapiro, 1973) and plants (Cohn, 1964; Reddy *et al.*, 1974).

The reaction of HA with DNA is very complex and depends upon many conditions: concentration of HA (Freese *et al.*, 1961), temperature of reactive mixture, pH, osmotic pressure, oxygen pressure, inhibitors, peroxides, and others (Freese and Freese, 1965; Freese *et al.*, 1966, see also reviews of Kochetkov and Budowsky, 1969 and Singer and Fraenkel-Conrat, 1969).

The basic reactions of high molar solutions of HA and OMHA (near 1 molarity and higher) with different bases in DNA and RNA were described (Schuster, 1961; Brown and Schell, 1965; Budowsky, 1968; Small and Gordon, 1968; Budowsky *et al.*, 1971a,b; Budowsky and Sverdlov, 1972) (see page 136).

The compound V was described by Tichonenko *et al.* (1971) who studied the reaction of S^D phage with methoxyamine. The authors supposed that compound V may interact with proteins.

Compound II is unstable and it converts quickly into 5,6-dihydro-4,6-hydroxyaminocytidine (compound III).

The compound IV may exist also in imino form:

(XII)

Besides reactions with cytosines HA and OMHA may replace amino groups of adenine. This reaction proceeds at maximally at pH 5, and the rate of this reaction in 50–100 times lower than total rate of modification of cytosines (Budowsky *et al.,* 1971a; Budowsky and Sverdlov, 1972).

At last, it should be mentioned that HA react with uridines in RNA and the rate of formation of compound IX is a maximal at pH 8–9.

At pH 4–7 compound IX (in medium without HA) converts quickly into uracil, and in alkaline medium (at pH = 10) compound IX decomposes and yields ribosylurea radicals.

Methoxyamine reacts considerably more slowly with cytosine than does HA, but it does not react with uracil.

The reaction of HA and OMHA with cytosine moieties in polynucleotide structure occurs at a lower rate than with free bases (Brown and Schell, 1965). It was shown (Sklyadneva *et al.,* 1970) that the interaction of OMHA with DNA, packed into phage S^D heads, takes place chiefly in the regions with destroyed secondary structure of DNA. Mutagenesis of bacteriophage by 1 and 5×10^{-3} M hydroxylamine and 5×10^{-3} M *N*-methylhydroxylamine was inhibited by bubbling of nitrogen through phage suspension or by catalase treatment (Chu *et al.,* 1973).

This is how it is accepted that HA induces transitions of GC → AT. The substitutions GC → AT induced by HA were described in experiments with phages, bacteria, and fungi (Phillips and Brown, 1968; Levisohn, 1970; Malling and DeSerres, 1968, and others). Hewlings and Brown (1968) and Grossman (1968) have shown that ATP molecules were inserted by the polymerase into the nascent copolymer (instead of GTP), opposite to each modified base, after the modification by HA of the cytosine residues in the poly C, which served as a template for the polymerase. The authors believed that this transition was the result of the formation of dihydrocytosine derivatives, hydrogen-bonded specifically with adenine and not with guanine.

Recently it was shown that one product of reaction of hydroxylamine with cytidine—*N*-4-hydrocytidine, as well as *N*-4-aminocytosine and *N*-4-amino-deoxycytidine are highly mutagenic to *E. coli* (Salganik *et al.,* 1973) and for phage ϕ80 (Chu *et al.,* 1974). They are incorporated into the DNA.

Although the molecular mechanisms of HA mutagenesis is incompletely understood, it is most probable that the maximal mutagenic activity has compound IV and its imino form (compound XII). Singer and Fraenkel-Conrat (1969) believe that 10% of mutagenicity belongs to compound IV and 90% to compound XII.

At the same time, it is very probable that compound II plays a primary role in inactivating action of HA (e.g., Budowsky and Sverdlov, 1972).

Since HA and OMHA interact with adenine (although this reaction proceeds at a 50–100-fold decreased rate), it is possible that product of reaction with adenine may be also involved in mutagenesis.

Besides, HA promotes the breakage of DNA strands (Freese *et al.*, 1967) and inhibits DNA replication in bacteria treated by HA (Soyfer and Yakovleva, 1974). We do not know the mechanism of this reaction yet, but it may be supposed that the decrease in the viscosity and the sedimentation constant of DNA treated with HA (Bendich *et al.*, 1964) is the result either of the direct or indirect breakage of the sugar-phosphate backbone of DNA, or of the destruction and removal of bases. This may help to explain the observations of many authors on induction of aberrations in mammalian cells (Somers and Hsu, 1962; Borenfreund *et al.*, 1964; Engel *et al.*, 1967).

At the high concentrations of HA any reactions with thymine derivatives, purine bases, and pseudouridilic acid were observed by Freese *et al.*, (1961), when the authors studied the decrease of UV-absorption in the range of 260–280 nm (Fig. 5).

With low concentrations of HA the effect of this agent is complex. Different peroxides and radicals are formed (Freese and Freese, 1965) and it is important that not only cytosines but also deoxythymidyl-, guanyl-, adenosil-monophosphates, and uridylmonophosphates may react with HA at low concentrations of agent. The reactions with dTMP, dGMP, and UMP are inhibited by 0.05 M Na_2PO_2. However, pyrophosphate did not influence the reaction of HA with dCMP and UMP at high concentrations of hydroxylamine.

It is established that the inactivation effect of HA in low concentra-

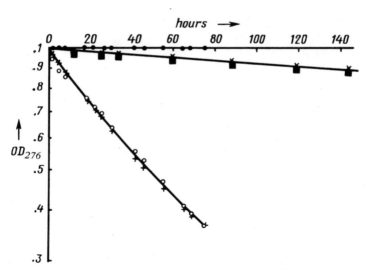

FIG. 5. Decrease of UV absorption at 276 nm of various nucleotides in 0.5 M HA at pH 7.5. ■—dHMP; ×—dHMP – glucose; +—dCMP; ○—UMP; ● —dTMP (from Freese *et al.*, 1961).

tions is stipulated by the radicals mentioned above. The inhibition of peroxide formation (through substitution of O_2 by N_2 or addition of EDTA, pyrophosphate, KCN, $FeCl_3$, catalase, peroxidase) coincided quantitatively with a decrease in the inactivation of transforming DNA (Freese and Freese, 1965) and with mutagenicity of bacteriophage (Chu et al., 1973).

At the same time, the mutagenic effect of 1 M HA solution in microorganisms did not depend on the oxygen, and was not inhibited by pyrophosphate. It depended only on the direct interaction of HA with cytosines. This interaction of HA at the low concentrations with the four bases in DNA can explain the results of Tessman et al. (1965), who described the mutations of all four bases in DNA of the single-stranded phage S13. In their experiments HA was added to the top layer of an agar medium in Petri dishes in a 2.5×10^{-2} M concentration. The authors noted under these conditions the following changes: $G \rightarrow A$, $A \rightarrow G$, $C \rightarrow T$, $T \rightarrow C$. It is interesting to note that in a preceding experiment (Tessman et al., 1964) when phage was treated by 1 M HA in vitro, there was observed only one type of transition: $GC \rightarrow AT$.

At the same time, induction of mutations by HA through reaction with cytosines in nucleic acids, discussed here, may be valid only for microorganisms and especially for viruses. In cells of higher organisms the mechanism of HA-induced mutagenesis may be strictly different from that found in microbes. First of all it should be emphasized that concentration of free HA inside of cells of higher organisms treated by HA is minimal. So Putrament et al. (1973) found that after treatment of yeast cells by 1 M HA during 1 hr the level of free HA within the cells quickly drops from 10^{-2} M to 2.5×10^{-5} M, i.e., decreases about 200–500 times. Such decrease of free HA in cells could be due to the reaction of HA with cellular components or diffusion from the cells.

On the other hand, HA (interacting with different aldehydes and ketones) acts on different enzymatic processes, and their action on cytosine and other bases in DNA may be minimal in comparison with the other effects of HA in cells. It is interesting that inhibition of enzymatic activity by HA as well as induction of chromosomal aberrations in animal and plant cells is observed at such low concentrations of HA as 0.1 M and lower. Thus, the direct reaction of HA with nucleotides in DNA may be negligible compared to other reactions of this agent inside of cells.

Hydrazines

The reaction of hydrazines with oxygen leads to the formation of peroxides which inactivate transforming DNA (Freese et al., 1967; Latarjet et al., 1958; Zamenhof et al., 1953). Therefore it is not surprising that the effects of hydrazine and of its derivatives are increased in the presence of traces of metals, and is inhibited by chelating compounds and by inhibitors of the radicals (Zeller *et al.*, 1963). Bubbling of nitrogen through $\phi80$ phage suspensions or addition of catalase inhibited mutagenic action of hydroxylamine but had no effect on the mutagenicity of 1 M hydrazine and completely inhibited mutagenicity of 5×10^{-3} M hydrazine (Chu *et al.*, 1973). Chromosome aberrations induced by some hydrazines were described in *Vicia faba* (Kihlman, 1956, *Oryza sativa* (Reddy *et al.*, 1974), and in ascites tumor cells of mice (Rurishauer and Bollag, 1963). The induction of gene mutations was studied by Goldfarb and Chubukoff (1963) in phage T2.

Hydrazine (N_2H_4, or diamid) similar to hydroxylamine, reacts mainly with pyrimidine bases. Hydrazine was tested in experiments with free bases in 1924 (Fosse *et al.*, 1924). According to Baron and Brown's experiments (1955), a concentrated solution of hydrazine can lead after the reaction with UMP and CMP to the ring opening of uracil and cytosine, and yield pyrasol derivatives: pyrasol for uracil and 3-aminopyrasol for cytosine and also ribosylurea:

Uracil $\xrightarrow{\text{hydrazine}}$ HNH—CO—NH$_2$ (Urea) + Pyrasol

Cytosine $\xrightarrow{\text{hydrazine}}$ HNH—CO—NH$_2$ (Urea) + 3-Aminopyrasol

Urea can be hydrolytically split from the sugar of ribosophosphate. By treating RNA with anhydrous hydrazine, pyrimidine bases are split off, and riboapyrimidilic acid is formed (Takemura, 1957). Following treatment of DNA (Takemura, 1959) cytosine bases are split off completely, but some thymine bases remain. However, Freese *et al.* (1961), using aqueous solution of hydrazine, arrived at the opposite conclusion. Observing a decrease in the UV absorption (at 276 nm) after treatment with 0.5 M hydrazine (at pH 8.5), the authors noted that the reaction with dCMP and deoxymethylcytidilate was negligible in comparison with reactions with dTMP, UMP, and 5-BDU. However, Brown *et al.* (1966) suggest that hydrazine reacts with cytosine to produce 5,6-dihydrocytosine, which should behave very much like thymine.

Hydrazine, similarly to hydroxylamine, inactivates phages more intensively at a low molarity (0.1 M) than at high concentrations (1 M and more). Its effect, similar to hydroxylamine, increased when pH increased. The mutagenic action of hydrazine in T4 phage was weaker in comparison to HA. Inactivating and mutagenic action of hydrazines is apparently caused by their reactions with thymine bases in DNA.

Methylhydrazine (CH_2NHNH_2) increases the sensitivity of bacteria and mammalian tissue culture cells to the action of X-rays. Using centrifugation in alkaline sucrose gradient for detection of single-strand breaks in DNA, Moroson and Furlan (1969) showed that methylhydrazine induced many single-strand breaks in DNA. It seems that such breaks stimulate chromosome breaks and aberration formation in *Vicia faba* (Gupta and Grover, 1970).

Certain hydrazines are used for treatment of depression and tuberculosis. Taking into account the inactivating and mutagenic activity of hydrazines, Freese *et al.* (1968) tested a set of hydrazines used in medical practice. The results obtained showed that both mono- and disubstituted hydrazines produce H_2O_2 and inactivate the transforming DNA of *Bacillus subtilis*. The strongest inactivation was observed with iproniazed phosphate, phenelzine sulfate, pheniprazine, and benzylhydrazine dihydrochlorid, while nialamide, isocarboxamid, and mebanzin oxalate had less activity. Isoniazid and pivalylbenzhydrazine had the least activity. The nonhydrazin antidepressant meprobamate showed no H_2O_2 production. Of course, on the basis of these experiments it is premature to make any conclusions about using these drugs in practice, since it is not clear how long peroxides induced by drugs exist in the cell. However, the authors were right to note that "nevertheless, it may be preferable to use a non-HZ antidepressant wherever feasible."

Recently isoniazid was used as a mutagen in a host-mediated assay with *Salmonella typhimurium* (Kehler *et al.*, 1973). This agent did not induce point mutation in bacteria or dominant lethal mutations in mice, although hydrazine sulfate was strongly mutagenic for mammals in these experiments.

Urethan and Hydroxyurethan

Urethan was one of the first compounds to be used as a chemical mutagen. Ohlkers (1943) found that urethan induces chromosome breaks and aberrations, such as deletions, translocations, inversions. Mutagenic and inactivating action of urethan has been described in many organisms—plants and insects, and in tissue cultures (see Freese, 1965 for review).

The important property associated with the action of urethan on biological systems is that low urethan concentrations (from 10^{-4} to 10^{-3} M) lead to lethality, or to inhibition of plant development, or to decreasing mitotic index. These low concentrations of urethan induce chromosome breakage and structural rearrangements. At the same time a mutagenic effect (for example, reverse mutations of *Neurospora*) was observed (Jensen et al., 1951) after the treatment of spores with high urethan concentrations (0.1 M).

This situation, similar to the action of hydroxylamine, was investigated in detail by Freese (1965). Having compared the formulas of urethan, hydroxyurethan, and hydroxylamine,

$$
\begin{array}{ll}
\underset{H}{\overset{H}{{>}}}N-\overset{\overset{\textstyle O}{\|}}{C}OC_2H_5 & \text{Urethan} \\[3ex]
\underset{HO}{\overset{H}{{>}}}N-\overset{\overset{\textstyle O}{\|}}{C}OC_2H_5 & \text{Hydroxyurethan} \\[3ex]
\underset{H}{\overset{H}{{>}}}N-\overset{\overset{\textstyle O}{\|}}{C}-NH_2 & \text{Urea} \\[3ex]
\underset{HO}{\overset{H}{{>}}}N-\overset{\overset{\textstyle O}{\|}}{C}-NH_2 & \text{Hydroxyurea} \\[3ex]
\underset{HO}{\overset{H}{{>}}}N-H & \text{Hydroxylamine} \\[3ex]
\underset{HO}{\overset{H}{{>}}}N-CH_3 & \text{N-Methyl-hydroxylamine} \\[3ex]
\underset{HO}{\overset{H}{{>}}}N-OCH_3 & \text{O-Methyl-hydroxylamine}
\end{array}
$$

Freese noted that mutagenic action must be connected with hydroxyurethan but not with urethan. Having examined this in experiments with the transforming DNA of *B. subtilis*, she confirmed the validity of her supposition. The curves for the induction of fluorescent mutations are shown in Fig. 6. One can see that the frequency of mutations decreases along with decreasing concentration of the agent. An increase in mutation frequency was observed in the pH range from 4.2 to 6.2, while a decrease was observed in the pH range from 7.5 to 9.0. At low concentrations the mutagenic effect of hydroxyurethan was changed to inactivation. Freese believes that the older data on the mutation induction with urethan can be explained by the activity of the hydroxy-derivative of urethan but not by urethan itself. She sees a verification of this hypothesis in the data of Miller *et al.* (1960), who demonstrated that amines are enzymatically transformed in higher organisms into hydroxylamine derivatives, and particularly in the data of Boyland and Nery (1965), who observed the change of urethan to *N*-hydroxyurethan. The chemical and functional similarities of hydroxylamine and hydroxyurethan are, in Freese's opinion, sufficient reason for accepting the validity of view that hydroxyurethan induces point mutations by transitions of $GC \rightarrow AT$.

At the end of this section on the activity of agents, containing or producing free NOH groups in the cell such as hydroxylamine, *N*-Methylhydroxylamine, hydroxyurea, hydroxy-urethan, and hydrazines, I must emphasize that the nature of the action of these agents is still insufficiently known, in spite of the many studies in this field.

Freese *et al.* (1967) emphasized that all these agents cause their lethal action mainly by peroxide formation. They suppose that those agents which contain NH_2-groups instead of NOH groups (urea, urethan) do not inactivate DNA. It is quite possible that this opinion is correct. I believe that the fact discovered by Rozenkranz *et al.* (1971) is very important. They describe the sharp inactivating action of formamidoxyme ($H_2N-CH=NOH$) on DNA which differs from urea by only one additional NOH group.

The Mutagenic Action of Low pH

The exposure of T4 phage to an acid medium (from pH 4.2 to 5.0, at 37° to 54°C) leads to an increased frequency of mutations (Freese, 1959a; Strack *et al.*, 1964). Freese looked for an explanation of this effect in the depurinization of the DNA bases, followed by repair rebuilding of the gaps by any of the four bases. The depurinization under the influence of the acidification was described by Tamm *et al.* (1952). In their experiments at pH

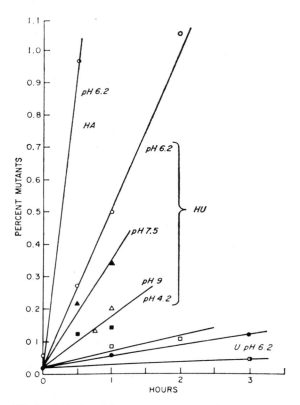

FIG. 6. Induction of fluorescent mutants by hydroxyurethane, urethane, ammonium chloride, and HA HU — 1 M: pH 6.2; ▲ pH 7.5; △ pH 9; ■ pH 4.2 HU — 10⁻² M: □ pH 6.2; HA — 1M: ◐; U — IM: ● NH Cl — IM: ◪ (Freese, 1965).

1.6 to 4.0, guanine and adenine were split off almost completely, and guanine was split off at a higher rate than adenine.

If a gap is not repaired before the replication of DNA, then four possible events can occur in the course of replication (Freese, 1961), three of which will lead to mutations (Fig. 7). According to the scheme of Freese, two-thirds of all mutations induced by low pH should be transversions. However, the results of his own experiments (Freese 1959a; 1961) did not confirm this supposition. Thus, 77% of all induced mutants were transitions and not transversions. It is impossible to explain this result by supposing that the acidification which leads to the formation of apurinic acid is accompanied by cleavage of the sugar-phosphate bonds. Such cleavage was

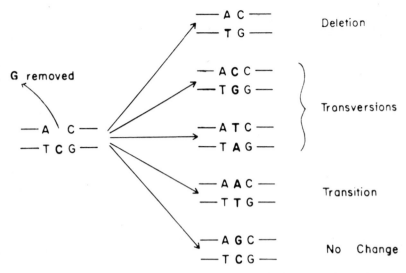

FIG. 7. Hypothetical mechanism by which the removal of guanine from DNA may occasionally produce a point mutation if replication occurs across the gap (from Freese, 1966).

described by Tamm *et al.* (1953) but, of course, it may lead to inactivation rather than to mutagenic consequences.

After the work of Shapiro and Klein (1966) it became possible to explain the high frequency of transitions appearing in DNA under the influence of acid conditions. These authors have described deamination of cytidine with acid buffer. According to their data the deamination reaction at pH 6.0 (although it has a low rate at 37°) can be represented as

As the substitution of a cytosine base by an uracil leads to transitions, the results obtained by Shapiro and Klein can explain the data obtained in the experiment by Freese. Finally, it must be mentioned that acidification can induce DNA denaturation (Marmur *et al.,* 1961). This offers a new possibility for the action of intracellular endogenous mutagenic compounds on DNA. However, first, this effect is manifested only at pH higher than 4.0 and, second, it will have more lethal than mutagenic action. It is possible that at low pH crosslinks can occur in DNA (Freese and Cashel, 1964).

The Mutagenic Action of High Temperatures

The increase of mutation frequency at high temperatures was noted in experiments with *Drosophila* (Plough, 1941) and *Tradescantia* (Sax, 1941). The mutagenic action of heat for microorganisms was described in the laboratory of Zamenhof (Zamenhof and Greer, 1958). *E. coli* cells were heated to 60°C at the neutral pH. In a subsequent study, the authors correlated the mutagenic effect of heat with the depurinization of DNA (Greer and Zamenhof, 1962).

Shapiro and Klein (1966) showed that when cytosine is heated in acetate or phosphate buffer at 50°C (at pH 4.5) for 19 h, a new spot appears on chromatograms. This spot may represent uracil. The appearance of a small quantity of uridine was observed also during heating of cytidine in distilled water or in sodium acetate at pH 7.2. The authors concluded that the mutagenesis described by Zamenhof and Greer may be explained by deamination of cytosine. Starting with his hypothesis of the mutagenic action of heating, Zamenhof (1967) proposed a new method for obtaining mutants of *B. subtilis.* Spores of *B. subtilis* were heated at 110°C after which the frequency of mutations of some genes increased.

Alkylating Agents

To generalize the data on the action of alkylating compounds on hereditary molecules is rather difficult. A large amount of data has accumulated on this subject, but it is rather heterogeneous because of the variety of objects (organisms, isolated DNA, free bases and nucleotides, etc.) and because of important differences in the experimental conditions.

An important review of the action of alkylating agents on different organisms has appeared (Loveless, 1966). In more detail, the molecular mechanisms of mutagenesis after the action of alkylating agents was surveyed by Freese (1964), Orgel (1965), Bartoshevich (1966), and Safonova (1973). The cytotoxic action of alkylating agents was discussed by Wheeler (1962). A short review of mutagenicity of nitro compound will be discussed in a separate section of this chapter.

The Types of Alkylating Compounds. In 1939, Rapoport established the extremely high mutagenic activity of ethylemimine. This work was published only after World War II (Rapoport, 1948*a*). In 1944, Auerbach and Robson discovered mutagenic action of another type of alkylating agent—mustard gas.

These pioneer works initiated a wide range of investigations which culminated in the discovery of the mutagenic action of dialkylsulfates (Rapoport, 1947), epoxides (Rapoport, 1948*a*), diazocompounds (Rapoport, 1948*b*), alkylalkanesulfonates (Bautz and Freese, 1960), β-propiolacton (Ross, 1962), and others. Detailed lists of compounds having alkylating activity are cited in the reviews of Freese (1964), Loveless (1966), and Bartoshevich (1966). A full analysis of the alkylating reactions (i.e., the addition of alkyl groups—CH_3, C_2H_5, and so forth, and some metals into the DNA structure) is hardly possible now because of the insufficient study of this problem. Nucleophile addition of alkyl groups may involve different radicals in DNA: phosphate groups, purine and pyrimidine bases. The alkylation of histones is genetically important, although it is poorly studied: this process may manifest itself in modification of genetic activity.

Some alkylating agents are listed below.

$$S \underset{CH_2-CH_2-Cl}{\overset{CH_2-CH_2-Cl}{<}}$$

Mustard (Di(2-chloroethyl) sulfide

$$CH_3-N \underset{CH_2-CH_2-Cl}{\overset{CH_2-CH_2-Cl}{<}}$$

Nitrogen mustard (Di-(2-chloroethyl) methylamine

$$CH_3-CH_2-O-\overset{O}{\underset{O}{\overset{\|}{\underset{\|}{S}}}}-O-CH_2-CH_3$$

Dialkylsulfate (Diethylsulfate)

$$CH_3-CH_2-O-\overset{O}{\underset{O}{\overset{\|}{\underset{\|}{S}}}}-CH_2-CH_3$$

Alkylalkanesulfonates (Ethyl-methanesulfonate)

Triethylenemelamine (TEM)

$$\underset{O=\underset{}{\overset{}{\bigcirc}}=O}{\overset{\overset{H}{\underset{}{\diagdown}}\,\underset{}{\overset{H}{\diagup}}}{N-N}}$$ Maleic hydrazide

$$\underset{O}{H_2C\!\!-\!\!CHR}$$ Epoxides

$$\underset{\underset{R}{|}}{\underset{N}{H_2C\!\!-\!\!CH_2}}$$ Ethyleneimine

$$\underset{O\!\!-\!\!CO}{H_2C\!\!-\!\!CH_2}$$ β-propiolactone

$^+N{=}N{-}R$ Diazocompounds (R-alkyl groups)

Reactions Due to Alkylation of Nucleic Acids. Only some alkylating agents have been studied in detail: in particular—dimethylsulfate (DMS), diethylsulfate (DES), methylmethanesulfonate (MMS), ethylmethanesulfonate (EMS), ethylenimine (EI) and its derivatives, alkylated derivatives of acridines, and some nitroso compounds. These achievements became possible primarily due to the work of two groups—laboratories of London Chester Beatty Research Institute (O. G. Fahmy, M. J. Fahmy, Lawley, Brookes, Loveless, and others), and Institute of Physical Chemistry of USSR Academy of Sciences (Moscow), headed by Professor I. A. Rapoport.

The alkylation reaction may be realized either by substitution of hydrogen atoms:

$$R{-}H + CH_3{-}CH_2{-}R_1 \rightarrow R{-}CH_2{-}CH_3 + R_1H$$

or by addition of an alkyl radical (e.g., an amino-alkyl group to a tertiary nitrogen atom of base ring, with the formation of a quaternary ion):

$$NR_3 + \underset{\underset{H}{|}}{\underset{N}{H_2C\!\!-\!\!CH_2}} \rightarrow NH_2{-}CH_2{-}CH_2N^+R$$

The rate of reaction is determined by the nature of nucleophilic groups and the concentration of hydrogen ions. In biological materials the degree of alkylation of separate atoms decreases in the following way (Ross, 1962):

$$S > N > O$$

TABLE I.   Reactivity of Different Groups in DNA

Functional group	pK	Possible alkylation at pH 7.5
The primary phosphoryl group	2.0	0.9999
The secondary phosphoryl group	6.0	0.96
The hydroxyl of aromatic type (uracil, thymine, guanine)	10.2	0.02
The amino group of aromatic type		
guanine	2.3	0.9999
adenine, cytosine	3.9	0.999
The hydroxyl of pentose	13.0	10^{-5}

The ionization of functional groups promotes their more active alkylation.

The study of alkylation of transforming DNA with various agents (Reiner and Zamenhof, 1957; Brookes and Lawley, 1960; 1961) opened the possibility of estimating the reactivity of different groups in DNA (according to Bartoshevich, 1966):

As we see from Table I phosphates, the nitrogen atoms of purine rings, and amino groups of purines have more reactivity.

Having used ^{14}C-labeled EMS, Lawley and Brookes (1963) established that the nitrogen atoms in bases in DNA and RNA are situated as follows:

in DNA: N-7-guanine > N-3-adenine > N-1-adenine > N-1-cytosine
in RNA: N-7-guanine > N-1-adenine > N-1-cytosine > N-3-adenine

In addition, the alkylation of purine bases in DNA molecules occurs more actively than that of free nucleotides and nucleosides (Lawley, 1957a), apparently as a result of the transfer of the alkyl radical from phosphate groups to nitrogen atoms of bases (mainly to N-7-guanine).

The Alkylation of Phosphate Groups.   The primary reaction with phosphate groups was described by Elmore *et al.* (1948); Reiner and Zamenhof (1957); Alexander (1952), Alexander and Stacey (1958); Lett *et al.* (1962).

As a result of reaction the phosphate triester is formed, where R_1 is the deoxyribose residue to which phosphate group is added:

$$
\begin{array}{ccc}
R_1 & & R_1 \\
| & & | \\
O & & O \\
| & & | \\
O{=}P{-}O^- + CH_3{-}CH_2{-}R_2 & \rightarrow & O{=}P{-}O{-}CH_2{-}CH_2{-}R_2 \\
| & & | \\
O & & O \\
| & & | \\
R_1 & & R_1
\end{array}
$$

The phosphate triester is unstable and as a rule is hydrolyzed either in a newly formed bond (the "a" reaction) with an alkyl radical splitting off and with the subsequent restoration of an unbroken DNA structure (Alexander and Stacey, 1958), or in an ester bond between carbohydrate and phosphate (the "b" reaction):

Futhermore, sometimes alkyl radical may be transferred to the N-7 atom of guanine (Lett *et al.,* 1962). The reaction of phosphate groups occurs more intensively in comparison with the reactions of alkylation of other functional groups in DNA (Alexander, 1952; Reiner and Zamenhof, 1957; Stacey *et al.,* 1958). Freese (1964) believes that if there are many unhydrolyzed alkyl groups in DNA at the moment of the beginning of replication, then they will repress the DNA replication process. The breakage of a bond between carbohydrate and phosphate may most likely become lethal, but it can also lead to structural rearrangements.

The Reactions of Base Alkylation. The different alkylated derivatives of purine bases and cytosine: 7-alkyl-guanine, 7-methyl-, and 7-ethylguanine (Reiner and Zamenhof, 1957; Bautz and Freese, 1960), 3-methyladenine, 1-methyladenine, and 1,3-dimethyladenine and cytosine derivatives also were described (Brookes and Lawley, 1960). The alkylation in N-7 atom of guanine occurs more actively as compared with other alkylation reactions (Reiner and Zamenhof, 1957; Loveless, 1959). Consider again the conclusion of Lawley and Brookes that the degree of alkylation falls in the following order: N-7-guanine > N-3-adenine > N-1-adenine > N-1-cytosine. Look at the picture below of intramolecular localization of these changeable atoms. We may come to the conclusion that mainly free nitrogen atoms, not included in the formation of hydrogen bonds and localized near phosphate groups (being outside in the twisted double-helix DNA), are subjected to the alkylation.

The atoms, subjected to alkylation, are circled in the scheme. It should be noted that the planes of bases in DNA are perpendicular to the axis of an ordered double-stranded molecule of DNA and therefore N-7-guanine atom and N-3-adenine atom project outside. This circumstance may condition the highest reaction with the mutagen.

The alkylation of guanine in the 7th nitrogen atom may lead to the following consequences: Since the existence of a quaternary N-7-nitrogen atom in the system of conjugated double bonds will lead to a thermodynamically unstable state, it will be the cause of breakage of one of the following bonds (Fig. 8):

(a) Splitting off the alkyl radical with restoration of the initial molecular configuration (reaction "a," the transition of structure I into structure II). The half-life of an alkylated deoxyguanine ribonucleotide equals about 20 hr at pH 7.0 (Lawley, 1957b).

(b) The cleavage of one of the double bonds in the ring (reactions b′ and b″ of structures I and II correspondingly) with the formation of a carbonium ion (structure III). The carbonium ion, interacting with water, forms the compound IV which partially lost the cyclic form.

(c) Bond cleavage between sugar (reaction "c," structure II) and the

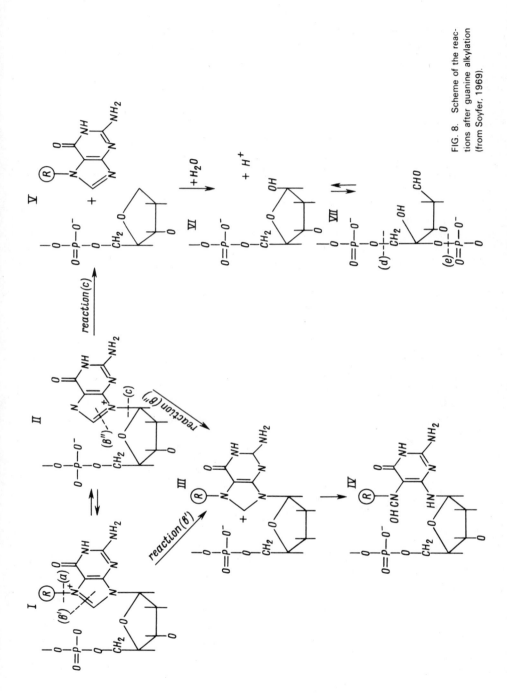

FIG. 8. Scheme of the reactions after guanine alkylation (from Soyfer, 1969).

base with the formation of 7-ethylguanine and deoxyribose (structure V).
The splitting off of alkylated purines from alkylated DNA (but not RNA)
was investigated by Lawley and Brookes (1963); 3-methyladenine is split off
quicker than 7-methylguanine. The furanose ring of deoxyribose ion (structure VI), reacting with water is broken with the formation of structure VII.

(d) The deoxyribose ion may be hydrolyzed with the scission of the
chain of DNA (reactions "d" and "e" of structure VII).

The reactions of other bases with alkylating compounds were studied in
less detail. Only the end products of alkylation are known, but schemes of
reactions and further fate of the alkylated products appearing in the DNA
structure are not clear.

Adenylic acid is alkylated with DMS at pH 7.0 in aqueous solution
with the formation of N-1 and N-3 derivatives (Brookes and Lawley, 1960):

1-Methyladenine   3-Methyladenine

However, Haynes et al. (1964) consider that only the nitrogen atom in the
1-position is subjected to methylation.

The methylation of cytosine by dimethylsulfate, dissolved in dimethylformamide at 37°C and 100°C, leads to the formation of the 1-methyl and
dimethyl derivatives of cytosine, transforming into methyluracil derivatives
later:

The alkylation of cytidine with diazomethane leads to the formation of
1-methylcytidine (Haynes et al., 1964).

Uridine and its derivatives are alkylated at the 1-nitrogen atom:

Ribose

The Alkylation Reactions Leading to Lethal Consequences.   Having considered the chemistry of reactions of alkylation, we can now review the biological consequences of alkylation and first of all of inactivation. It is true that the cleavage of the sugar-phosphate backbone will lead mainly to lethal consequences (and only rarely will induce structural rearrangements), while the alkylation of bases and depurinization may lead to lethal and mutagenic results.

The methylating agents.   The treatment of bacteria with EMS and DMS leads to rapid methylation of purine bases and phosphates in DNA (Reiner and Zamenhof, 1957). The methylation results in significant depolymerization of DNA at the expense of the hydrolysis of the sugar-phosphate backbone of the molecule (Lett *et al.,* 1962), accompanied by a sharp rise of viscosity. Loveless (1959) has demonstrated that removal of MMS activity by dilution, or by addition to the medium of sodium thiosulfate, does not stop phage inactivation. This effect may be connected with DNA hydrolysis although it is impossible to exclude depurinization of DNA. The removal of 3-methyladenine from T4 phage DNA was observed by Lawley and Brookes (1963). The methylating agents inactivate phages and bacteria more intensively than do the ethylating compounds, but the former have less expressed mutagenic properties.

The ethylating agents.   EMS, EES, EI, and other ethylating compounds have less expressed inactivating properties than methylating agents (Freese, 1964). The reactions of scission of the backbone of DNA are observed rarely (Lett *et al.,* 1962). This conclusion is connected with the observation that the esterification of phosphate groups by means of ethylating and methylating compounds occurs with the same rate. However, the former agents induce the mutations more frequently than do the latter (Alexander and Stacey, 1958).

Bifunctional agents.   The bifunctional compounds (such as nitrogen and sulfur mustards) have a clearly expressed cytotoxic action, much greater than single-armed compounds (Ross, 1962) have. The supposition of Goldacre *et al.* (1949) that the bifunctional compounds link two chains of DNA was confirmed by Brookes and Lawley (1961) and Lawley and

Brookes (1963), who discovered that a part of the mustard molecules is involved in the diguanyl-mustard compound. Interstrand crosslinks in DNA induced by mustard treatment have been described by Geiduschek (1961), Kohn et al. (1965), Rutman et al. (1969). It was shown that diepoxybutane (Verby et al., 1971) also induces interstrand links. The reaction between guanines in neighboring chains in DNA and the mustards (in particular, the sulfur mustard $S(CH_2CH_2Cl)_2$) may be imagined as shown:

Chromosome aberrations in *Vicia faba* root tips were induced by a bifunctional alkylating agent maleic hydrazide (Swietlinska and Zuk, 1974).

Now we should emphasize that the dark repair of damage induced by bifunctional compounds such as mustards, which was described in the literature (Papirmeister and Davison, 1964; Kohn et al., 1965; Hanawalt and Haynes, 1965), may be explained only with difficulty because such repair needs consecutive excision-resynthesis first from one strand of DNA and only later from the other. However, this repair has been proved convincingly, and evidently such repair can explain the nonlethality of crosslinks in transforming DNA (Freese and Cashel, 1964).

Trifunctional agents. Trifunctional compounds are studied only poorly. As shown, they have mutagenic and inactivating action (Michaelis et al., 1962; Rieger and Michaelis, 1964; Akifiev et al., 1965; Kimball, 1965). So, triethylene melamine (TEM) induces point mutations (Kimball, 1965) apparently at the expense of alkylating of guanine. Besides, it was established (Doscočil, 1965; Lawley, Brookes, 1967a,b) that TEM induced interstrand links. Doscočil (1965) established formation of diguanyl derivative of triethylenemelamine. It was observed that trifunctional ethyleneimino derivate—Trenimon—is a very potent mutagen. In experiments with human lymphocytes the threshold value of this mutagen was estimated (Kaufmann et al., 1973). This examination revealed that a concentration of 2.8×10^{-10} M was minimal for chromosome aberrations. At higher concentrations the agent caused a marked dose-dependent increase of chromosome aberrations.

The Alkylation Reactions Leading to Mutation Production. The consequences of alkylation leading to mutation production are shown in Figs. 9 and 10. One may suppose that alkylation of bases will lead to point mutations mainly because (1) alkylating bases will have the possibility of pairing with a wrong partner and (2) after depurinization a gap will be left in one strand of DNA in which only one base may be inserted (if only this gap would not be enlarged by exonucleases). The first supposition on the appearance of tautomeric forms under the alkylation of guanines was studied and confirmed by some authors (Freese, 1959b; Lawley and Brookes, 1961; Nagata et al., 1963). The reaction of such a kind will lead only to transitional substitutions (Freese, 1963).

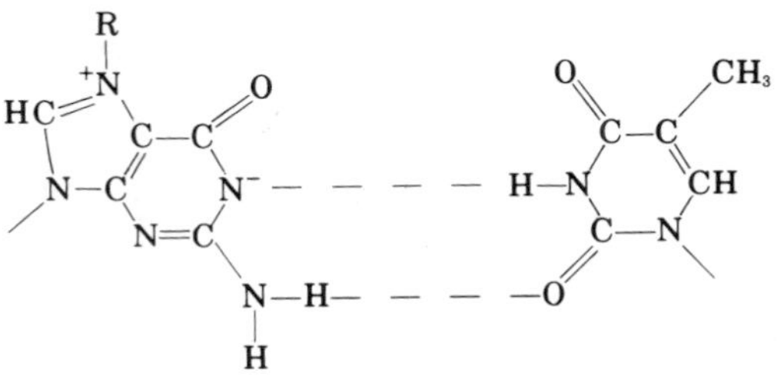

The conclusive evidence of validity of this idea was obtained in the experiments of Green and Krieg (1962). The basic advantage of their work was that they have studied the progeny of an individual bacterium infected by phages (so-called single-burst experiments) and not simply the effects of the mutagen on populations.

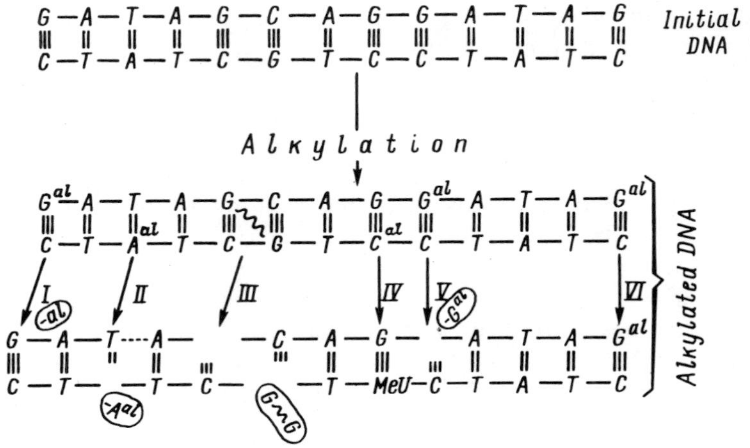

FIG. 9. The mutagenic consequences of alkylation of double-stranded DNA in the absence of replication. G^{al}, A^{al} and C^{al}—alkylated bases; G~G—crosslinks between guanines; MeU—methyluracil. I—the restoration of initial base pair after splitting off of alkyl-group; II—single or double-strand break after depurinization; III—double-strand break after excision; IV—formation of MeU from methylcytosine; V—the formation of a gap after depurinization; VI—the formation of alkylated guanine (from Soyfer, 1969).

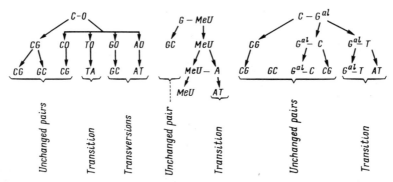

FIG. 10.   The consequences of alkylation of bases in DNA after its replication. O—gap in DNA after depurinization. See for other indications the legend to Fig. 9 (from Soyfer, 1969).

In this experiment the suspension of infected and treated bacterial cells was dissolved so that in each tube there was, on the average, a single infected cell. In this case all plaques forming from each probe (after plating the content of each tube on a Petri dish) will be formed by progeny of only one phage particle, propagating in one bacterium.

Utilizing this method, Green and Krieg first showed that 7-alkylguanine in the structure of phage DNA is a long-living compound. As a consequence, it was possible to prove (Green and Krieg 1962; Krieg 1963a) that alkylated DNA is heterozygous: one of its strands is normal and after replication, yields wild-type phage particles, and the second strand is mutated and yields either mutant or inactivated DNA. In accordance with this, 1% of the mutants was found with inactivation as low as 50%.

Krieg (1963a) tried to explain also why after the influence of EMS on phage, the inactivation curve has double-component character. He believed that inactivation occurs mainly at the expense of depurinization and therefore two reactions must be realized: alkylation of bases and destruction of DNA due to depurinization.

It is evident that depurinization can result not only in inactivation but in mutation as well. Clearly expressed mutagenic effects of depurinization of DNA had been described by many authors (Lawley and Brookes, 1961; Freese, 1961; Ross, 1962; Krieg, 1963a; Verly, 1974). Splitting off of purine occurs even at neutral pH. The loss of bases with the formation of gaps in the template can lead to insertion of any bases into these gaps during the next replication. This situation is analogous to mutagenesis resulting from acidification and heating, and, as it takes place in the latter case, it can lead to the transitions and transversions. If depurinization as a result of alkylation is similar to depurinization after the acidification and the heating, then both kinds of treatment will lead to one and the same set of mutations.

Having checked this hypothesis experimentally, Freese (1961) confirmed it. The appearance of transitions and transversions was recorded in both cases.

The occurrence of gaps in the sequence of bases in DNA, with preservation of unbroken sugar-phosphate backbone, makes possible mutations without replication of DNA. An increased frequency of mutations was found after incubation of an extracellular phage in a medium with alkylating agents (Bautz and Freese, 1960) and also in experiments with bacteria (Strauss, 1962). Although Alikhanian and Mkrtumian (1964) decided that experiments with extracellular phages can easily prove the existence of process of mutation without replication, there is no doubt that Alikhanian's argument was erroneous.

Besides the point mutations, the alkylating compounds may induce chromosomal rearrangements (Rieger and Michaelis, 1965; Jenkins, 1967) or mutants of *Drosophila*. Some small deletions were observed in the experiments with *Drosophila* (Fahmy and Fahmy, 1959). It is interesting that the bifunctional compounds have proved more active in this respect than their single-armed analogues. The converse dependence has been noted in experiments with phages (Löbbecke, 1963).

Thus far we have dealt only with 7-alkylguanines and their part in the mutagenesis. However, not only 7-alkylguanines lead to mutations. Krieg (1963a) believes that the alkylation of adenine can also lead to mutations. According to the data of Tessman et al. (1964), EMS interacted not only with three, as expected, but with all four bases of the single-stranded DNA of the phage S13, and all alkylated bases gave rise to mutations. Because the highest frequency of mutations was associated with alkylation of one purine (7-alkylguanine) and one pyrimidine (cytosine), mutations which arose owing to alkylation of other bases were masked. This point of view is evidently correct, for many studies showed that although the main portion of the alkylations is connected with 7th atom of guanine, other bases are also subject to alkylation.

It should be mentioned that monofunctional alkylating agents are more active mutagens than are bifunctional compounds and the former less toxic for cells and multicellular organisms (Fahmy and Fahmy, 1957; Loveless, 1959; Freese and Freese, 1961b; Sree, 1970) than the bifunctional compounds. If we speak of monofunctional compounds, we have to note, as a rule, much less mutagenic activity of methylating agent as compared with ethylating ones (Westergaard, 1957; Strauss, 1962). Both types of agents can cause transitions by alkylation of bases (mainly GC → AT), and also the transitions and transversions by depurinization. The reactions leading to inactivation and mutation owing to the action of alkylating agents depend to a great extent not only on the concentration of substance and the property of solvent (Bhatia, 1967), but on pH (Wagner et al., 1968) and intracellular enzyme activity (Kimball, 1965; Corran, 1968) and also on the anionic residue reaction *in vivo* and *in vitro*. The mutagenic and inactivating property of various methylating and propylating (e.g., propyliodide) agents in transforming DNA *in vitro* and *in vivo* has been studied in *B. subtilis* by Zamenhof and Arikawa (1970). The authors showed that "at the same molarity ethylating agents are less active than methylating, in all the above phenomena; at the same mutagenicity, ethylating agents may be less active (ethyl methanesulfonate) or more active (diethylsulfate) than methylating (dimethylsulfate, V.S.) in cell killing, depending on anionic residue (p. 141)." There is an indication that to some extent molecular events leading to mutations are different in bacteria and phages (Strauss, 1961).

The reaction of alkylation may take place also with the help of mitomycin C (Iyer and Szybalski, 1964), its derivatives, and with porfiromycin (Weissbach and Lisio, 1965). The latter can alkylate not only DNA, but also RNA and polysaccharides. Weissbach and Lisio found, in contradiction to Iyer and Szybalski, that mitomycin did not link the two strands of DNA, but acted as a monofunctional agent leading to alkylation of only one strand. For every 500 bases in DNA, one base turns out to be alkylated.

To conclude, it should be emphasized that the results of recent experiments support the hypothesis that the level of mutability (as well as inactivation) induced by alkylating agents is influenced by many factors and first of all by the capability of repair (Howell-Saxton *et al.*, 1973; Zadrazil *et al.*, 1973, 1974), a physiological state of cells, and additive action of some compounds (e.g., caffeine; see Roberts and Ward, 1973).

Induction of Mutations at the Time of Replication and Recombination of Nucleic Acids

Incorporation of Base Analogues. Like the analogues of naturally occurring bases, such analogues of pyrimidine and purine bases as halogen-derivatives of uracil (5-bromo-, 5-chloro-, 5-iodo-, 5-fluoro-uracil) and purine analogues (2-aminopurine, 2,6-diaminopurine), and different dyes (especially, those belonging to acridine series), may be incorporated into DNA and RNA. All have more or less mutagenic action, though the molecular mechanisms of mutagenesis for many are not clear.

5-Bromouracil and 5-Bromodeoxyuridine

Thymine in DNA can be replaced by halogenated compounds of uracil and uridine in bacteria (Weygand *et al.*, 1952; Zamenhof and Griboff, 1954;

Wacker *et al.*, 1960; Rudner, 1961; Strelzoff, 1961, 1962), phages (Dunn and Smith, 1954, Litman and Pardee, 1956; Benzer and Freese, 1958), and tissue culture cells (Eidinoff *et al.* 1959; Hakala, 1959; Szybalski and Djordjevic, 1959, Stark and Littlefield, 1974). The incorporation of bromouracil (BU) and bromodeoxyuridine (BDU) is strongly intensified with the inhibition of thymine synthesis (or without thymine in the medium). When the organisms with BU are cultivated in the light, the mass death of cells and phages can be observed. No such lethal action can be seen when phages (Stahl *et al.*, 1961) and transforming DNA (Szybalski *et al.*, 1960) are kept in the dark. BUDR induces sex-linked mutations in *Drosophila,* but it does not give chromosome aberrations (Kaufman and Gay, 1970). The authors suppose that a large majority of mutations are not connected with direct interaction of BUDR with DNA, but more probably result from secondary effects of BUDR action. The mutagenic effect of BUDR was studied in diploid human fibroblasts, and comparative investigation of BUDR and ethyl methanesulfonate showed that the average rates of induced mutations for EMS and BUDR were 7.8×10^{-6} and 6.3×10^{-6} cell/generation, respectively (Stark and Littlefield, 1974). It seems that the mutagenic activity of BU in microorganisms is connected with the change in density of the electrons of the molecule. Though the bromine atom and methyl group are nearly the same size, bromine is more electronegative and, due to this, the diffusion of electrons from the electron cloud of pyrimidine ring can occur. Various authors give different meaning of the pKs for BU and thymine—7.83 and 8.1 for BU, and 9.45 and 9.8 for thymine (Katritzky and Waring, 1962; Lawley and Brookes, 1962), but in all determinations, the pK of BU is less than the pK for thymine.

The change of electronic density of pyrimidine favors the keto-enol tautomerization of 5-bromouridine and 5-bromodeoxyuridine molecules:

As a result, 5-bromouracil can pair not only with adenine but with guanine as well. The evidence of the possibility of pairing of BU with guanine was obtained in experiments by Trautner *et al.* (1962). When polydeoxythymidilic acid was used in a cell-free system for DNA synthesis, the synthesized product was pure polydeoxyadenilic acid, even in the presence of deoxyguanine triphosphate. However, polydeoxy-5-bromouridilic acid gave a mixed copolymer of deoxyguanylic and deoxyadenylic acids.

Adenine   5-bromouracil   Guanine   5-bromouracil
           (Normal keto-form)   (Rare enol-form)

BU may cause mutations both as a result of pairing mistakes and from replication mistakes (Fig. 11), and in both cases, it causes point mutations of the transitional types (Benzer and Freese, 1958; Freese, 1959b). Terzaghi *et al.* (1962), studying the reversions of T4 phages containing BU, tried to define how many mutants, appearing after the replication and pairing mistakes, are among all induced mutants. Four mutants of the GC → AT type (mistakes of pairing) and two mutants of the AT → GC type (replication mistake) were found among six selected mutants of the *e* locus, responsible for synthesis of phage lysozyme. Howard and Tessman (1964a) believe that sometimes BU can result in substitution of thymine by cytosine.

5-Fluorouracil

Along with BU, incorporating into nucleic acids of another analogue of pyrimidine bases, 5-fluorouracil (FU), which is involved in RNA, and not in DNA, instead of uracil, was found. Naono and Gros (1960) have shown that cells of *Escherichia coli* and *Bacillus megatherium* incorporate FU into RNA and subsequently synthesize protein modified in their amino acid composition. An analogue of purine bases, 8-azaguanine, has the analogous property.

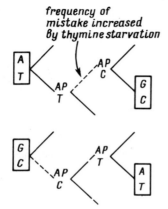

FIG. 11. Induction of transition mutations by incorporation of 2-aminopurine into DNA. Mistakes in replication (upper scheme); mistakes in incorporation (lower scheme) (from Freese, 1969).

Using the ability of FU to incorporate into RNA, Champe and Benzer (1962) have used it for the reversion of mutations on phenotypic level: certain rII mutants of phage T4 began to multiplicate in *E. coli* KB in the presence and not in the absence of FU. Although the mutation in DNA exists, synthesizing mRNA has been corrected at the expense of incorporation of FU (Fig. 12).

Barnett and Brockman (1962), using the property of FU and 8-azaguanine to induce phenotypic reversions, showed that approximately 50% of the nitrous acid-induced ad-3 mutants of *Neurospora crassa* exhibit partial phenotypic reversions when grown in the presence of either FU or 8-azaguanine.

The possibility of change of RNA molecules with FU was ingeniously used by Cooper (1964). He induced the mutation in RNA-containing virus of poliomyelitis with 5-fluorouracil. Using 5-FU Herrington and Takahachi (1973) induced mutation in bacteriophage PBS 2 (*in vivo*) at a high frequency (0.6%).

2-Aminopurine and 2,6-Diaminopurine

The compounds 2-aminopurine (AP) and 2,6-diaminopurine are potent mutagens for phages (Freese, 1959b; Herrington and Takahachi, 1973) and weak ones for bacteria (Wacker *et al.*, 1960). The mutagenic action of 2-aminopurine in *Saccharomyces cerevisiae* can be made evident only by the

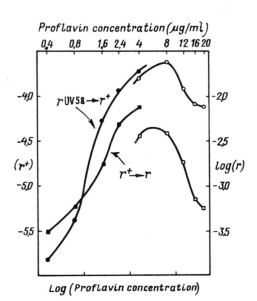

FIG. 12. The concentration dependence of proflavin mutagenesis in bacteriophage T4 (from Drake, 1970).

cells after adenine starvation (Sora *et al.*, 1973). AP in normal tautomeric form may pair both with thymine and with cytosine, giving transitions of both types. In experiments with yeasts it was found that the absolute majority of mutational events induced by 2-AP was due to transitions AT → GC (Sora *et al.*, 1973).

2-Aminopurine   Thymine

2-Aminopurine   Cytosine

Aminopurine incorporates into DNA weakly (Wacker *et al.*, 1960; Rudner, 1961*a*; Freese, 1959*b*), and the definition of points of its incorporation is a difficult task. Nevertheless, Freese believes that aminopurine is an analogue for adenine, though it will be included into DNA rarely (as compared with 5-bromouracil), but after incorporation it will act as a strong mutagen. As BU, aminpurine can induce replication and pairing mistakes (Fig. 11). Induced mutations can revert under the influence of the same agent (Tsugita, 1962).

Acridines

The mutagenic action of acridine and its derivatives (the structural formulas of acridines are presented below in the section "Photodynamic Action of Dyes," Fig. 25) was demonstrated in poliomyelitis virus (Dulbecco and Vogt, 1958), phages (De Mars, 1953; Brenner *et al.*, 1958; Orgel and Brenner, 1961; Brenner *et al.*, 1961; Drake, 1964; Lotz *et al.*, 1968, and so forth), microorganisms—*E. coli, B. subtilis, S. typhimurium, Pneumococcus* (Witkin, 1947; Eisenstark and Rosner, 1964; Zampieri and Greenberg, 1965; Sicard, 1964; Stewart, 1968; Sesnowitz-Horn and Adelberg, 1968), fungi (Ball and Roper, 1966; Magni, 1963; Magni *et al.*, 1964; Magni and

Puglisi, 1966), and also in insects (*Drosophila*—Alderson and Khan, 1968; silkworm, *Bombyx mori*—Murakami, 1973). Moreover, addition of acridine to cells infected by phages decreases phage yield from cells and interferes with phage maturation (Foster, 1948; De Mars, 1953; Lotz *et al.*, 1968).

The mechanism of action of acridine dyes is very complex and different for various methods of treatment. First, it is different for treatment in the dark and in the light. In the latter case the mutations are induced by photodynamic action. Second, acridines can induce mutations in bacteria only at certain times of their life. So, proflavin is usually not an effective mutagen for bacteria (Lerman, 1963; Orgel, 1965), but it may become very mutagenic during sexual conjugation (Sesnowitz-Horn and Adelberg, 1968). Third, under normal growth conditions it is possible to induce mutations by acridines in bacteria, but drugs must be used at very high concentrations which kill most of the bacterial cells (Witkin, 1947; Zampiery and Greenberg, 1965). However, a reverse relationship was noted for bacteriophage T4 (Drake, 1970). The frequency of forward and back mutations decreases at supraoptimal dye concentrations (Fig. 12).

In 1961 Crick and his collaborators showed that acridine mutagenesis is connected with deletion and insertion of one or more nucleotides. Drake (1966) has shown that some of the largest frame-shift mutations were deletions of about 20 base pairs.

Several hypotheses were proposed to explain the mechanism of acridine mutagenesis. All of these assume that the mutagenicity of acridines results from intercalations into DNA molecules. The insertion of acridines into the DNA structure was proved in experiments of Lerman (1961). The intercalation of acridines into DNA has been proved with different methods: the increase of DNA viscosity following the adsorption of acridine molecules on DNA, increased rigidity of the DNA, the decreased molecular mass along the length of molecules, by the protection of DNA from deamination with nitrous acid after the formation of the complex of DNA with acridines, and so forth. The inhibition and activation of polynucleotide phosphorylase during the formation of DNA-acridine complex has been shown before (Beers *et al.*, 1958). Lerman in his experiments established that acridine, complexing with DNA, intercalates between two bases of one strand of the DNA helix:

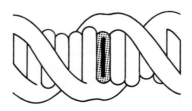

stretching of the normal 3.4 Å distance between the bases to 6.8 Å. The lengthening of the molecule was equal according to Lerman to the size of one nucleotide. This explains the appearance of frame-shift mutations (Brenner *et al.,* 1961; Lerman, 1961; Orgel, 1965). This hypothesis of acridine mutagenesis may be called the hypothesis of mistakes of replication. According to this supposition, any "redundant" base incorporates into a newly synthesized chain of DNA at the time of the replication, opposite to an acridine molecule inserted between the twists or gyres of DNA strands. At the next round of replication, this "redundant" base will pair with a complementary partner, and the process of the insertion of one base pair and the shift in reading on one sign is completed. If at the time of the first act of replication acridine is inserted into the new chain of DNA, and at the time of the second act of replication it is lost, then such a molecule will have one base less than the template, and a point deletion will appear (Fig. 13). However, some data (Drake, 1964; Streisinger *et al.,* 1966) are at variance with the opinion of the existence of mere mistakes of replication.

The second hypothesis concerning the nature of frame-shift mutations induced by acridines was proposed by Lerman (1963). He concluded that the occurrence of mutations in this case is connected with unequal crossing over (or unequal recombination). According to his model a correlation should exist between recombination by crossing-over and frame-shift mutagenesis. Such a correlation was found experimentally. Magni and von Borstel (1962), studying the frequency of spontaneous reversions in *Saccharomyces,* found that the frequency of mutations at meiosis increases 7–12 times. Since crossing-over occurs at meiosis, Magni decided to check if the increase of the mutagenesis rate was connected with crossing-over, and found a correlation between the rate of mutation and the frequency of recombination of closely linked markers (Magni, 1965).

Having obtained such an effect, Magni and his collaborators decided to examine the possibility of increase of the mutational process at the time of recombination. The authors used 5-aminoacridine. This agent can actually increase the frequency of direct and back mutations after the treatment of yeasts in the dark at the time of meiosis (Magni *et al.,* 1964). At mitosis such an increased frequency of mutations has not been observed. The agent, 5-aminoacridine even has an antimutagenic effect on the same system at mitosis. A similar antimutagenic effect of acridine orange had previously been reported for vegetatively growing bacteria (Webb and Kubitschek, 1963) in visible light. An association between acridine mutagenesis and recombination was shown also in experiments with bacteria (Sesnowitz-Horn and Adelberg, 1968), when these authors showed that proflavine induces mutations mainly at the time of conjugation. The mutagenicity of acridine was recorded also at the time of transduction (Boyer, 1966). The

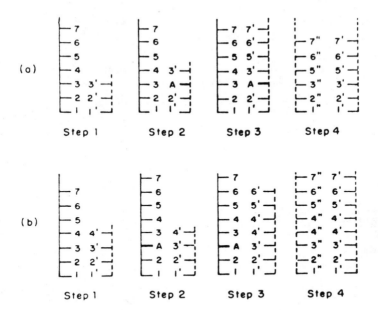

FIG. 13. Hypothetical scheme of replication mistakes after acridine incorporation into DNA, according to Orgel (1965). a—point deletion; b—point insertion.

experiments with bacteriophages are also in agreement with the scheme of recombinational mutagenesis. However, many data are completely contradicted or do not fully agree with the supposition of Lerman and Orgel. These data were obtained in experiments with both phages (Drake, 1967; Lindstrom and Drake, 1970) and with higher organisms. As mentioned above, acridine mutagenesis in bacteria was found in some cases when recombination was apparently absent, because of high acridine concentrations, which caused rapid killing of the bacterial population (Witkin, 1947; Zampiery and Greenberg, 1965). However, Stewart (1968) described mutagenesis in *B. subtilis* by acridine yellow in the dark, at concentrations that cause no loss of viability. The author failed to observe any association between mutagenesis and recombination of linked markers. A similar result was described in the experiments with higher organisms (Alderson and Khan, 1968). However, I want to call attention to the fact that their experimental evidence cannot definitively decide whether acridine-induced mutagenesis in *Drosophila* requires a recombinational event for its expression.

The recombinational hypothesis was developed further, making more precise the principle of recombination mistakes. The appearance of mutations can be connected with mistakes in the DNA-polymerase, induced by acridines at the time of normal replication (Orgel, 1965), or with mistakes in enzymes inducing unequal crossing-over (Lerman, 1963), or with the repair of single-stranded breaks in the course of recombination (see a scheme of recombination in the last section of this paper).

A more detailed formulation of the hypothesis of participation of acridines at the time of recombination belongs to Streisinger *et al.* (1966). Supposing that "there is now a great deal of evidence that recombination in phage T4 and other bacteriophages occurs through the formation of internal heterozygotes (Sechaud *et al.*, 1965)" (p. 81), they offered the following scheme of recombination (Fig. 14): (1) the regions of parent molecules are connected by hydrogen bonds in the heterozygous areas; this implies that these regions are wholly or partly complementary; (2) the regions remaining single-stranded are rebuilt in such a way as was supposed under the repair replication.

Frame-shift mutations can accordingly appear when one or more bases in the synthesizing region have no time for pairing with a complementary partner and "hang" free. If the next base joins with the former by a sugar-phosphate chain and pairs with the complementary partner, then a frame shift has occurred (Fig. 15).

Streisinger *et al.*, 1966 wrote that "the insertion would be most likely to occur in a region of repeating bases or base doublets through the pairing of a set of bases in one chain of the DNA molecule with the wrong, but complementary, set in the other chain" (p. 81). The authors saw evidence of the correctness of their conclusion in the following observation. In mutants analyzed by them, the shift of reading happened only in those regions where the repeating sequences of bases existed. In the authors' opinion, the role of acridine consists in the following: acridine stabilizes the DNA molecule

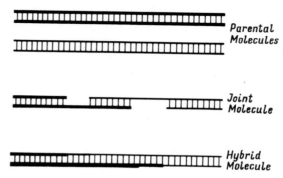

FIG. 14.   Scheme of recombination of two molecules of DNA (from Streisinger *et al.,* 1966).

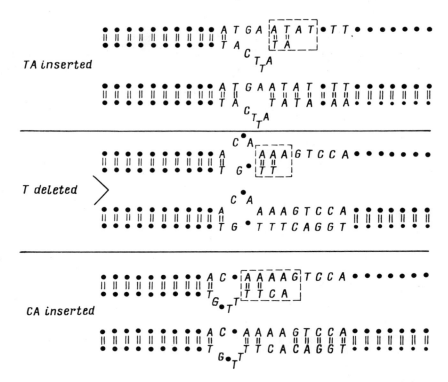

FIG. 15.   Possible mode of origin of some frame-shift mutations (from Streisinger *et al.*, 1966).

rather than disjoins it. Owing to the stabilization, the life-span of a nonpair-ing base increases, and the possibility of complementary pairing of the next bases also increases. The hypothesis of Streisinger and co-workers answered two very important questions. First, it explained why acridine mutagenesis was observed with recombination and at the time of repair synthesis, and, second, why the frame-shift mutations appeared mainly with the insertion of acridines into DNA.

We may note that Streisinger *et al.* (1966) proposed an hypothesis for the origin of acridine hot-spots. If the regions with repeating bases favor the occurrence of frame-shift mutations, then such regions can be hot-spots for acridine mutagens. The mutagens which do not induce frame shifts interact with the same efficiency with both separate bases, and with combinations of them, and such mutagens cannot induce hot-spots. Mn^{2+} ions are an example of such mutagens. The map of "manganous" mutations was published by Orgel and Orgel (1965). This map is rather primitive, but nevertheless it showed the absence of hot-spots for this mutagen (Fig. 16).

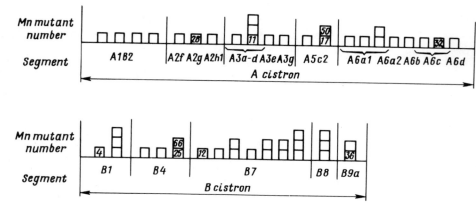

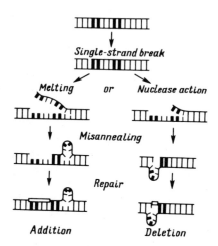

FIG. 16. The genetic map of rII-mutants of T4 bacteriophage induced by KMnO₄ (from Orgel and Orgel, 1965).

Streisinger's hypothesis concerning the role of repair mistakes in the appearance of frame-shift mutations was further developed by Drake (1970). Drake showed that cases when some bases are incorporated or deleted, may be explained within the framework of this hypothesis (Fig. 17).

A new, interesting although speculative hypothesis on the nature of frame-shift mutations proposed recently by Schreiber and Danne (1974) is based on the data obtained from a study of fluorescence and fluorescence quenching of acridine complexes with synthetic polydeoxyribonucleotides.

In conclusion, note that acridine dyes can eliminate the F-factor in bacteria (Hirota, 1960). I believe that this can occur also at the time of recom-

FIG. 17. Hypothetical mechanism of production of frame-shift mutations by mispairing (from Drake, 1970).

bination, when the episome is inserted into the genome of a bacterium. It is true that data on the elimination of episomes by other mutagens—with nitrosoguanidines (which, it would seem, can manifest its activity also at the time of recombination), and with ethyl-methanesulfonate—contradict this point of view to some extent. Nitrosoguanidine was more active in elimination of episomes than EMS, and both agents were weaker than acridines (Willets, 1967).

Alkylated Derivatives of Acridine

At the beginning of the 1960s at the Institute of Cancer Research in Philadelphia, in the laboratory of H. Creech, there were synthesized alkylated derivatives of acridines which were called, in accordance with the first letters of the name of this Institute—ICR. Ames and Whitfield (1966) tested more than 50 similar compounds. Later the mutagenic action of ICR compounds was studied in many organisms including phages (Sederoff, 1966), bacteria (Brammar *et al.,* 1967; Martin, 1967), fungi (Brockman and Goben, 1965; Malling, 1967; Munz and Leuppold, 1970; Brusick, 1970), and *Drosophila* (Carlson and Oster, 1962; Carlson *et al.,* 1967).

It was shown (Chan and Ball, 1971) that mutagenic alkylated acridines (having both mutagenic and carcinogenic activity) bonds to DNA.

Among ICR-compounds ICR-170 was encountered most frequently.

ICR-170

Brockman and Goben (1965) showed that ICR-170 is a potent mutagen in *Neurospora* and came to the conclusion that the characteristics of ad-3 mutants induced by this agent are different from those induced by X-rays, nitrous acid, 2-aminopurine, and of spontaneous origin. A detailed comparison of mutagenicity and specificity of ICR-170 and related mutagens was made by Malling (1967). Among eight ad-3B mutants of *Neurospora crassa* four mutants would appear to revert only by base-pair substitution (revertible by nitrous acid and/or hydroxylamine and usually with EMS); two revert only spontaneously, and two only by ICR-170. The two mutants which are revertible only after treatment with ICR-170 were classified as frame-shift mutants. Now the conclusion is thoroughly augmented that the alkylated derivatives of acridine induce mainly the frame-shift mutations (Hartman *et al.*, 1973). Though ICR mutants induce base substitutions, the proportion of these mutants is relatively small in comparison with frame-shift mutants.

It is established that both the acridine ring and the alkyl chain (for ICR-170–nitrogen mustard group) of the ICR molecules are essential for their mutagenic activity. An hypothesis may be put forward on the nature of high mutagenicity of these compounds. It is possible that an alkyl-radical began mutagenic reaction, and while repair enzymes remove this initial lesion, forming a gap in DNA structure, the frequency of mutations greatly increases at the time of rebuilding of this gap because of the work of acridine nuclei.

Nitrosoguanidine

In 1960 Mandel and Greenberg described the mutagenic action of *N*-nitro-*N'*-nitrosoguanidine:

$$O_2N-NH-\underset{\underset{NH}{\parallel}}{C}-NH-NO$$

It was quickly established that NG is a very potent mutagen despite its small inactivation ability.

The mutagenic action of NG was observed in experiments *in vivo* for bacteriophages propagating in bacterial cells (Goldfarb *et al.*, 1966; Zampieri *et al.*, 1968), *E. coli* (Mandel and Greenberg, 1960; Adelberg *et al.*, 1965), *Salmonella typhimurium* (Eisenstark *et al.*, 1965), *Schizosaccharomyces pombe* (Loprieno and Clarke, 1965), *Streptomyces coelicolor* (Delit *et al.*, 1970) and even for higher organisms, *Arabidopsis thaliana* (Muller and Gichner, 1964). Singer and Fraenkel-Conrat (1967) observed the mutagenic action of *N*-methyl-*N'*-nitro-*N*-nitrosoguanidine in RNA-containing virus (TMV), although *in vitro* neither inactivation, nor mutation induction was observed upon the treatment of RNA-containing phage particles (Herrington and Takahachi, 1973).

Initially it was thought that NG has a mutagenic action only *in vivo*. However, it was shown that NG had mutagenic ability *in vitro*, although with rather less efficiency. Inactivation of transforming DNA of *B. subtilis* was found under the influence of NG (Terawaki and Greenberg, 1965).

Singer and Fraenkel-Conrat (1967) noted strong inactivation of RNA of TMV in aqueous solutions of NG (99%), in phosphate buffer, and in foramide (97%), a smaller inactivation in dimethylsulfate and dimethylformamide (30–60%), but, in all cases, mutagenesis *in vitro* was so weak that it was 5–10 times less than with nitrous acid, hydroxylamine, and the photodynamic action of dyes. A similar result was obtained also with the treatment of DNA by NG. The frequency of mutations decreases in 10–100 times with the treatment of DNA of *B. subtilis* in comparison with the treatment of intact cells (Kolb and Kaudewitz, 1970). The influence of NG on the template activity of synthetic polyribonucleotides and some transfer RNA was studied by Chandra *et al.*, (1967).

In addition to mutagenic activity, NG induces antitumor action in mice (Leiter and Schneiderman, 1959; Goldin *et al.*, 1959; Skinner *et al.*, 1960; Greene and Greenberg, 1960) and has radiomimetic action in plants, inducing the chromosome aberrations (Gichner *et al.*, 1963; Sax and Sax, 1966), and it can also induce prophage (Allan and McCalla, 1966) and induces destruction of *Euglena gracilis* chloroplasts (McCalla, 1965).

The Mechanism of Mutagenic Action of Nitrosoguanidine. The detailed study of mechanism of action of NG and its methylated derivative:

$$HN = C - NH - NO_2$$
$$\underset{\displaystyle ON - N - CH_3}{|}$$

was started by Adelberg *et al.*, (1965). The authors tested the mutagenic activity of MNG in *E. coli* K12 under different conditions of treatment. First, they showed that the most active concentrations of MNG lie in the range of 100–300 $\mu g/ml$ (but later it was shown on a model of the mammalian virus SV40 (Tegtmeyer *et al.*, 1970) that NG at high concentrations—5000 $\mu g/ml$—does not induce mutations at all, but has only inactivating ability).

The highest activity of MNG was observed at pH 5.0–6.0. The maximum yield of mutations was noted after 15–20 minutes of treatment. Adelberg and his collaborators noted a very important circumstance—that survival under the NG treatment mostly depends on the medium composition. The survival of *E. coli* cells suspended in buffer amounted to 66% with 2.2×10^{-3} mutation frequency (in relation to survivals), while survival of cells suspended in minimal medium was 200 times less (0.3% at the 3.1 mutants among 10^3 survivals). But, perhaps, a more interesting result of their work (to which the authors paid no attention, however) was that the highest mutagenic effect of MNG was observed in cells at logarithmic stages of growth. The frequency of mutations was 6–10 times less in cells at the stationary stage. A little later it was proved that NG, like acridine and some other mutagens, has the highest mutagenic activity in replicating DNA.

Altenbern (1966) noted that the frequency of mutations in *Staphylococ-*

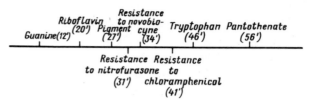

FIG. 18. The genetic map of *Staphylococcus aureus*. The position of genes on the map was marked after the treatment of synchronously dividing cells by nitrosoguanidine (from Altenbern, 1966).

cus aureus, treated with NG was increased at times of replication of cellular DNA. He has succeeded in determining more exactly the time of appearance of the mutations induced by NG. Having synchronized the initiation of replication of the whole population of cells, this author noted that during one cycle of DNA replication the highest yield of mutants per gene appeared at the time of replication of this gene. So, the mutants of the guanine locus appeared with the highest frequency at the 12th minute, of riboflavin locus in the 20th minute, and so on. The whole cycle of DNA replication was completed within 60 min and the author noted the localization of eight genes on the genetic map. Thus, using NG, it was possible to reproduce the genetic map of *Staphylococcus aureus* (Fig. 18).* This new method of genetic mapping allows one to obtain the maps even for organisms in which the process of conjugation is either unknown or suppressed. Cerda-Olmedo *et al.* (1968) used this method for mapping of *E. coli* TAU bar. The map obtained coincides almost completely with the map of *E. coli* K12, made on the basis of numerous conjugation experiments (Fig. 19). Ward and Glaser (1969) used NG and, noting the time of appearance of prototrophs by mutation of different auxotrophs of *E. coli* B/r, could determine not only the marker which replicates first (the initial point of replication of DNA), but could also determine the direction of synthesis of DNA in cells of this strain. Kimball and Setlow (1974) found that the time course of mutation fixation after treatment of *Hemophilus influenzae* (in transformation experiments) coincided with the time course of DNA replication.

The exact coincidence of the time of replication and the time of highest yield of mutations after the treatment with NG was noted by Kimball (1970) in synchronized populations of *Paramecium* (Fig. 20).

The molecular mechanism of mutagenic and inactivating action of NG

---

* It should be noted that in 1973 Altenbern found that nitrosoguanidine is not a preferential "growing point" mutagen in *Staphylococcal aureus*.

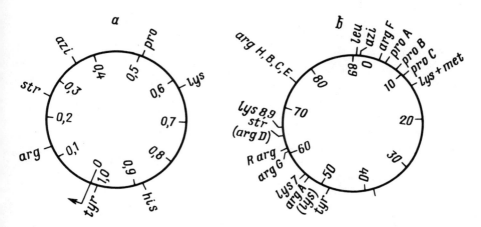

FIG. 19. The genetic maps of *E. coli* K12 and *E. coli* TAU bar a) the map of *E. coli* TAU bar which was made on the base of experiments with nitrosoguanidine; b) genetic map of *E. coli* K12 after conjugational experiments. The beginning and the direction of the replication indicated by an arrow (from Cerda-Olmedo *et al.,* 1968).

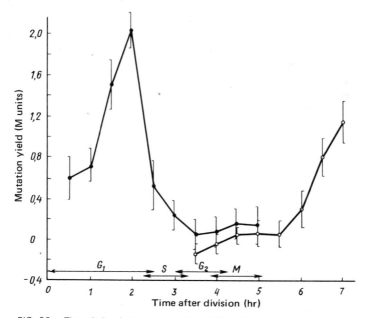

FIG. 20. The relation between mutation yield and the time of treatment by MNG. In the experiment shown by open circles the treatment was given late in the first cell circle, and continuing into the next. The cell cycle stages are shown at the bottom of the graph (from Kimball, 1970).

has not yet been studied sufficiently. NG can induce the both types of transitions: AT → GC and GC → AT (Eisenstark *et al.,* 1965; Whitfield *et al.,* 1966; Baker and Tessman, 1968), and also transversions (Eisenstark *et al.,* 1965), but gives few, if any, frame-shift mutations (Whitfield *et al.,* 1966; Dhillon and Dhillon, 1974). All the NG *in vivo* induced mutations in the rII region of T4 phage were reverted by 2-aminopurine.

The study of base substitutions in the ochre-triplet of alkaline phosphatase of *E. coli,* treated with NG, showed that NG may be considered the most versatile mutagen in its molecular action, because all possible changes have been recorded. All transitional and transversional substitutions have been observed (Weigert *et al.,* 1967). After treatment with NG two modified bases were found: 7-alkylguanine and 3-methyladenine (Lawley, 1968), and no other modified bases formed with alkylated agents were observed.

Mutagenic and lethal actions of NG strongly depend on the conditions of treatment and, first of all, on pH. NG is stable only at pH 5.0, and it decays in alkaline conditions to diazomethane and nitrocyanamide. In acid medium it gives nitrous acid (McKay and Wright, 1947; Zimmermenn *et al.,* 1965). Each of these compounds has its own mutagenic and lethal actions.

The inactivating action of NG increases with the pH from 5.5 to 7.0 (Kolb and Kaudewitz, 1970) and this effect is apparently connected with the DNA depurinization. At pH 5.0, NG strongly suppressed the synthesis of inducible proteins, and to some extent it suppressed mRNA synthesis (Cerda-Olmedo and Hanawalt, 1968). The authors believe that the lethal action of NG is connected not only with damage to DNA, but also with the level of translation, and the translation effect is connected with a change of ribosomes. It is more difficult to understand the basis of the mutagenic effect of NG at pH 5.0. Cerda-Olmedo and Hanawalt believe that it is not connected with NG itself, but rather with some compound such as diazomethane. They also showed that the mutagenic effect of NG is connected with nitrous acid in conditions in the decrease of pH to acidity. The mutagenic effect of diazomethane was demonstrated in 1948 by I. A. Rapoport (see also Rapoport, 1963), and since then mutagenic action of diazomethane has been described many times (Jensen *et al.,* 1949; 1951; Marquardt *et al.,* 1964), as well as its carcinogenic effect (Schoenthal, 1960). The influence of diazomethane on DNA results in methylation of bases. Since an alkylation of nucleic acids leads to clearly expressed mutagenesis (see previous section on "Alkylating Agents"), and in these conditions guanine is attacked most, it is not surprising that NG induces splitting of guanine from nucleic acid, as was shown by Singer and Fraenkel-Conrat (1967).

In concluding the section on the mutagenic action of substances which manifest their effect at the time of replication, it must be mentioned that the high activity of acridines and NG may be due to the fact that both agents do not interact directly with DNA, but rather with structures (possibly with membrane structures) involved in DNA replication, or with the enzymes controlling the process of DNA replication. Inasmuch as these enzymes may be located on bacterial (or nuclear, in higher organisms and other eukaryotes) membranes, it is reasonable to suppose that NG has an affinity to lipids; acridines which can bind to the cellular surface as well as to DNA (Silver et al., 1968) may have an influence on lipopolysaccharide structures of the cellular and nuclear membranes, and this may cause modification of the activity of enzymes controlling the DNA replication process.

Mutagens with Indeterminate Time of Activity

Thymine Deprivation

In thymine-deficient bacteria growing in a medium without thymine, some changes are observed. Among them are exponentially increasing lethality (called "thymineless. death"), induction of prophages, increase of the frequency of recombination, change in time of chromosome replication, structural changes in DNA (see Holmes and Eisenstark, 1968, for references).

The first communications on mutations in bacteria grown without thymine were rejected by Lederberg (1950). He believed that these data were incorrect and that the increased number of mutants might be explained by selection of preformed mutants. However, the following experiments showed that thymine deprivation is not only a selective, but also a potent mutagenic factor. Starvation led to the appearance of mutants of E. coli defective in synthesis of some amino acids (Coughlin and Adelberg, 1956; Kanazir, 1958), and resistant to streptomycin (Weinberg and Latham, 1956). Only AT pairs are affected under thymine deprivation (Smith et al., 1973; Bresler et al., 1973). Thymine starvation produces mostly AT → GC transitions (Pauling, 1968) and possibly transversions (Smith et al., 1973). Thus thymine deprivation is the first example of an AT specific mutagen treatment.

A detailed study of the mutants appearing as a result of thymine starvation, was made by Holmes and Eisenstark (1968). These authors obtained 70 mutants of S. typhimurium defective in some of the essential amino acids, purine and pyrimidine bases, and some vitamins. These mutants were treated with DES, NG, 2-aminopurine, BUDR, and β-propiolactone for the study of the process of reversions with these agents. Spontaneously or after treatment with at least two mutagens 77.2% of all mutants revert to wild type; 17.1% do not revert either spontaneously or after the action of any of the tested mutagens. These mutants were scored as deletions. Four mutants (5.7% from all selected mutants) can revert spontaneously, but cannot revert after the treatment with mutagens. Therefore, thymine starvation is a mutagenic factor with a very wide range of mutagenic activity.

The study of the rate of the mutation process showed that 5 hours of thymine deprivation results in the appearance of 1.7 to 4.8% of mutants among the survivors. The authors proved that the mutants obtained have nothing to do with selection of preformed mutants. Bresler *et al.* (1970) described almost 100% mutation among surviving bacterial cells grown on a solid medium without thymine. The effect was not observed when bacteria were grown in liquid media.

Steroid Diamines

Mahler and Baylor (1967) described a mutagenic action of irehdiamine A (IDA, pregn-5-ene-3β, 20α-diamine) and malouetine (Mal, 5α-pregnan-3β, 20α-ylenebis[trimethyl-ammoniumiodide]) in the DNA-phages T4D and T2H and the RNA-phage MS2 at the stage of replication. The authors postulate DNA to be the primary target of diamine action *in vivo*. Earlier, strong and peculiar chemical interactions between nucleic acids and certain steroid diamines had very strong inhibitory action on the propagation of bacteria and bacteriophages. IDA (in concentration 1.2 $\times$ 10^{-4} M) effectively suppresses the reproduction of phages T2 and T4. The replication of the DNA-phage T2 is more effectively inhibited than is that of MS2. IDA (di-primary amine) is much more effective than is Mal (di-quaternary amine). The effect is dependent on the anionic strength and on the incubation temperature. The inactivation appears to approximate single-hit kinetics.

Diamines have no influence on recombination although the mean yields of phages (burst sizes) in single-burst experiments were strongly reduced. Both agents are potent mutagens. Apparently these mutagens are the first agents that can give both base substitutions (either transitions or transversions), and frame-shift mutations.

The Mutagenic Action of Some Biopolymers

In 1939, S. M. Gershenson found that a partly depolymerized preparation of calf thymus DNA induces numerous mutations in *Drosophila melanogaster*. However, Rapoport (1940) and Muller (1941), and later Di Paolo (1952) and Theleu and Shoffner (1970) obtained negative results when using DNA preparations as mutagenic factors. Numerous experiments, undertaken in the Laboratory of Gershenson in Zabolotny Institute of Microbiology in Kiev (Gershenson *et al.*, 1948; Gershenson, 1965*a,b*; 1966; 1969), as in other laboratories (Fahmy and Fahmy, 1961; 1964; Fox and Yoon, 1966) confirmed the validity of the original conclusion of Gershenson concerning the possibility of induction of mutations with DNA preparations. It was found that the DNA preparations induced many visible mutations but scarcely any lethals in the X chromosome, whereas with other chemical mutagens the situation is just the reverse. No gross chromosome rearrangements are induced by DNA. A study of visible mutations both in the X-chromosomes and autosomes showed that certain genes mutate with a high frequency under the influence of exogenous DNA while the mutability of others remains as low as in the control. The second peculiarity of the mutagenic action of exogenous DNA is its delayed character (Gershenson, 1965*b*).

Another example of mutagenic action of biopolymers is induction of mutations by viruses (Alikhanian and Ilyna, 1958; Shein and Enders, 1962; Nichols, 1963; Helstrom *et al.*, 1963; Vogt and Dulbecco, 1963; Stolier and Collman, 1965; Boiron *et al.*, 1966; Kerkis *et al.*, 1967; Stich *et al.*, 1968). In these experiments mutagenic action was shown on human cells and on laboratory mammals and their tissue cultures with such viruses as herpes simplex, measles, chicken pox, rubella, infectious hepatitis, parainfluenza, adenoviruses, simian vacuolizing virus SV40, polyoma, Shope's papilloma, Rous sarcoma virus, and others. Mutations were induced in insects and also in plants (Baumiller, 1966; Sprague *et al.*, 1963). In all these cases viruses induced mutations in those organisms in which they can normally reproduce.

Another case of mutation induction by viruses and their nucleic acids is observed in experiments with noninfectious viruses, which cannot multiplicate in a given host. The first data of this kind were obtained by Fahmy and Fahmy (1964), who showed that mutations in *Drosophila* can be induced by mouse leukemia viruses. Burdette and Yoon (1967) induced mutations in *Drosophila* with Rous sarcoma virus. Gershenson *et al.* (1971) studied in detail mutagenic action of nuclear polyhedrosis virus, *Tipula* iridescent virus, poliomyelitis virus, and influenza virus in *Drosophila*.

Undoubtedly, it is hard to distinguish the mutagenic action of DNA from transforming action in some cases. The first attempts to induce DNA-mediated transformation in eukaryotes did not achieve their objective, but the first specific transformations in *Drosophila* were brought about by Fox and Yoon (1966) after treatment of *Drosophila* embryos of particular genotypes with "homologous" DNA prepared from adults of the same genotype.

Successful attempts were made at genotypical modifications in human tissue cells in culture under the action of exogenous DNA (Szybalska and Szybalski, 1962; Merill *et al.*, 1973).

The positive results in transformational experiments with higher plants were obtained by Hess (1969), who studied genetic modification of *Petunia* after treatment of seeds with DNA preparation, by Ledoux *et al.* (1971, 1974) with experiments with *Arabidopsis thaliana* and by Turbin *et al.*, (1973, 1975) (see also Soyfer and Turbin, 1974) in experiments with waxy mutant of barley injected by DNA isolated from wild type plants.

Caffeine and Related N-Methylated Purine Bases

Besides inhibition of repair activity (mainly post-replicative repair—Trosco *et al.*, 1973; Roberts and Ward, 1973; Lehman and Kirk-Bell, 1974), caffeine has a marked mutagenic action (Fries, 1950; Gezelius and Fries, 1952; Novick, 1956). The effect is inhibited by adding guanine to the medium (Novick and Szillard, 1952). In higher organisms the mutagenic action of caffeine was noted both in plants (Kihlman and Levan, 1949; Schoneich *et al.*, 1970; Kaul and Zutschi, 1973), and in ani-mals—*Drosophila* (Andrew, 1959; Alderson and Khan, 1967; Clark and Clark, 1968; Mittler *et al.*, 1967), mice (Lyon *et al.*, 1962; Adler, 1966; Adler and Schöneich, 1967; Cattenach, 1962), and human cells (Ostertag, 1966). A detailed review of the data on the mutagenic action of caffeine was published by Schöneich *et al.* (1970).

The use of caffeine in combination with alcohol and sucaryl leads to the production of aberrations in the onion root tips greater than the sum of the effects of alcohol and sucaryl and caffeine each used alone. The synergistic effect was about twice the cumulative effect (Sax and Sax, 1968). In addition, a cytotoxic action of caffeine in tissue culture cells of man (HeLa cells) was observed by Goroshkina and Soyfer (1971).

6-Mercaptopurine

The analogue of adenine and antileukemic drug 6-mercaptopurine destroys purine pathways in cells and thus influences nucleic acid synthesis. An elevated level of chromosome breakage was shown by Holden *et al.* (1973) in mouse bone marrow cells within 12 hours after oral or parenteral administration of drug.

Manganese Ions

The mutagenic action of manganous salts in bacteria was carefully studied in the laboratory of M. Demerec. In 1951, Demerec and Hanson published an investigation of mutagenic action of $MnCl_2$ in *E. coli* B/r. The authors studied mainly the mutations from streptomycin dependence to independence, and found a striking peculiarity of Mn^{2+} mutagenesis. The frequency of mutations varies 10,000-fold, depending upon culture conditions. Manganese chloride is a potent mutagen with little inactivating effect (see also Böhme, 1961). Orgel and Orgel (1965) showed that Mn^{2+} ions produced mutations mainly at the time of replication of the bacteriophage T4. Mutations are distributed more or less equally along the map (see Fig. 16). It is possible that mutations induced by Mn^{2+} ions belong to the transitional type. The permanganate in solution interacts basically with pyrimidine bases and mainly with thymine (Iida and Hayatsu, 1971). From 60% to 70% of the thymine is oxidized under these conditions. The reaction rate is higher at pH 8.6. The products shown below are formed in this reaction:

The manganese ions change the mutagenic activity of ethyleneimine and propane sultone in experiments with plants (Kak and Kaul, 1973).

4-Nitroquinoline-1-oxide

Kondo and Kato (1966) showed that 4-nitroquinoline-1-oxide (4NQO) is a potent mutagen for bacteria. The damage induced by this agent is subject to repair by the dark repair enzyme system. The agent, 4NQO manifests an inactivating effect similar to ultraviolet light in *Bacillus subtilis* (Felkner and Kadlubar, 1968). Malkin and Zahalsky (1966) stated that 4NQO is bonded to the adenine in DNA. In other experiments GC base pair substitutions were demonstrated after 4NQO treatment (Prakash *et al.*, 1974). The authors showed that 4NQO induced GC → AT transitions, GC → TA transversions and possibly GC → CG transversions. Hartman *et al.*, (1973) found that 4NQO causes frame-shift mutations in repeating $\frac{GGGGGG}{CCCCCC}$ or $\frac{GCGCGC}{CGCGCG}$ sequences of DNA in *Salmonella*.

Glycosides and Plant Toxins

The mutagenic action of glycosides and plant toxins isolated from plants and fungi was studied in *Salmonella, Drosophila, Phaseolus vulgaris* (Moh, 1970), and *Neurospora crassa* (Ong, 1970). From seeds of *Vicia faba*, catechins and saponin were isolated which had an antimutagenic action (Winkler and Smisslova, 1967). A review of mutagenicity of some alkaloids and toxins was made by Tazima (1974).

Polyimines

The high mutagenicity of some polyimines used as sterilizing agents in the control of insect populations was demonstrated in experiments with phages (Drake, 1963), *E. coli* (Szybalski, 1958) and *Neurospora* (Kaney and Atwood, 1964). Although the rates of inactivation of T4B phage by these compounds varied considerably in Drake's experiments, the number of mutations per lethal hit was approximately constant for the studied compounds. Because the compounds used are very similar to some alkylating agents (e.g., TEM), Drake supposed that these compounds are likely to function as alkylating agents. The author notes that the high mutagenicity of APO, MAPO, and EGPC (ethylene glycol-bis((1,2)-propylene)-carboxamide) indicates that their use as insect sterilizing agents should be confined to systems avoiding their free distribution in nature, especially where transmission to human beings could occur.

Tris(1-aziridinyl)
phosphine oxide
(APO)

Tris(1-(2-methyl)
aziridinyl) phosphine oxide
(MAPO)

Triethylene melamine
(TEM)

Nitroso Compounds

A large group of substances belonging to the nitroso compounds and having mutagenic effects in all living organisms was studied. These include: dimethyl- and diethylnitrosoamines (Malling, 1966*b*), methyl-, ethyl-, and propyl-*N*-nitrosourea (Rapoport, 1966; Veleminsky and Gichner, 1970; Veleminsky *et al.*, 1970; Magee, 1971; Lawley *et al.*, 1973; Gichner and Veleminsky, 1973), *N*-methyl-*N*-nitrosoaniline (Shanbulingappa, 1968), *N*-nitroso-*N*-methyl-urethane (Clarke, 1969; Gecheff and Nicoloff, 1970).

The extremely high mutagenicity of dilute solutions of these mutagens *in vivo* is accompanied by weak, inactivating action. The mutagenicity of these compounds is due to derivatives of these substances formed in cells, radicals, and alkylating derivatives (Schoenthal, 1960; Druckrey *et al.*, 1963).

The following compounds 3-methylthymidine, O^4-methylthymidine, O^6-methylguanidine, and some other products of reaction of DNA and synthetic polydeoxyribonucleotides with nitrosoureas were isolated (Lawley, 1971; Lawley *et al.*, 1973). Lawley (1971) proposed that the ability of nitrosocompounds to methylate the O^6 atom of guanine could account for their mutagenic potency, since this could be expected to cause mispairing (transversions).

It is possible that nitrosocompounds manifest their mutagenic activity through interaction with proteins and other molecules but not only with DNA (Veleminsky *et al.*, 1970). This class of substances also has carcinogenic action (Magee, 1971). In addition, the inactivation of transforming DNA was recorded after the treatment of DNA with nitrofurasone (Terawaki and Greenberg, 1965).

Indazole Analogues

Indazole and chloro-, sulfonate-, and other derivates (hycanthone, lucanthone, niridazole, and others) are often used as schistosomicidal drugs. It was found that hycanthone methanesulfonate depressed survival and induced mutations in bacteria and phages (Hartman *et al.*, 1973), yeast (Meadows *et al.*, 1973), *Drosophila* (Knaap and Kramers, 1974), rat bone marrow (Sauro and Green, 1973) and in cultured mammalian cells (Clive, 1974).

$$NH(CH_2)_2NE_2^+$$

$$CH_2OSO_2Me$$

Hycanthone methanesulfonate

Hycanthone methanesulfonate injected into *Drosophila melanogaster* adults or fed to larvae, produced a 10-fold increase of sex-linked recessive lethals (Knapp and Kramers, 1974). Mutagenic activity of hycanthone and a number of structural analogues of hycanthone have been examined in mouse lymphoma cells. Mutagenicity of these agents as a function of the molar concentration of the mutagen had a complex mode of action compared to other analogues tested (Clive, 1974). For example, 6-chloro-indazole derivative of lucanthone (IA-3) was nonmutagenic at concentrations leading to up 90% inhibition of growth, while the corresponding des-chloro-indazole derivative (IA-5) is 10–100-fold more mutagenic than IA-3. Some indazoles (chloro-indazole thioxanthene, IA-4, and niridazole) failed to elicit a statistically significant mutagenic response with regard to the production of grossly visible chromosomal damages in rat (Sauro and Green, 1973). In *Saccharomyces cerevisiae* IA-4 depressed survival slightly but did not cause genetic reversions (Meadows *et al.*, 1973), while in *Salmonella* IA-4 and two other 2-hydroxymethyl substituted derivatives of thioxanthene IA-6 were weakly mutagenic and IA-4 and IA-5 were mutagenic for T4 phage.

The strong mutagenicity was observed when hycanthone was used. The markers reverted to wild type by hycanthone in yeast were of such a nature as to suggest that it is a general mutagen which can induce both base substitutions and additions or deletions (Meadows *et al.*, 1973). In *Salmonella typhimurium* (as a model) induction of frame-shift mutations by hycanthone is strictly proportional to dose over a greater than 250-fold range.

Hartman *et al.* (1973) found that hycanthone and analogues of lucanthone (IA-4 and IA-6) cause frame-shift mutations in repeating $\frac{GGGGGG}{CCCCCC}$ or $\frac{GCGCGC}{CGCGCG}$ sequences in *Salmonella*.

The effects of four analogues of hycanthone and lucanthol (e.g., miracil D) on the sedimentation characteristics of phage PM2 DNA were investigated by Waring (1973). The author found that all four compounds bind to DNA by intercalation and that they remove and reverse the supercoiling of the closed circular duplex DNA molecules of phage *in vitro*.

Carcinogenic Benz[a]anthracenes

The mutagenicity of some carcinogenic derivatives of anthracene was studied in experiments with *E. coli* (see, e.g., Tarmy *et al.*, 1973), *Drosophila* (Fahmy and Fahmy, 1973) and cultured mouse embryo cells (Baird and Brookes, 1973). The important role of repair enzymes in mutation production by 7-bromomethyl-benz[a]anthracene was revealed in experiments with different repair-deficient mutants of *E. coli* (Tarmy *et al.*, 1973). The authors found that the wild type strain of *E. coli* was more resistant to the cytotoxic effect than was uvr⁻, exr⁻, or double mutants. An intact exr-function was required for the induction of mutation, and excision-repair enzymes were efficient in removing premutation lesions induced by benz[a]anthracene.

The molecular structure of benz[a]anthracenes and their epoxides plays an important role in mutagenicity of these compounds. K-region epoxides were mutagenic on the euchromatic genes of *Drosophila* and were less effective on the ribosomal RNA loci than the corresponding parent hydrocarbons. The parent compounds were inactive on the euchromatic parts of the genome but were effective on certain heterochromatic sites, especially on the rRNA genes that yield bobbed mutations (Fahmy and Fahmy, 1973).

Baird and Brookes (1973) revealed modified bases in DNA of some mammalian cultivated cells after their treatment by 7-methyl-benz[a]anthracene. DNA from the treated cells was isolated and enzymatically degraded to deoxyribonucleotides and the products obtained were separated by column chromatography. The changed nucleosides eluted as 2 peaks. These experiments demonstrated that benz[a]anthracenes interacted with cellular DNA.

Polycyclic Carcinogens

Polycyclic carcinogens (e.g., 2-nitrofluorene) induce frame-shift mutations in *Salmonella* (Isono and Yourno, 1974) by deleting a $_{GC}^{CG}$ or $_{GG}^{CC}$ doublet from the DNA sequences with repetite GC pairs.

Nitrofuran Derivatives

A number of 2-nitrofuran derivatives were mutagenic for *E. coli* (McCalla and Voutsinos, 1974). The presence of the nitro group in nitrofurans was essential for mutagenic activity of these compounds and strains lacking the enzymes involved in reduction of the nitro group were resistant to the nitrofurans.

Ethidium Bromide

It was shown that ethidium bromide induced rough variants of *Streptococcus* mutants from a mucoid type strain (Higuchi *et al.*, 1973).

Ethidium bromide

Free Radicals

Free radicals induce different types of mutations (see review of Fournier, 1973) on the role of free radicals in intoxications, mutagenesis, carcinogenesis, and senescence).

Environmental Mutagenesis

The mutagenic role of many naturally occurring substances is discussed with enthusiasm in both trade and scientific literature. Recently some interesting reviews of environmental mutagenesis appeared (Bergel, 1973; Tazima, 1974; and others). Mutagenic activity of some chemosteriliants, herbicides, insecticides, pyrrolizidine and other alkaloids, cycasin in cycads, bracken toxins in ferns, mycotoxins, natural and man-made cytotoxic pollutants have been discussed in the literature. Below I shall dwell on only a few examples, which obviously do not cover the whole complex and ever-growing problem of environmental effects on heredity.

Namdeo and Dube (1973) demonstrated that such herbicides as dalapon and paraquat at so small a concentration as 500 ppm inhibited growth of rhizobia isolated from soil and acted as mutagenic factors. The rhizobia did not decompose the herbicides. Dalapon at 2000 ppm and paraquat at 200 ppm induced morphological and pantothenate-requiring mutants, and reversions are possible only with paraquat-induced mutants.

Mutagenic activity of some organophosphorus insecticides (dichlorvos, oxydemetonmethyl, dimethoate, and Bidrin) was investigated by Mohn (1973). The author found that all studied compounds induced mutations in *E. coli* and that the lowest concentrations of these compounds needed for marked mutagenic action were 2×10^{-3} M–3.3×10^{-4} M.

Mutagenic activity of some chemosteriliants belonging to alkylamino-bis(1-aziridinyl) phosphine oxides was studied using the dominant lethal test on mice (Šram and Zudowa, 1974). The frequency of chromosome aberrations induced by a dose of 6×10^{-3} mol per kilogram of body weight were higher for methyl-, ethyl-, and isopropyl derivatives than for the propyl and butyl derivatives. The effectiveness of equimolar doses decreased in the following order: methyl > ethyl > propyl > butyl derivatives.

Frame-shift mutations were demonstrated in *Salmonella* treated by a fungistatic agent mixin (Hartman *et al.*, 1973).

Potential hazard of pesticides for man was demonstrated by Yoder *et al.* (1973). Lymphocyte cultures from 42 pesticide applicators and 16 controls were examined. Cultures prepared from the individuals exposed to pesticides showed a marked increase in the frequency of chromatid lesions.

Some naturally occurring compounds may be transformed into a mutagenic state under the action of some enzymes of mammalian metabolism. Endo and Takahashi (1973) showed that nitrosation of methylguanidine by real or simulated human gastric juice at pH 1.2–1.4 caused a potent mutagencity toward *Salmonella typhimurium*. The observed mutation frequency was 1000 times greater than the spontaneous rate. The authors came to the conclusion that nitrosation of methylguanidine, present in several kinds of feed, may cause spontaneous gastric cancer.

And finally it should be noted that some medicines also have a weak mutagenic activity. For example isoniazid is transformed into hydrazine in *Salmonella* cells as host and induces mutations (Kehler *et al.*, 1973). Although mammalian metabolism is accounted for in the host-mediated assay, the mutations are actually produced in microorganisms and gave indications, but no real proof, that isoniazid is mutagenic in mammals.

Azaserine

Azaserine, suppressing some enzymes, controlling purine base biosynthesis, has a weak mutagenic activity (Iyer and Szybalski, 1969).

D_2O

A weak mutagenic action of deuterium was observed in experiments with phages. The induced mutations may belong to the transition class and to frame-shift mutations. They can be reverted by analogues of bases and by acridine (Giovanni, 1960; Konrad, 1950).

Peroxides

There are communications that these agents have mutagenic action on *Drosophila* (Sobels, 1956), ascite tumor cells of mice (Schöneich, 1967), and other organisms (cf. Wyss *et al.*, 1963).

Formaldehyde

The mutagenic action of formaldehyde was described in experiments with *Drosophila* (Alderson, 1961; Sobels, 1963). Alderson believes that formaldehyde induces mutations interacting with the 6-amino groups of adenines.

Inorganic Salts and Commonly Used Chemicals

The results of numerous investigations of Newell and Maxwell who studied a possibility of mutation production by various inorganic salts and commonly used chemicals were published in U.S. National Technics Information Service Bulletins in 1971–1973 (for references see *Chemical Abstracts*).

RADIATION MUTAGENESIS

Although there are very many studies on the biological action caused by radiation, the chemical principles of radiation mutagenesis are to a great extent obscure in comparison to chemical mutagenesis.

Ionizing Radiation

Inactivation and mutagenesis under the influence of ionizing radiation on living cells result mainly in lesions of nuclear structures of cells and their DNA (Astaurov, 1947; Zirkle and Bloom, 1953; Mortimer, 1958; Szybalski and Lorkiewich, 1962; Kaplan and Moses, 1964; Sparrow, 1965; Baetcke *et al.*, 1967; Blok and Loman, 1973).

One of the first effects of irradiation on DNA and nucleoproteins in solution noted in the literature was a decrease in molecular weight (Butler, 1949; Koenig and Perrings, 1953; Scholes and Weiss, 1954; Cavaliery *et al.*, 1955), accompanied by a change in viscosity (Sparrow and Rosenfeld, 1946; Taylor *et al.*, 1948; Butler and Smith, 1950). These effects were due to the appearance in DNA of single- and double-strand breaks (Taylor *et al.*, 1948; Alexander and Stacey, 1955; Harrington, 1966; Munson and Bridges, 1969). The single- and double-strand breaks after γ and X irradiation were found in DNA, protein-DNA complexes, isolated chromosomes, in cultivated cells and in whole plants and animals (see, e.g., Gilot-Delhalle *et al.*, 1973; El-Metainy *et al.*, 1973; Van der Schans *et al.*, 1973, and others). In the presence of oxygen the frequency of DNA breakage is 4 to 5-fold higher than under nitrogen anoxia (Johansen, *et al.*, 1974).

Another important consequence of ionizing irradiation is the appearance of crosslinks in DNA (Kaplan, 1955; Shugar and Baranowska, 1960, and others) and links between DNA and proteins (Alexander and Stacey, 1958; Smith, 1966).

The secondary effect of irradiation is apparently a denaturation of DNA (Cox *et al.*, 1958; Peacocke and Preston, 1960; Hagen and Wild, 1964; Collyns *et al.*, 1965). There are indications in the literature that ionizing irradiation leads to deamination and depurinization (Scholes and Weiss,

1960), to a destruction of purine and pyrimidine bases, to a liberation of inorganic phosphate, ^3H_2O from tritiated DNA, and to formation of phosphomonoester groups (Ward, 1972; El-Metainy *et al.,* 1973; Olast and Bertinchamps, 1973).

The analysis of inactivation of biologically-active double-stranded circular DNA of phage PM2 by ionizing radiations revealed that $4.5 \pm 0.5\%$ of inactivation is a consequence of double-stranded breaks, $8.5 \pm 4.2\%$ of single-strand breaks, and $87.0 \pm 4.2\%$ of nucleotide damages (Van der Schans *et al.,* 1973). It is important to underline that only about 2% of the single-strand breaks is lethal (other breaks either lead to inactivation or may be repaired), whereas the efficiency of inactivation due to nucleotide damage is more than 30%.

Recently many attempts have been made to understand the physico-chemical changes leading to strand breaking in DNA. Ward (1972) showed that release of inorganic phosphate from 5′-deoxynucleotides is caused by an hydroxyl radical abstracting a hydrogen atom from the deoxyribose moiety. The important role of transport of electrons from purines toward pyrimidines was revealed by an electron-spin resonance study by Olast and Bertinchamps (1973).

In the course of the investigation of products formed after the irradiation of aqueous solutions of purine and pyrimidine bases there were isolated some oxy- and peroxy-derivatives. Among the pyrimidine bases the highest destruction of rings was observed with cytosines (Scholes and Weiss, 1960). The formation of oxides and peroxides of pyrimidine bases begins with the cleavage of a 5,6-double band. Ekert and Monier (1960) separated out three products of irradiation of an aqueous solution of cytosine in an oxygen medium and suggested the scheme shown below for their conversions:

$$\begin{array}{c}
\text{NH}_2 \quad \text{H} \\
\text{N} \quad\!\!\!-\text{OH} \\
\text{HO} \quad \text{N} \quad\!\!\!-\text{H} \\
\text{OH}
\end{array}$$

$$-\text{NH}_2 \swarrow \qquad\qquad \searrow -\text{H}_2\text{O}$$

$$\begin{array}{c}
\text{OH} \quad \text{H} \\
\text{N} \quad\!\!\!-\text{OH} \\
\text{N} \quad\!\!\!-\text{OH} \\
\text{HO} \quad \text{N} \quad \text{H}
\end{array}
\qquad\qquad
\begin{array}{c}
\text{NH}_2 \\
\text{N} \quad\!\!\!\text{OH} \\
\text{HO} \quad \text{N}
\end{array}$$

$$\searrow -\text{H}_2\text{O} \qquad\qquad \nearrow \text{NH}_2$$

$$\begin{array}{c}
\text{OH} \\
\text{N} \quad\!\!\!\text{OH} \\
\text{HO} \quad \text{N}
\end{array}$$

Pletikha-Lansky (1972) showed that as a result of γ-irradiation of cytosine and uracil in water solutions, isobarbituric, 4,4′-diisobarbituric, isodialuric acids and alloxan appeared. As a result of their further hydrolysis parabanic and oxaluric acids were formed.

The irradiation of aqueous solutions of thymine leads to the formation of organic peroxide compounds, also appearing after 5,6-bond cleavage (Scholes et al., 1956; Ekert and Monier, 1959):

$$\begin{array}{c}
\text{O} \quad \text{CH}_3 \\
\text{HN} \quad\!\!\!-\text{O}_2\text{H} \\
\text{O} \quad \text{N} \quad\!\!\!-\text{H} \\
\text{H} \quad \text{OH}
\end{array}
\qquad\qquad
\begin{array}{c}
\text{O} \quad \text{CH}_3 \\
\text{HN} \quad\!\!\!-\text{OH} \\
\text{O} \quad \text{N} \quad\!\!\!-\text{H} \\
\text{H} \quad \text{O}_2\text{H}
\end{array}$$

5-peroxy-6-hydroxythymine   5-hydroxy-6-peroxythymine

The product formed is a mixture of 80%-cis and 20%-trans isomers (Latarjet et al., 1961). The formation of ring-saturated radiation products of thymine 6-hydroxy-6-hydroperoxy dihydrothymine type was supported by Roti et al., 1974).

Besides, the γ-ray induced formation of tritiated water from [Me-^{3}H]thymine was observed.

These two reactions [H_2O liberation and 6-(hydroxy and hydroperoxy)-dihydrothymine formation] were studied not only in mononucleotide mixture but in single-stranded φX174 phage DNA and double-stranded E. coli DNA (Swinehart et al., 1974). It was revealed that for native DNA of E. coli the initial rates of ^3H_2O and hydroxylated dihydrothymine formation were lower by factors 9 and 12, respectively, relative to φX174 DNA.

In addition, while studying the role of the chromosomal proteins

(histones and nonhistone proteins) it was found that the chromosomal proteins effectively protected the thymine residues from destruction by radiation
(Roti *et al.*, 1974).

Purine bases are more stable under irradiation than pyrimidines
(Scholes and Weiss, 1960). The irradiation of adenine, according to Scholes
(1963), is accompanied by a 5,6-bond cleavage:

Conlay (1963) believes that the main product of adenine irradiation by
X-rays is adenine oxide:

The destruction of adenine can be observed together with the formation
of oxides:

The other products
of degradation(?)

With the irradiation of guanine one may observe the scission of the five-membered ring of the molecule, followed by the depurinization of DNA.

The mutagenic action of ionizing irradiation was described in practically all representatives of the living world, beginning with extracellular bacteriophages (Kaplan *et al.*, 1960; Ardashnikov *et al.*, 1963; 1964; Krivisky and Solovieva, 1963; Tessman, 1965; Brown, 1966, and others) to higher organisms.

The nature of the changes in the genetic code following the action of ionizing irradiation is, in a large measure, obscure. Grossman and his collaborators (Ono *et al.*, 1965) believe that the highest mutagenic action is connected with a transition of cytosine to uracil, whereupon the substitution of GC-pair by AT-pair is completed. Bridges and Munson (1968) believe that the transition AT → GC is as likely as the other type of substitution. Bridges *et al.* (1968) also believe that single-stranded breaks appear to be enlarged into single-stranded gaps under the influence of exonucleases (Emmerson and Howard-Flanders, 1965), and can give rise to mutations in the course of their repair. We must note that in the last case both transitions and transversions and also frame-shift mutations can occur.

The study of DNA-polymerase I activity in a cell-free system using γ-irradiated polydeoxyribonucleotides as a template reveals that the γ-irradiation of poly(dA) strand leads to the incorporation of dG into the complementary strand of newly synthesized DNA, irradiation of poly(dC) and poly(dG) leads to incorporation of dA, whereas irradiation of poly(dT) does not lead to incorporation of any wrong bases (Saffhill, 1974).

And finally, it should be noted that frame-shift mutations induced by X irradiation of *Salmonella* was revealed (Ino and Yourno, 1974). These mutations were shown to be a deletion of $\frac{GC}{CG}$ or $\frac{CC}{GG}$ pairs from DNA sequences $\frac{GCGCGC}{CGCGCG}$ or $\frac{CCCCCC}{GGGGGG}$ lying in the histidinol dehydrogenase gene of *Salmonella*.

Ultraviolet Irradiation

In 1958 Beukers *et al.* were the first to isolate chromatographically a substance formed after UV-irradiation of cold, aqueous thymine solutions. In 1960 this substance was characterized (Beukers and Berends, 1960), and is called a dimer of thymine:

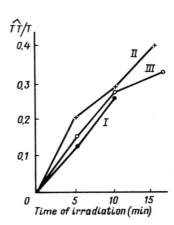

Dimers of thymine were found after UV-irradiation of polynucleotides (Shugar and Baranowska, 1960) and of native DNA of bacteria (Wacker *et al.*, 1960). The dimerization of the nearest pyrimidine bases in DNA was described in cells of higher organisms—in DNA of calf thymus (Salganik *et al.*, 1967), in tissue culture cells of mammals (Trosco *et al.*, 1965; Steward and Humphrey, 1966) and of man (Regan *et al.*, 1968; Horikawa *et al.*, 1968; Setlow *et al.*, 1969; Soyfer *et al.*, 1970), in diploid cells of human embryos (Soyfer and Yakovleva, 1972), in isolated plant DNA (Soyfer and Cieminis, 1974), in plant tissue cultures (Trosco and Mansour, 1969), and in whole plant seedlings (Soyfer and Cieminis, 1974). It was shown that both in bacteria and in human tissue culture cell dimerization follows single-hit kinetics (Fig. 21).

It was shown that in DNA at least four (Wulff and Frankel, 1961) or five (Blackburn and Davies, 1965; Weinblum and Johns, 1966) basic isomeric forms of thymine dimers can appear, and it was surmised that those most frequently appearing in DNA are *cis*-isomers of dimers. The isomeric forms of dimers obtained (Fig. 22) have different properties (Weinblum and Johns, 1966; Varghese and Wang, 1968; Stafford and Donnelan, 1968; Ben Hur and Rosenthal, 1970). Thus the III and the IV-isomers, indicated in Fig. 22 (*cis*-anti and *trans*-anti) are unstable in an acidic medium,

FIG. 21.   The kinetics of thymine dimerization in HeLa cells by UV-irradiation (from Soyfer *et al.*, 1970).

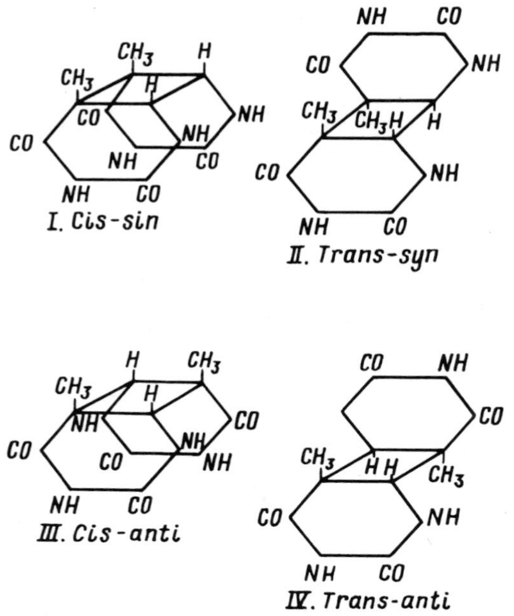

FIG. 22.   The isomeric forms of cyclobutane type of thymine dimers.

though others are stable enough, and all four isomers have different coefficients of chromatographic separation (R_f).

Beside the dimers of thymine, there were described cytosine dimers
(Sauerbier, 1964; Setlow, 1966; Regan *et al.*, 1968), and mixed dimers of cytosine and thymine, which can transform into a dimer of cytosine-uracil
after cytosine deamination (Weinblum, 1967), and also into uracil-uracil
dimers appearing in RNA molecules (Wacker, 1965):

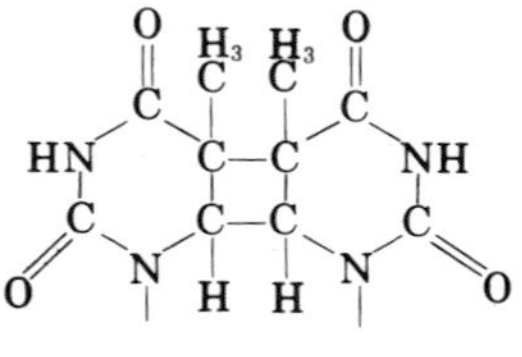

Thymine-thymine cyclobutane dimer 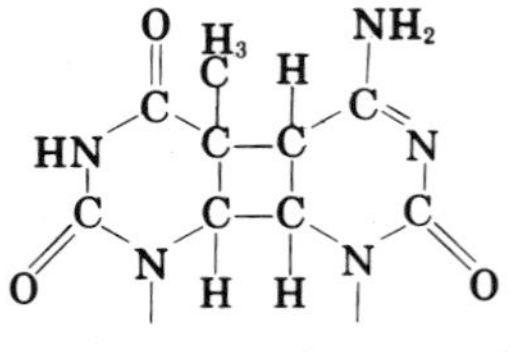   Thymine-cytosine cyclobutane dimer

TNC-dimer   TOT-dimer

It is found that $\widehat{CC}$, $\widehat{CT}$, and $\widehat{TT}$ dimers were formed in the approxi-mate ratios of $1:1:2$ at doses of UV light below 2000 ergs/mm². Dimers are formed in native DNA at twice the rate they are formed in cells (Unrau *et al.,* 1973; Soyfer and Cieminis, 1974). Cyclobutane dimers of $\widehat{TT}$ and $\widehat{TC}$ have been studied in detail. It was shown that they play a great role in mutagenesis and inactivation.

Dimers of the TNC type which appear between thymine residues in op-posite strands of DNA (Wang and Varghese, 1967) and dimers of the TOT type (Pearson *et al.,* 1965) which form with the help of an oxygen bridge, have not been thoroughly investigated and their role is still obscure.

Another widespread and well studied type of photochemical lesion in DNA is the hydratation of pyrimidine bases (Sinsheimer and Hastings, 1949; Moore, 1958; Wang, 1958; Shugar, 1962; McLaren and Shugar, 1964):

Cytosine-hydrate

A cytosine-hydrate can transfer quickly into uracil-hydrate or even uracil, which produces transitions (Schuster, 1964; Ono, *et al.,* 1965; Johns *et al.,* 1965).

Thymine hydrate probably reverses very rapidly under ordinary condi-tions. Besides, 5,6-dihydrothymine may appear (Yamane *et al.,* 1967):

5,6-dihydrothymine

The formation of hydrates and dimers was also noted during the study of irradiated RNA (Small *et al.,* 1969). These authors found five photoproducts—uridine hydrate, two isomeric forms of cyclobutane-type dimers, and two unknown photoproducts. Pyrimidine bases can, in addition, produce links with proteins (Alexander and Moroson, 1962; Smith, 1964). One example of such was found by Smith (1964). They isolated and described 5-hydro-6-*S*-hydrouracil, which was obtained from cytosine.

Ultraviolet irradiation as well as ionizing radiation induces mutations in a whole set of living organisms, beginning with extracellular phage (Kaplan *et al.,* 1960; Folsome and Levine, 1961; Krivisky, 1961, Soyfer, 1965 and others) to higher organisms.

The general scheme of damage induced by ultraviolet and ionizing irradiation in DNA is presented in Fig. 23. The highest mutagenic activity is inherent in I, II, and V structures. Structures IV, VI, and VIII have high inactivation properties. The study of mutagenic specificity of UV irradiation has shown that all types of substitutions (both transitions and transversions), as well as frame-shift mutations, can be induced by UV-irradiation. The first work in this field was done by Folsome and Levine (1961). They could revert a part of the obtained mutants with analogues of bases, and came to the conclusion that this class of mutants can be transitions.

A more detailed quantitative investigation of the relation between transitions and other types of mutations was made by Drake (1963; see also Drake, 1966). Drake obtained 329 rII–mutants of T4 phage and showed (together with Benzer) that they fall into 99 sites of rII region, and that the distribution of UV-induced mutants differs essentially from the distribution of spontaneous mutants (Fig. 24). Half the mutants were scored as frameshift mutants, others could be reverted by analogues of bases and by hydroxylamine. Drake supposes that a majority of them belong to transition mutations of GC → AT type. Folsome (1962) also showed that UV-induced rII mutants of T4 phage are divisible into three classes—transitions, transversions, and frame-shift mutants, but their distribution on the genetic map was different from the data obtained by Drake.

The study of the consequences of UV irradiation of S13 phage with single-stranded DNA (Howard and Tessman, 1964*b*) showed that more

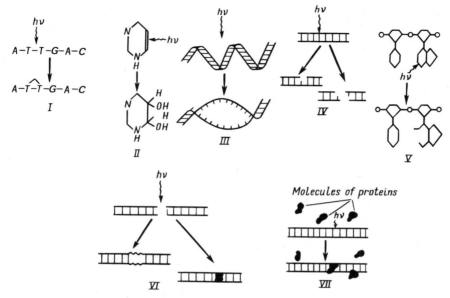

FIG. 23. Types of molecular changes after irradiation. I—dimers of pyrimidine bases; II—hydrates of pyrimidine bases; III—local denaturation; IV—single and double-stranded breaks; V-destruction of bases; VI—intra and intermolecular cross-links; VII—cross-links of DNA and proteins (from Soyfer, 1969).

than a half of the mutants are transitions (cytosine was substituted by thymine), the remaining 5 of 16 studied mutants were either transitions AT → GC, or transversions. The existence of transitional changes of GC → AT type in *E. coli* was proved by Yanofsky *et al.*, (1966). In their experiments it was found that about 9% of all mutants were of the frame-shift type (Berger *et al.*, 1968).

The Photodynamical Action of Dyes

In 1963, Böhme and Wacker described the mutagenic action of dyes (thiopyronine and methylene blue) in combination with visible light. This work was done on *Proteus mirabilis,* which was first used in genetic experiments by Böhme. Several mutant lines of *P. mirabilis* (single and double-auxotrophs) were isolated by Böhme. He treated the cells with thiopyronine and methylene blue, and after that the suspension of the cells was illuminated with a daylight lamp. The dyes did not induce mutations without irradiation. The irradiation of cells with visible light led to an

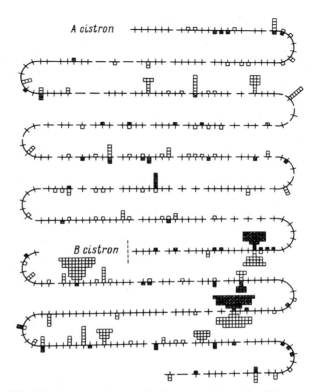

FIG. 24. The genetic map of rII-mutants induced by UV-rays (open squares) and spontaneous mutants (hatched squares) (from Drake, 1963).

increased mutation frequency from 5×10^{-7} to 10^{-3} or 10^{-5}. The frequency of mutation increased linearly with the increased time of irradiation, and the inactivation curves were approximately exponential.

The mutagenic photodynamic action of dyes was widely described (see Table II): microorganisms (Nakai and Saeki, 1964; Singer and Fraenkel-Conrat, 1966; Drake and McGuire, 1967; Brendel, 1968; Böhme and Geissler, 1968; Böhme, 1968) and fungi (Kaplan, 1950b; Alderson and Scott, 1970; Singht, 1973). A detailed review of this kind of mutagenesis is presented in Drake's book (1970, p. 171–176).

In most experiments a linear kinetics for the mutation induction was found. Brendel (1968) noted, however, that in his experiments with treatment of the phage T4 with methylene blue and visible light, the mutation frequency followed a nonlinear, multiple-hit kinetics. Brendel showed that the observed mutation frequency increase was not connected with selection of some spontaneous preformed mutants more resistant to photodynamic inactivation.

The molecular nature of photodynamic action of dyes at first seemed simple. After the studies of Simon and Van Vunakis (1962, 1964) and Wacker *et al.* (1963), which proved that the photodynamic action may be explained by destruction of purine bases both in solution and in deoxypoly-nucleotides (guanine was mainly attacked, according to these authors), a majority of researchers believed that mutational changes are induced at guanine sites. However, using the lines of rII mutants of T4 phage exactly mapped by Benzer, Brendel (1968) could show that the mutagenic action of dyes is manifested also in the AT pair (see also Ritchie, 1964; 1965). This conclusion was conclusively proven by the work of Drake and McGuire

TABLE II.   Photodynamic Mutagenesis (from Drake, 1970)

Organism	Mutation	Dye[a]	Reference
Serratia marcescens	Aberrant colonies	ER	Kaplan, 1949, 1956
Penicillum notatum	Aberrant colonies	ER	Kaplan, 1950b
Sarcina lutea	Penicillin resistant	MP, TB	Mathews, 1963
Proteus mirabilis	phe$^-$ → phe$^+$	MP, TB	Böhme and Wacker, 1963
Escherichia coli	Resistance to phage T7	ER	Kaplan, 1950a
	Resistance to phage T5	AO	Webb, Kubitschek, 1963
	try$^-$ → try$^+$	AO, AY, MB, TB	Nakai, Saeki, 1964
Phage kappa of *Serratia marcescens*	Plaque morphology	MB	Brendel, Kaplan, 1967
Phage T4	r$^+$ → r	PF	Ritchie, 1964, 1965
	r$^+$ → r	PS, TP	Drake, McGuire, 1967
	rII → r$^+$	MB	Baricelli, DelZoppo, 1968
	rII → r$^+$	MB	Brendel, 1968

[a] AO—acridine orange, AY—acridine yellow; ER—erythrosine; MB—methylene blue; MP—8-methoxypsoralen; PF—proflavine; PS—psoralen; TB—toluidine blue; TP—thiopyronine.

(1967). The authors tested some thousands of T4r mutants obtained by them after the treatment of extracellular phage with thiopyronine and psoralen (Fig. 25). The mutants were distributed among the rI, rII, and rIII regions of the T4 genome. After that the rII mutants were localized on the genetic map (Fig. 26), and used in reversion tests with base analogues, proflavine, and hydroxylamine. As a result, the authors were able to define the quantitative characteristics of the relationships between transitions and transversions. No mutants of the reading frame-shift type were observed. About a third of the induced mutants could not be reverted by base analogues, and they presumably represent transversions. The majority of the others were scored as transitions, because they could be reverted by base analogues. The authors found that only half of all reverting mutants can be reverted by hydroxylamine. On this basis they decided that transitions are connected with substitution of GC-pairs by AT-pairs, and *vice versa*.

The exact molecular mechanisms of mutagenic and photodynamic action of dyes is obscure. Though there has been some success in investigating the photochemistry of this process, we are still far from an adequate understanding of the situation.

Combined Action of Mutagens

A synergistic mutagenic effect was noted frequently in experiments with microorganisms, plants, and insects with combinations of several mutagens. The use of a combination of several mutagens leads to a rise of the mutation frequency greater than the sum of the mutagenic effects of individual mutagens, used separately. The first communication on the synergistic character of the origin of mutations was made by Auerbach and Kølmark (1959), who used UV rays and diepoxybutane (DEB) in *Neurospora* deficient in inositol and adenine synthesis. Having studied the reverse mutations, the authors found that if cells were treated with DEB before UV irradiation, an additive effect was noted for the inositol locus, and a synergistic effect for the adenine locus. If UV irradiation was preceded by DEB treatment, the synergistic effect was not noted.

The synergistic effect was noted by several authors using different combinations of mutagens: e.g., with irradiation and nitrogen mustard treatment of bacteria (Haynes and Inch, 1963), irradiation and hydroxylamine treatment (Salganik, 1963; Shankel, 1963, and others), ultraviolet and X-ray irradiation (Baptist and Haynes, 1972; Martignoni and Smith, 1973) and also in plants (Ahund-Zade and Hvostova, 1966; Sax and Sax, 1968; Reddy *et al.*, 1974). The synergistic effect was about twice the cumulative effect.

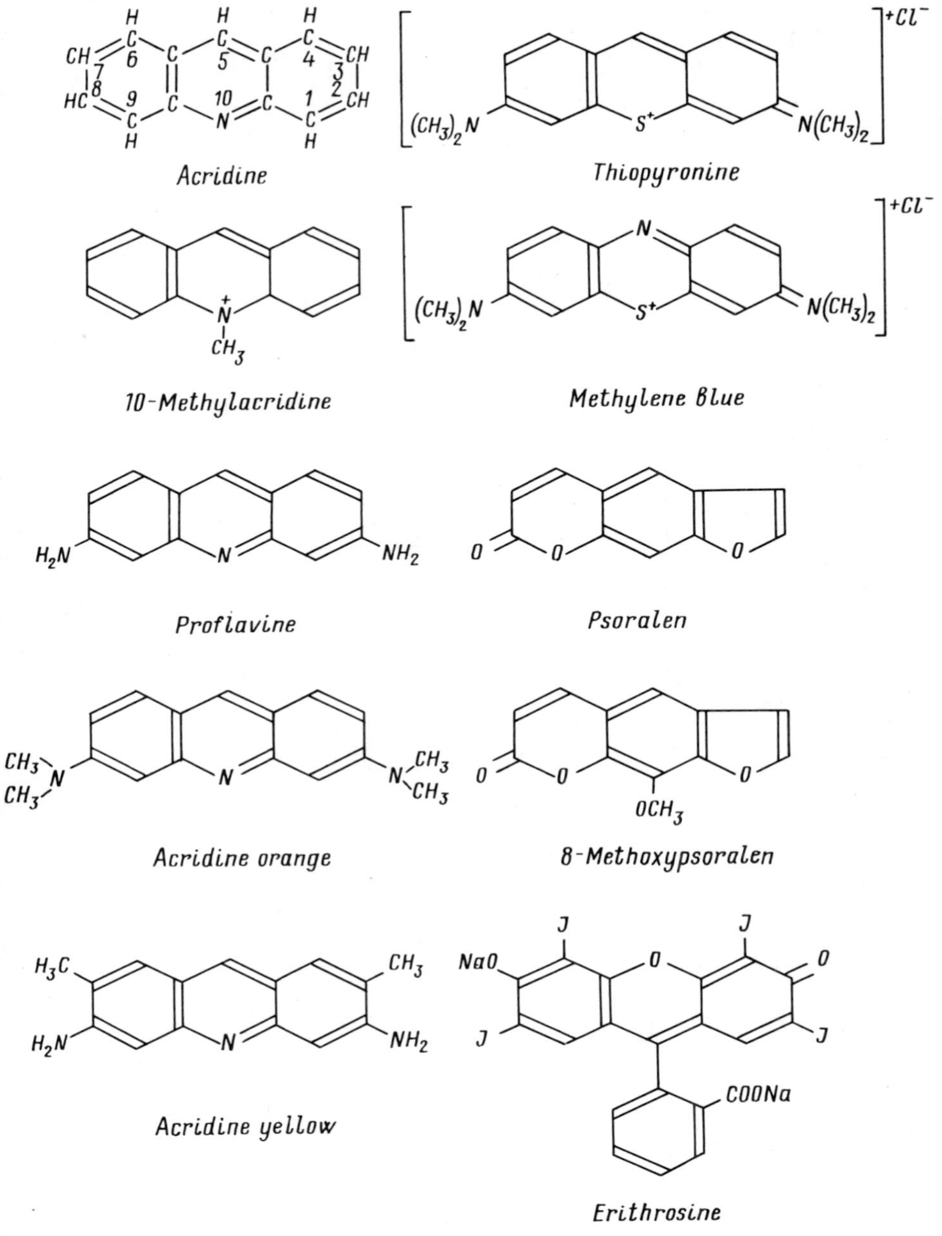

FIG. 25. Chemical structures of acridines and dyes.

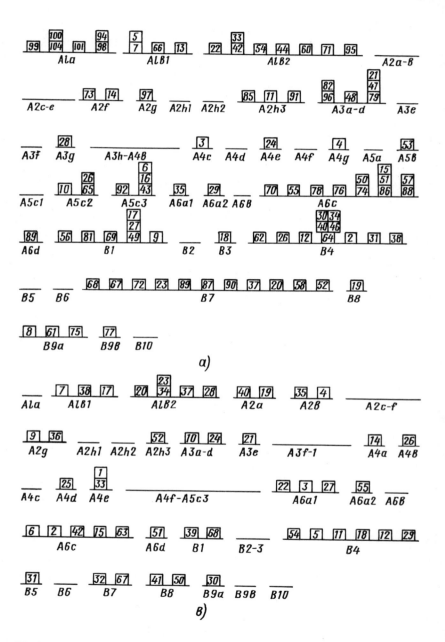

FIG. 26.   The maps of rII-mutants of T4 phage photodynamically induced by psoralen (a) and thiopyronine (b) (from Drake and McGuire, 1967).

A detailed analysis of the synergistic effect under combined treatment by mutagens was made in the Laboratory of R. I. Salganik in Novosibirsk Institute of Cytology and Genetics (Panfilova *et al.*, 1966; Salganik, 1969), and by D. M. Goldfarb in Gamaleya Institute (Goldfarb and Chubukoff, 1963; Goldfarb and Belyaev, 1966). Salganik and collaborators noted a synergistic effect for some loci (leu and thr) of *E. coli*, but the absence of this effect, or even a decrease of the mutation frequency, in other loci (for example, B_{12} vitamin synthesis). Salganik and his co-workers believed that the possible explanation of a synergistic effect is connected with local denaturation of DNA induced by some mutagens (possibly by hydroxylamine). After the denaturation, interaction of DNA with photons of UV rays may occur more intensively. The important role of repair enzymes in the nature of the synergistic effect was discussed by Goldfarb. However, the first experiments in this field gave negative results (Soyfer, 1970*b*). Krivisky and Tichomirova (1970), in experiments with T4 and S_D phages, did not obtain a synergistic effect having used hydroxylamine followed by UV irradiation. In *E. coli* B/r, but not in *E. coli* B_{s-1}, 5-bromouracil incorporation and purine starvation almost completely abolish UV and following X-ray synergism (Baptist and Haynes, 1972).

The synergistic effect was noted in experiments of Sharma and Glover (1970) in males of *Drosophila* when ethylmethanesulfonate and X-rays were used, and in experiments of Shankel (1963) when UV and X rays were combined. The molecular mechanisms of the synergistic effect of combined action of chemical or radiation mutagens are obscure.

PARTICIPATION OF THE INTRACELLULAR ENZYMES IN THE CONTROL OF THE MUTATION PROCESS

Enzymatic Mistakes as a Cause of Mutation Production

Before the discovery of induced mutagenesis, geneticists, in their efforts to understand the reasons for spontaneous mutability, restricted themselves to consideration of only intrinsic factors as causes of mutagenesis.

With the finding of external (mutagenic) factors speeding up a mutational process dozens, hundreds, and thousands of times, geneticists' attention was centered only on them as the unique cause of induced hereditary variability. However, this is a one-sided view considering only external factors as causes of the mutational process. The reasons for hereditary variability should also be sought in the immanent properties of living matter. The advances of biochemistry, the enzymology of genetic processes, in particular, allow, at present, the singling out of several essential biochemical processes, underlying the vital functions and the evolution of living matter and, at the same time, responsible for errors in the structure of the hereditary molecules, in other words, leading to mutations (Soyfer, 1970a, pp. 200–220). Replication, recombination, and repair are the processes in the course of which mutations may be induced owing to the mistakes of the enzymes (Borstel, 1969; Witkin, 1968; Soyfer, 1971b; 1972).

Besides the direct action of environmental factors (exposure to radiation or chemical agents) on DNA, they apparently may speed up mutagenesis playing the role of triggers in the enzymatic processes. The mutability of a living cell or organism is a result of the internal and external mutagenic factors.

Mutation Induction as a Result of Mistakes of Replicating Enzymes

Until recently it was thought that no mutation could arise in the structure of the synthesized DNA strand during replication. However, it was experimentally supported (and it has a direct bearing on the problem considered) that the process of replication proceeds with errors. Hall and Lehman (1968) reported that using DNA-polymerase of phage T4 in the experiments with DNA synthesis *in vitro* with polydeoxycytidilic acid as a template, they observed wrong bases (thymines) appearing in a newly synthesized product, polydeoxyguanilic acid with a frequency of 10^{-5} to 10^{-6}. It is remarkable that when the enzyme from t_s-mutant (just for the polymerase gene) phage T4 was used, the mutant enzyme made, on the average, four times more errors than the normal one. Finally, the frequency of errors made by both enzymes increased with the concentration of deoxythymidine triphosphates in the environment and it may enhance 5–20 fold if Mg^{2+} ions are replaced by Mn^{2+}. Similar data on the mistakes of DNA-polymerases were obtained by Springgate and Loeb (1973) with enzymes isolated from human leukemic cells.

Thus, the DNA-synthesizing machine itself without exposure to any external factors would introduce errors in the genetic template while the environment would lead to speeding up of the mutational process. It is also quite significant that quantitatively the frequency of errors of the replicating enzyme (10^{-4} to 10^{-6} per replication event) fully coincides with the overall statistic frequency of spontaneous mutagenesis (Dobzhansky, 1941).

This coincidence is hardly accidental. It is highly likely that in the course of evolution the error frequency was fixed by the natural selection and all the organisms whose DNA-polymerase showed a higher or a lower error frequency were eliminated by natural selection.

Thus, errors of replicating enzymes may constitute one of the causes of mutagenesis. The general scheme of mutational events resulting from the mistakes of these enzymes is presented at Fig. 27.

The frequency of errors made by the replicating enzymes is still higher when a cell or organism is exposed to chemical agents or radiation affecting the efficiency of these enzymes. It refers, for example, to dyes of the acridine series and nitrosoguanidine (see appropriate sections above). Streisinger *et al.* (1966), who studied lysozyme mutations of phage T4 showed that acridines did induce frame-shift mutations. Similar data were obtained by many authors studying the histidinol dehydrogenase gene of *Salmonella* after mutagenic treatment by several mutagens (Hartman *et al.*, 1973; Ino and Yourno, 1974, and so forth). Their findings may be treated as evidence of the errors of replicating enzymes. It is remarkable that many authors found frame-shift mutations in those regions of a gene where thymine nucleotides would repeat many times. This clustering of the same bases may be one of the reasons for the appearance of hotpoints (i.e., points of the highest mutation frequency in a genome).

Mutation Induction as a Result of Mistakes of Repair Enzymes

During the last two decades several modes of repair of the normal DNA structure, damaged by physical or chemical agents, were discovered. It was found that in some cases repair enzymes eliminated the damage completely, causing no structural changes in the DNA helix, while in others the process of repair considerably affected the DNA structure (Witkin, 1968; Pons, 1973; Verly, 1974, and others). We shall consider the scheme of dark repair, dwelling on the stages at which mutations may be generated as a result of the errors of repair enzymes.

First of all, the errors may arise when the first nuclease is cutting the DNA strand at the site of the damage. If this endonuclease makes mistakes with the same insignificant frequency as "precision" enzymes, such as

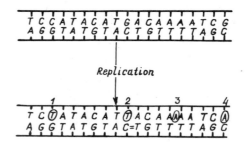

FIG. 27. Mutations arising from the mistakes of replication. 1—point mutation due to wrong replication (DNA-polymerase mistake); 2—wrong pairing with damaged region (for example, opposite to CT dimer region); 3—the insertion of extra base opposite to region of repeating bases; 4—mistake on the end of nascent chain.

DNA-polymerase (i.e., 10^{-5} to 10^{-6}), the results may be quite meaningful. The thing is that the two events—cutting the damaged strand itself or the opposite one—are not at all equal in their consequences. According to a number of authors (Rupp and Howard-Flanders, 1968; Bremer *et al.,* 1968), the gap opposite the thymine dimer will not be resynthesized for a long time, the delay of the synthesis for each dimer amounting to 10 seconds. Since the rates of the DNA syntheses in bacteria and phages and chromosomes of higher organisms (Cairns, 1966) are the same and equal to 1000–1500 nucleotides per second, the speed of resynthesis of the gap opposite the dimer drops 10^4-fold. It leaves sufficient time for the second stage of cutting the damaged strand by endonuclease with resulting fragmentation of the DNA molecule. If replication occurs at this region, then a wrong base will be inserted opposite the damaged one and hence complete mutation will be produced (pure mutant clones). This mechanism and its role in mutagenesis and recombination were considered by Soyfer (1969, pp. 417–425) and Dubinin and Soyfer (1969). Experimental evidence of the role of repair enzymes in the induction of complete mutation was obtained (Soyfer and Dorohov, 1972; Roberts *et al.,* 1974; Makino and Okada, 1974).

The mechanism discussed above allows therefore the explanation of the nature of pure and mixed mutant clones and the quantitative laws of the production of complete and mosaic mutations.

It should be noted that at the stage of resynthesis of the gap during DNA repair the generation of errors in the DNA structure is governed by the rules discussed in connection with normal semiconservative replication (i.e., by way of erroneous resynthesis of bases). Besides, such errors as insertion of extra bases in a resynthesized chain, especially in the region of recurrent nucleotide sequences in irradiated DNA may also lead to mutations for, according to the reports of many authors (Streisinger *et al.,* 1968; Hartman *et al.,* 1973 and Isono and Yourno, 1974), it is the region where frame-shift mutations occur most often.

Thus, mutations may result from the errors of a number of repair enzymes (Fig. 28).

We shall point out one more reason for mutation production in the course of erroneous enzymatic repair. If dimers or other lesions were "ignored" by repair enzymes and preserved in DNA structure after repair, then these damages produced mutations in the next act of replication. The persistence of thymine dimers in the DNA structure throughout, at least, four replication cycles was reported by Bridges and Munson (1968).

The persistence of damaged regions in DNA during its replication may lead to mutations for at least three reasons (Fig. 29):
1.   Wrong bases will arise opposite the damaged regions;
2.   Unstable lesions (e.g., cytosine or thymine-cytosine dimers) will

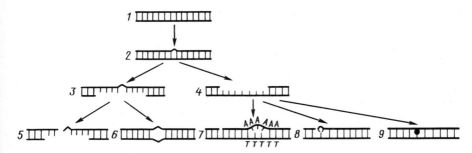

FIG. 28. Mutations arising from the errors of repair enzymes. 1—original DNA; 2—DNA with the damage recognized by repair enzymes; 3—the error in excision causing splitting out of the region opposite to the damaged one; 4—normal repair process at the stage of gap formation; 5—double-stranded DNA breakage along the pathways of a long-lived gap; 6—the complete mutation; 7—the insertion of an extra base in reparative replication; 8—the insertion of an extra base at the end of repaired fragment in reparative replication; 9—the error of the repairing DNA-polymerase resulting in the substitution of the wrong base in a repaired fragment (from Soyfer, 1970a).

bring about "mutagenic" bases in the DNA structure and substitution of a base in a pair;

3. A long-lived gap will appear opposite the lesion, hence mispairing or molecular fragmentation may take place.

Mistakes Arising During Recombination

Mutation production during recombination was previously postulated by Witkin (1968). Based on the above principles, we can point out some other possibilities of the mutation production during recombination. As in the case of repair and replication errors, we shall consider mutation production in the course of recombination as a result of enzymatic errors (Fig. 30):

1. If there is injury in one of the recombining molecules in the region that serves as a template for the recombination synthesis, it will be erroneously replicated and will give rise to a mutation.

2. During the recombination synthesis, mutations may arise from the errors of recombinational DNA polymerase.

3. An injury at the end of a fragment of a broken single strand of one of the recombining molecules may give rise to a mutation during joining with the single-stranded end of another molecule.

4. After the synapse of the recombining molecules (i.e., formation of a heteroduplex molecule) certain single-stranded regions will be resynthesized by so-called additional synthesis; an injury in the maternal strand will be fixed in the course of the additional synthesis.

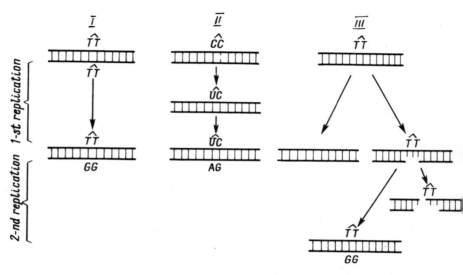

FIG. 29. Molecular mechanism production of the mutation resulting from the errors of the repair system on passing of the initial damage over the succeeding replication cycles (from Soyfer, 1970a).

5. Finally, at the sites of the additional syntheses, while a resynthesized fragment is joining to an earlier formed one, the insertion of an extra nucleotide or union of the unfinished strand with the fragment may give rise to frame-shift mutations.

We have dwelled upon several external and internal conditions leading to mutations. Such an examination has become possible with the recognition of a complicated nature of the reactions underlying the vital functions, in particular, those which were very aptly described as "enzymology of the genetic processes."

Some authors have already advanced certain ideas as regards possible participation of repair, recombination, and replication in the appearance of spontaneous mutations, although a sufficiently comprehensive analysis has not yet been made (see, e.g., von Borstel, 1969, Witkin, 1968, etc.).

The present study is also incomplete, because we have just started to investigate the phenomenon connected with the storage, hereditary transmission, and realization in ontogenesis of the genetic information. But even at the outset it allows us to outline new approaches to the investigation of mutations and, what is of paramount importance, to the control over this process. Also, it allows us to point to some previously disregarded causes of spontaneous and induced mutagenesis, which are the main point of the present work.

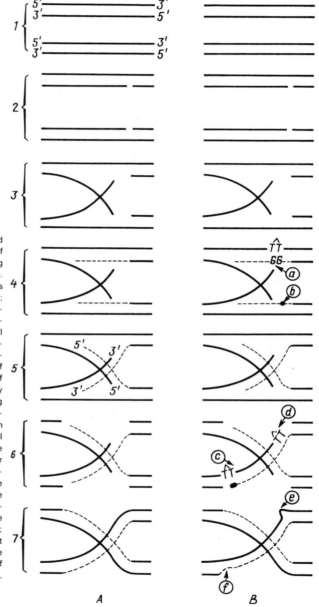

FIG. 30. Mutation induction caused by the errors in the work of recombining enzymes considering Whitehouse's recombination scheme. A—normal recombination process after Whitehouse (to the left); B—the same scheme marking the alternative points of mutation production (to the right). 1—original molecules; 2—single-stranded cutting; 3—scission of strands; 4—additional synthesis; 5—the formation of synapse between the strands of original molecules and those of newly synthesized ones; 6—the damaging of single-stranded regions; 7—joining of molecule ends; a—mutation arising in the region of the additional synthesis opposite the changes in the maternal strand (e.g., opposite dimer TT); b—a DNA-polymerase error during the additional synthesis; c—the insertion of the wrong base opposite dimer TT at the final point of the additional synthesis; d—changes at the molecule terminus (near dimer TT); e—the insertion of an extra base at the end of rejoining molecules; f—the union of fragment with the deletion of the terminal base (from Soyfer, 1970a).

A   B

REFERENCES

Adelberg, E. A., Mandel, M., and Chen, G. C. C., 1965, Optimal conditions for mutagenesis by N-methyl-N-nitro-nitrosoguanidine in *Escherichia coli* K12, *Biochem. Biophys. Res. Commun.* **18**:788–795.

Adler, I. D., 1966, Cytogenetic investigations of mutagenic action of caffeine in premeiotic spermatogenesis in mice, *Humangenetik* **3**:82–83.

Adler, I. D., and Schöneich, G., 1967, Mutagenic action of caffeine in an ascites tumor strain of mice: cytogenetic investigation, *Humangenetik* **4**:374–376.

Ahund-Zade, A. I., and Khvostova, V. V., 1966, Cytogenetical analysis of mutagenic effect of ionized radiation and alkylated compound in *Pisum, Genetika* **No. 6**, 47–54 (in Russian).

Akifiev, A. P., Makarov, V. B., Polunovsky, V. A., and Jurchenko, V. V., 1965, The study of mutagenic after-effect in cultivated L-cells. *Genetika,* **No. 3,** 19–35 (in Russian).

Alberts, B. M., 1968, Characterization of a naturally occuring, crosslinked fraction of DNA. II. Origin of the cross-linkage, *J. Mol. Biol.* **32**:405–421.

Alderson, T., 1961, Mechanism of mutagenesis induced by formaldehide, *Nature* **191**:251–253.

Alderson, T., and Khan, A. H., 1967, Caffeine-induced mutagenesis in *Drosophila, Nature* **215**:1080–1081.

Alderson, T., and Khan, A. H., 1968, Acridine mutagenesis (in dark) in *Drosophila, Mutation Res.* **5**:147–154.

Alderson, T., and Scott, B. R., 1970, The photosensitising effect of 8-methoxypsoralen on the inactivation and mutation of *Aspergillus conidia* by near ultraviolet light, *Mutation Res.* **9**:569–578.

Alexander, P., 1952, Interference with formation of a nucleoprotein complex by radiomimetic compounds, *Nature* **169**:226–227.

Alexander, P., and Moroson, H., 1962, Cross-linking of DNA to protein following ultraviolet irradiation of different cells, *Nature* **194**:882–883.

Alexander, P., and Stacey, K. A., 1955, Production of "hidden" breaks in DNA by the direct action of ionized radiation, *In Progress in Radiobiology, Internat. Conf. Radiobiol.,* Cambridge, p. 105.

Alexander, P., and Stacey, K. A., 1958, Comparison of the changes produced by ionizing radiation and by the alkylating agents: evidence for a similar mechanism at the molecular level, *Ann. N.Y. Acad. Sci.* **68**:1225–1237.

Alikhanian, S. I., and Ilyina, T. S., 1958, Mutagenic action of actinophage, *J. Gen. Biol.* **19**:348–356.

Alikhanian, S. I., and Mkrtumian, N. M., 1964, The induction of host mutations in extracellular phage without preliminary replication in sensitive cells, *Mutation Res.* **1**:355–361.

Allan, R. K., and McCalla, D. R., 1966, Prophage induction by N-methyl-N′-nitro-N-nitrosoguanidine, *Canad. J. Microbiol.* **12**:202–204.

Altenbern, R. A., 1966, Apparent genomic mapping of *Staphylococcus aureus* by a new method, *Biochem. Biophys. Res. Commun.* **25**:346–353.

Altenbern, R. A., 1973, Nitrosoguanidine mutagenesis and chromosomal replication in *Staphylococcal* [sic!] *aureus. Microbial. Genet. Bull.* **35**:6.

Ames, B. N., and Whitfield, H. J., 1966, Frameshift mutagenesis in *Salmonella, Cold Spring Harbor Symp. Quant. Biol.* **31**:221–225.

Anagnostopolous, C., and Crawford, I. P., 1961, Transformation studies on the linkage of markers in the tryptophan pathway in *Bacillus subtilis, Proc. Natl. Acad. Sci.* **47**:378–390.

Andrew, L. E., 1959, The mutagenic activity of caffeine in *Drosophila, Amer. Natur.* **43**:135–138.

Ardashnikov, S. N., Soyfer, V. N., and Goldfarb, D. M., 1964, Induction of h-mutations in extracellular phage T2 with γ-irradiation, *Biochem. Biophys. Res. Commun.* 16:455–459; See also: Ardashnikov, S. N., Soyfer, V. N., and Goldfarb, D. M., 1963, *Proc. U.S.S.R. Acad. Medical Sci.* No 12, p. 43–50 (In Russian).

Astaurov, B. L., 1947, The experimental proof of direct injury of cytoplasm of living cell at X-irradiation. *Proc. U.S.S.R. Acad. Sci.* 58:887–890 (In Russian).

Auerbach, Ch., 1959, Specificity on the functional level, In Böhme, H., ed., *Chemical Mutagenese, Erwin Baur Gedächtnisvorlesungen,* 1:121–123, Akademie Verlag, Berlin.

Auerbach, Ch., 1967, The chemical production of mutations, *Science* 158:1141–1147.

Auerbach, Ch., 1971, History of research on chemical mutagenesis, *In* A. Hollaender, ed., *Chemical mutagens: Principles and Methods of their Detection,* 3:1–19, Plenum Press, N.Y.

Auerbach, Ch., and Kølmark, G., 1959, See Auerbach, 1959.

Auerbach, Ch., and Robson, J. M., 1941, Production of mutations by alkyl isothiocyanate, *Nature* 154:80–81.

Baetcke, K. P., and Sparrow, A. H., 1967, The relationship of DNA content to nuclear and chromosome volumes and to radiosensitivity, *Proc. Natl. Acad. Sci.* 58:533–540.

Baird, W., and Brookes, P., 1973, Isolation of the hydrocarbon-deoxyribonucleoside products from the DNA mouse embryo cells treated in culture with tritium-labeled 7-methylbenz[a]-antracene, *Cancer Res.* 33:2378–2385.

Baker, R., and Tessman, I., 1968, Different mutagenic specificities in phage S13 and T4: in vivo treatment with N-methyl-N′-nitro-N-nitrosoguanidine, *J. Mol. Biol.* 35:439–448.

Ball, C., and Roper, I. A., 1966, Studies on the inhibition and mutation of *Aspergillus nidulans* by acridines, *Genetical Res.* 7:207–221.

Baptist, J. E., and Haynes, R. H., 1972, The UV-X-ray synergism in *Escherichia coli B/r-1.* Inhibition by the incorporation of 5-bromouracil and by purine starvation, *Photochem. Photobiol.* 16:459–463.

Barnett, W. E., and Brockman, H. E., 1962, Induced phenotypic reversion by 8-azaguanine and 5-fluorouracil, *Biochem. Biophys. Res. Commun.* 7:199–203.

Baron, F., and Brown, D. M., 1955, Nucleotides. Part XXXIII. The structure of cytidylic acid *a* and *b, J. Chem. Soc.* 2855–2860.

Barricelli, N. A., and Del Zoppo, G., 1958, Genotypic reversion by methylene blue: the orientation of guanine-hydroxymethylcytosine at mutated sites in rII mutants of phage T4, *Molec. Gen. Genetics* 101:51–58.

Bartoshevich, J. E., 1966, Ethylenimines and mutational process, *In* Rapoport, I. A., ed., *Supermutagens,* 211–268 (in Russian).

Baumiller, R. C., 1966, Virus induced point mutations, *Genetics* 54:322.

Baur, E., 1916, See Gustafsson, A., 1959.

Bautz, E., and Freese, E., 1960, On the mutagenic effect of alkylating agents, *Proc. Natl. Acad. Sci.* 46:1585–1594.

Baylor, M. B., and Mahler, H., 1962, Effects of nitrous acid on intracellular T2 bacteriophage, *Virology* 16:444–451.

Becker, E. F., Zimmerman, B. K., and Geiduschek, E. P., 1964, Structure and function of crosslinked DNA. I. Reversible denaturation and *Bacillus subtilis* transformation, *J. Mol. Biol.* 8:377–391.

Beers, R. F., Hendley, D. D., Steiner, R. F., 1958, Inhibition and inactivation of polynucleotide phosphorylase through the formation of complexes between acridine orange and polynucleotides, *Nature,* 182:242–244.

Bendich, A., Borenfreund, E., Korngold, G. C., Krim, M., and M. E. Balis, 1964, Amino acids or small peptides as punctuation in the genetic code of DNA. Acidi nucleici e loro funzione biologica, *Istituto Lombardo di scienze e lettere, Pavia,* 214–237.

Ben Hur, E., and Rosenthal, J., 1970, Photosensitized splitting of pyrimidine dimers, *Photochem. and Photobiol.* **11**:163–168.

Benzer, S., and Freese, E., 1958, Induction of specific mutations with 5-bromouracil, *Proc. Natl. Acad. Sci.* **44**:112–119.

Bergel, F., 1973, Carcinogenic hazards in natural and manmade environments, *Proc. Roy. Soc., Lond.* (B) **185**:165–181.

Berger, H., Brammar, W. J., and Yanofsky, C., 1968, See J. M. Drake, 1970.

Beukers, R., and Berends, W., 1960, Isolation and identification of the irradiation product of thymine, *Biochim. Biophys. Acta* **41**:550–551.

Beukers, R., Ijlstra, J., and Berends, W., 1958, The effect of ultraviolet light on some components of the nucleic acids. II. In rapidly solid frosen solutions, *Recueil trav. chim. Pays-Bas* **77**:729–737.

Bhatia, C. R., 1967, Increased mutagenic effect of ethyl methanesulfonate when dissolved in dimethyl sulfoxide, *Mutation Res.* **4**:375–376.

Blackburn, G. M., and Davies, R. J. H., 1965, The structure of thymine photodimer, *Chem. Communs.* No. **1**:215.

Blok, J., and Loman, H., 1973, The effect of γ-irradiation in DNA, *Curr. Top. Radiat. Res. Q.* **9**:165–245.

Böhme, H., 1961, Streptomycin-abhängige Mutanten von *Proteus mirabilis* und ihre Verwendung in Mutationsversuchung mit Manganchlorid, *Biol. Zentralbl.* **80**:5–32.

Böhme, H., 1968, Absence of repair of photodynamically induced damage in two mutants of *Proteus mirabilis* with increased sensitivity to mono functional alkylating agents, *Mutation Res.* **6**:166–168.

Böhme, H., and Geissler, E., 1968, Repair of lesions induced by photodynamic action and by ethyl methanesulphonate in *Escherichia coli, Molec. Gen. Genetics* **103**:228–232.

Böhme, H., and Wacker, A., 1963, Mutagenic activity of thiopyronine and methylenblue in combination with visible light, *Biochem. Biophys. Res. Commun.* **12**:137–139.

Boiron, M., Tanzer, J., Thomas, M., and Hampe, A., 1966, Early diffuse chromosome alterations in monkey kidney cells infected in vitro with herpes simplex virus, *Nature* **209**:737–738.

Bollum, F. J., and Setlow, R. B., 1963, Ultraviolet inactivation of DNA primer activity. I. Effects of differing wavelengths and doses, *Biochim. Biophys. Acta* **68**:599–607.

Borenfreund, E., Krim, M., and Bendich, A., 1964, Chromosome aberrations induced by hyponitrite and hydroxylamine derivatives, *J. Natl. Cancer Inst.* **32**:667–679.

Borstel von, R., 1969, On the origin of spontaneous mutation, *Jap. J. Genet.* **44**, Suppl. 1:102.

Boyer, H., See Ames, B. N., and Whitfield, H. J., 1966.

Boyland, E., and Nery, R., 1965, The metabolism of urethan and related compounds, *Biochem. J.* **94**:198–208.

Brammar, W. J., Berger, H., and Yanofsky, C., 1967, Altered amino acid sequences produced by reversion of frameshift mutants of tryptophan synthetase A gene of *E. coli, Proc. Natl. Acad. Sci.* **58**:1499–1506.

Bremer, H., Michalke, H., and Mueller, K., 1968, Messenger RNA synthesis in vivo and in vitro, *In Annual Report of the Biology Division of Southwest Center for Advanced Studies for 1967 year.* Dallas, p. 6.

Brendel, M., 1968, Induction of mutation in phage T4 by extracellular treatment with methylene blue and visible light, *Molec. Gen. Genetics* **101**:111–115.

Brendel, M., and Kaplan, R. W., 1967, Photodynamische Mutationsanlösung und Inaktivierung bein Serratia-Phagen Kappa durch Methylenblau und Licht, *Molec. Gen. Genetics* **99**:181–190.

Brenner, S., Benzer, S., and Barnett, L., 1958, Distribution of proflavine-induced mutations in the genetic fine structure, *Nature* **182**:983–985.

Brenner, S., Barnett, L., Crick, F. H. C., and Orgel, A., 1961, The theory of mutagenesis, *J. Mol. Biol.* **3**:121–124.

Bresler, S., Mosevitsky, M., and Vyacheslavov, L., 1970, Complete mutagenesis in a bacterial population induced by thymine starvation on solid media, *Nature* **225**:764–766.

Bresler, S. E., Mosevitsky, M. T., and Vyacheslavov, L. G., 1973, Mutations as possible replication errors in bacteria growing under conditions of thymine deficiency, *Mutation Res.* **19**:281–293.

Bridges, B. A., and Munson, R. J., 1968, Mutagenesis in *Escherichia coli*: evidence for mechanism of base change mutation by ultraviolet radiation in a strain deficient in excision-repair, *Proc. Roy. Soc. London (B)* **171**:213–226.

Bridges, B. A., Law, J., and Munson, R. J., 1968, Mutagenesis in *Escherichia coli*. II. Evidence for a common pathway for mutagenesis by ultraviolet light, ionizing radiation and thymine deprivation, *Molec. Gen. Genetics* **103**:266–273.

Brockman, H. E., and Goben, W., 1965, Mutagenicity of a monofunctional alkylating derivative of acridine in *Neurospora, Science* **147**:750–751.

Brookes, P., and Lawley, P. D., 1960, The methylation of adenosine and adenilic acid, *J. Chem. Soc.* 539–545.

Brookes, P., and Lawley, P. D., 1961, Reaction of mono- and di-functional alkylating agents with nucleic acids, *Biochem. J.* **80**:496–503.

Brookes, P., and Lawley, P. D., 1964, Reaction of some mutagenic and carcinogenic compounds with nucleic acids, *J. Cell. Comp. Phys.* **64**:11–127.

Brown, D. F., 1966, X-ray-induced mutation in extracellular bacteriophage T4, *Mutation Res.* **3**:365–373.

Brown, D. M., and Schell, P., 1965, Nucleotides, 68. The reaction of hydroxylamine with cytosine and related compounds, *J. Chem. Soc.,* 208–215.

Brown, D. M., McNaught, A. D., and Schell, P., 1966, The chemical basis of hydrasine mutagenesis, *Biochem. Biophys. Res. Commun.* **24**:967–971.

Brusick, D. J., 1970, The mutagenic activity of ICR-170 in *Saccharomyces cerevisiae, Mutation Res.,* **10**:1–19.

Budowski, E. I., 1968, The modification of nucleic acid by hydroxylamine and its derivatives. *Thesis of Doct. Diss., Moscow* (In Russian).

Budowsky, E. I., and Sverdlov, E. D., 1972, Mechanism of mutagenic action of hydroxylamine, *In Molecular mechanisms of genetic processes, Moscow,* 81–88 (in Russian).

Budowsky, E. I., Sverdlov, E. D., Shibaeva, R. P., Monastyrskaya, G. S., and Kochetkov, N. K., 1971a, Mechanism of the mutagenic action of hydroxylamine. III. Reaction of hydroxylamine and O-methylhydroxylamine with the cytosine nucleus, *Biochem. Biophys. Acta,* **246**:300–319.

Budowsky, E. I., Sverdlov, E. D., and Monastyrskaya, G. S., 1971b, Mechanism of the mutagenic action of hydroxylamine. IV. Reaction of hydroxylamine and O-methylhydroxylamine with adenine nucleus, *Biochim. Biophys. Acta,* **246**:320–328.

Budowsky, E. I., Krivisky, A. S., Klebanova, L. M., Metlitskaya, A. Z., Turchinsky, M. F., and Savin, F. A., 1974, The action of mutagens on MS2 phage and its infective RNA, *V, Mutation Res.,* **24**:245–258.

Burdette, W. Y., and Yoon, J. S., 1967, Mutations, chromosome aberrations and tumors in insects treated with oncogenic viruses, *Science,* **155**:340–341.

Butler, G. C., 1949, The effects of X-radiation on sodium thymus nucleate, *Canad. J. Res.,* **27** B:972–987.

Butler, J. A. V., and Smith, K. A., 1950, The action of ionizing radiation and of radiomimetic substances on deoxyribonucleic acid. Part I. The action of some compounds of the "mustard" type, *J. Chem. Soc.* 3411–3418.

Cairns, J., 1966, Autoradiography of HeLa cells DNA. *J. Mol. Biol.* **15**:372–373.

Carbon, I., and Curry, I. B., 1968, A change in the specificity of transfer RNA after partial deamination with nitrous acid, *Proc. Natl. Acad. Sci.* **59**:467–474.

Carlson, E. A., and Oster, I. I., 1962, Comparative mutagenesis of the dumpy locus in *Drosophila melanogaster*. II. Mutational mosaicism induced with out apparent breakage by a mono functional alkylating agent, *Genetics* **47**:561–576.

Carlson, E. A., Sederoff, R., and Cogan, M., 1967, Evidence favoring a frameshift mechanism for ICR-170 induced mutations in *Drosophila melanogaster, Genetics* **55**:295–313.

Cattanach, B. M., 1962, Genetical effects of caffeine in mice. *Z. Vererbungslehre* **93**:211–219.

Cavaliery, L. F., Baturay, O., Eidinoff, L. M., Sharpe, J. S., and Laughlin, J. S., 1955, Irradiation of solutions of deoxyribonucleic acid, *Radiation Res.* **3**:218.

Cerda-Olmedo, E., and Hanawalt, P. C., 1968, Diazomethane as the active agent in nitroso-guanidine mutagenesis and lethality, *Molec. Gen. Genetics* **101**:191–202.

Cerda-Olmedo, E., Hanawalt, P. C., and Guerola, N., 1968, Mutagenesis of the replication point by nitrosoguanidine: map and pattern of replication of the *Escherichia coli* chromosome, *J. Mol. Biol.* **33**:705–719.

Champe, S. P., and Benzer, S., 1962, Reversal of mutant phenotypes by 5-fluorouracil: an approach to nucleotide sequence in messenger RNA, *Proc. Natl. Acad. Sci.* **48**:532–546.

Chan, E. W., and Ball, J. K., 1971, Interaction of DNA with three dimethyl derivatives of benz[c]acridines, *Biochim. Biophys. Acta* **238**:31–45.

Chandra, P., Wacker, A., Sussmuth, R., and Lingens, F., 1967, Wirkung von 1-Methyl-guanidin auf die Matrizenaktivität der Polynucleotide bei der Zellfreien Proteinsynthese, *Z. Naturforsch.* **22b**:512–517.

Chu, B. C. F., Brown, D. M., and Burdon, M., 1973, Effect of nitrogen and catalase on hydroxylamine and hydrazine mutagenesis, *Mutation Res.* **20**:265–270.

Chu, B. C. F., Brown, D. M., and Burdon, M. G., 1974, N(4)-amino and N(4)-hydroxycytosines as base analogue mutagens, *Mutation Res.,* **23**:267–273.

Clark, A. M., and Clark, E. J., 1968, The genetic effects of caffeine in *Drosophila melanogaster, Mutation Res.* **6**:227–234.

Clarke, C. H., 1969, Chemical mutagenesis in *Escherichia coli* B/r. The influence of repair system for UV damage, *Mutation Res.* **8**:35–41.

Clive, D., 1974, Mutagenicity of thioxanthenes (hycantone, Lucantone and four indazole derivatives) at TK locus in cultured mammalian cells, *Mutation Res.* **26**:307–318.

Cohn, N. S., 1964, Similar cytological effects of hydroxylamine and 5-BUDR, agents with different modes of action, *Experientia (Basel)*, **20**:158.

Collyns, B., Okada, S., Scholes, G., Weiss, J. J., and Wheeler, C. M., 1965, Chain scission and hydrogen bond breakage on irradiation of DNA, *Radiat. Res.* **25**:526–536.

Conlay, J. J., 1963, Effect of ionizing radiation on adenine in aerated and de-aerated aqueous solutions, *Nature* **197**:555–557.

Cooper, P. D., 1964, The mutation of poliovirus by 5-fluorouracil, *Virology* **22**:186–192.

Corran, J., 1968, The induction of super supressor mutants of *Bacillus subtilis* by ethyl methanesulfonate and the posttreatment modification of mutation yield, *Molec. Gen.* **103**:42–57.

Coughlin, C. A., and Adelberg, E. A., 1956, Bacterial mutation induced by thymine starvation, *Nature* **178**:531–532.

Cox, A., Overend, W. D., Peacocke, A. R., and Wilson, S., 1958, The action of γ-rays on sodium deoxyribonucleate in solution, *Proc. Roy. Soc. London (B)*, **149**:511–534.

Crick, F. H. C., Barnett, L., Brenner, S., and Watts-Tobin, R. J., 1961, The general nature of the genetic code for proteins, *Nature* **192**:1227–1231.

Delič, V., Hopwood, D. A., and Friend, E. J., 1970, Mutagenesis by N-methyl-N′-nitro-N-nitrosoguanidine (NTG) in *Streptomyces coelicolor, Mutation Res.* **9:**167–182.

De Mars, R. I., 1953, Chemical mutagenesis in bacteriophage T2, *Nature* **172:**964.

Demerec, M., and Hanson, J., 1951, Mutagenic action of manganeous chloride, *Cold Spring Harbor Symp. Quant. Biol.* **16:**215–228.

Dhillon, E. K. S., and Dhillon, T. S., 1974, N-Methyl-N′-nitro-N-nitrosoguanidine and hydroxylamine induced mutants of the rII-region of phage, *Mutation Res.* **22:**222–233.

Di Paolo, I. A., 1952, Studies on chemical mutagenesis utilising nucleic acid components, urethane and hydrogen peroxide, *Amer. Nat.* **86:**49.

Dobzhansky, Th., 1941, *Genetics and the origin of species,* Columbia University Press, New York.

Doskočil, J., 1965, Reaction of deoxyribonucleic acid with triethylenemelamine, *Collection Czech. Chem. Commun.* **30:**2434–2444.

Drake, J. W., 1963, Polyimines mutagenic for bacteriophage T4B, *Nature* **197:**1028.

Drake, J. W., 1964, Studies on the induction of mutations in bacteriophage T4 by ultraviolet irradiation and by proflavin, *J. Cell. Comp. Physiol.* **64:**19–31.

Drake, J. W., 1966, Ultraviolet mutagenesis in bacteriophage T4. 2. Photoreversal of mutational lesions, *J. Bacteriol.* **92:**144–147.

Drake, J. W., 1967, The length of the homologous pairing region for genetic recombination in bacteriophage T4, *Proc. Natl. Acad. Sci.* **58:**962–966.

Drake, J. W., 1969, Mutagenic mechanisms, *Ann. Rev. Genet.* **3:**247–268.

Drake, J. W., 1970, *The molecular basis of mutation,* Holden-day, San Francisco, London, Cambridge, Amsterdam.

Drake, J. W., and McGuire, J., 1967, Properties of r-mutants of bacteriophage T4 photodynamically induced in the presence of thiopyronine and psoralen, *J. Virology* **1:**260–267.

Druckrey, H., Steinhoft, D., Preussman, R., and Ivankovic, S., 1963, Krebserzeugung durch einmalige von Methylnitrosoharnstoff und verschiedenen Dialkyl-nitrosaminen, *Naturwiss.* **50:**735.

Dubinin, N. P., and Soyfer, V. N., 1969, Chromosome breakage and complete genic mutation production in molecular terms, *Mutation Res.* **8:**353–365.

Dulbecco, R., and Vogt, M., 1958, Studies on the induction of mutations in poliovirus by proflavin, *Virology* **5:**236–243.

Dunn, D. B., and Smith, J. D., 1954, Incorporation of halogenated pyrimidines into deoxyribonucleic acids of *Bacterium coli* and its bacteriophages, *Nature* **174:**305–306.

Eidinoff, M. L., Cheong, L., and Rich, M. A., 1959, Incorporation of unnatural pyrimidine bases into deoxyribonucleic acid of mammalian cells, *Science* **129:**1550–1551.

Eisenstark, A., and Rosner, I. L., 1964, Chemically induced reversions in the cys C region of *Salmonella typhimurium, Genetics* **49:**343–355.

Eisenstark, A., Eisenstark, R., and Sicklle, van R., 1965, Mutation of *Salmonella typhimurium* by nitrosoguanidine, *Mutation Res.* **2:**1–10.

Ekert, B., and Monier, R., 1959, Structure of thymine hydroperoxide produced by X-irradiation, *Nature* **184:**58–59.

Ekert, B., and Monier, R., 1960, Effect of X-ray on cytosine in aerated aqueous solution, *Nature* **188:**309–310.

El-Metainy, Tano, S., Yano, K., and Yamaguchi, H., 1973, Chemical nature of radiation-induced single-strand breaks in the DNA of dormant barley seeds *in vivo, Radiation Res.* **55:**324–333.

Elmore, D. T., Gulland, J. M., Jordan, J. O., and Taylor, H. F. W., 1948, Reactions of nucleic acids with mustard gas, *Biochem. J.* **42:**308–316.

Emmerson, P. T., and Howard-Flanders, P., 1965, Post-irradiation degradation of DNA following exposure of ultraviolet sensitive and resistant bacteria to X-ray, *Biochem. Biophys. Res. Commun.* **18**:24–29.

Engel, W., Krone, W., and Wolf, U., 1967, The action of thioguanine, hydroxylamine and 5-BDU on human chromosomes in vitro, *Mutation Res.* **4**:353–368.

Endo, H., and Takahashi, K., 1973, Methylguanidine, a naturally occuring compound showing mutagenicity after nitrosation in gastric juice, *Nature* **245**:325–326.

Fahmy, O. G., and Fahmy, M. J., 1957, Mutagenic response to the alkylmethanesulfonates during spermatogenesis in *Drosophila melanogaster, Nature* **180**:31–34.

Fahmy, O. G., and Fahmy, M. J., 1959, Small deficiencies and deletions under the monofunctional alkylating mutagens, *Brit. Cancer Comp. Ann. Reports* **37**:116–171.

Fahmy, O. G., and Fahmy, M. J., 1961, Induction of mutations by deoxyribonucleic acid in *Drosophila melanogaster, Nature* **191**:776.

Fahmy, O. G., and Fahmy, M. J., 1964, Mutagenicity of ionic polymers and leucomogenic virus relative to nucleic acid in *Drosophila melanogaster, Nature* **204**:46.

Fahmy, O. G., and Fahmy, M. J., 1970, Induction of bobbed (bb) mutations by polycyclic aromatic carcinogens in *Drosophila, Mutation Res.,* **9**:239–243.

Fahmy, O. G., and Fahmy, M. J., 1973, Oxidative activation of benz[a]antracene and methylated derivatives in mutagenesis and carcinogenesis, *Cancer Res.* **33**:2354–2369.

Felkner, I. C., and Kadlubar, F., 1968, Parallel between ultraviolet light and 4-Nitroquinoline-1-oxide sensitivity in *Bacillus subtilis, J. Bacteriol.* **96**:1448–1449.

Folsome, C. E., 1962, Specifity of induction of T4 rII mutants by ultraviolet irradiation of extracellular phage, *Genetics* **47**:611–622.

Folsome, C. E., and Levine, D., 1961, Detecting reversions in T4 rII bacteriophage to r⁺ induced by ultraviolet irradiation, *Nature* **192**:1306.

Fosse, R., Hieulle, A., and Bass, L. W., 1924, Action de l'hydrazine sur l'uracyle et la thymine, *Compt. Rend. Acad. Sci.* **178**:811–813.

Foster, R. A. C., 1948, An analysis of the action of proflavine on bacteriophage growth, *J. Bacteriol.* **56**:795–809.

Fournier, Et., 1973, Free radicals and toxicology, *J. Eur. Toxicology* **6**:109–122.

Fox, A. S., and Yoon, S. B., 1966, Specific genetic effects of DNA in *Drosophila melanogaster, Genetics* **53**:897–911.

Freese, E., 1959a, On the molecular explanation of spontaneous and induced mutations, *Brookhaven Symp. Biol.* **12**:63–75.

Freese, E., 1959b, The specific mutagenic effects of base analogues on phage T4, *J. Mol. Biol.* **1**:87–105.

Freese, E., 1963, The molecular mechanisms of mutations, *In* J. H. Taylor (ed), *Molecular Genetics, part 1,* Acad. Press. N.Y., London.

Freese, E., 1964, The influence of DNA structure and base composition on mutagenesis, *In Genetics Today,* 1964, p. 297–306.

Freese, E., and Cashel, M., 1964, Crosslinking of DNA by exposure to low pH, *Biochim. Biophys. Acta* **91**:67–77.

Freese, E., and Freese, E. B., 1965, The oxygen effect on deoxyribonucleic acid inactivation by hydroxylamines, *Biochemistry* **4**:2419–2433.

Freese, E., Bautz, E., and Freese, E. B., 1961, The chemical and mutagenic specificity of hydroxylamine, *Proc. Natl. Acad. Sci.* **47**:845–855.

Freese, E., Freese, E. B., and Graham, S., 1966, The oxygen-dependent reaction of hydroxylamine with nucleotides and DNA, *Biochim. Biophys. Acta* **123**:17–25.

Freese, E., Sklarow, S., and Freese, E. B., 1968, DNA damage caused by antidepressant hydrazines and related drugs, *Mutation Res.* **5**:343–348.

Freese, E. B., 1961, Transitions and transversions induced by depurination agents, *Proc. Natl. Acad. Sci.* **47**:540–545.

Freese, E. B., 1965, The effects of urethan and hydroxyurethan on transforming DNA, *Genetics* **51**:953–960.

Freese, E. B., and Freese, E., 1961*a*, Induction of reverse mutations and cross reactivation of nitrous acid-treated phage T4, *Virology* **13**:19–30.

Freese, E. B., and Freese, E., 1961*b*, On the mutagenic effect of alkylating agents, *Proc. Natl. Acad. Sci.* **46**:540–545.

Freese, E. B., Gerson, J., Taber, H., Rhaese, H., and Freese, E., 1967, Inactivating DNA alterations induced by peroxides and peroxide-producing agents, *Mutation Res.*, **4**:517–531.

Fries, N., 1950, The production of mutations by caffeine, *Hereditas* **36**:134–149.

Gamow, G., 1954, Possible relation between deoxyribonucleic acid and protein structure, *Nature* **173**:318.

Gecheff, K., and Nicoloff, Hr., 1970, The activity of N-nitroso-N-methyl-urethane on chromosomes of barley, *Mutation Res.* **10**:119–123.

Geiduschek, E. P., 1961, "Reversible" DNA, *Proc.. Natl. Acad. Sci.* **47**:950–955.

Gershenson, S., 1934, Effect of ether on gene changes, *Drosophila Information Service*, **1**:54.

Gershenson, S., 1939, Induction of directed mutations in *Drosophila melanogaster*, *Compt. Rend. (Doklady) Acad. Sci. U.S.S.R.* **25**:224–227.

Gershenson, S., 1965*a*, Induction of lethal mutations in *Drosophila melanogaster* by DNA, *Genet. Res. (Cambr.)* **6**:157–162.

Gershenson, S. M., 1965*b*, Delayed mutagenic effect of DNA in *Drosophila*, *In Mechanism of mutation and inducing factors*, 291–293, Praha, Academia.

Gershenson, S. M., 1966, Altering genes to order, *New Scientist* 29 Dec., 724–725.

Gershenson, S. M., 1969, Mutagenic action of some biopolymers in *Drosophila*, *Japan J. Genetics* **44**:114–119.

Gershenson, S. M., Zilberman, R. A., Liovochkina, O. A., Sytko, P. O., and Tarnavsky, N. D., 1948, Induction of mutations in *Drosophila* by thymonucleic acid, *Gen. Biol.* **9**:69–88.

Gershenson, S. M., Alexandrov, Y. N., and Maliuta, S. S., 1971, Production of recessive lethals in *Drosophila* by viruses noninfectious for the host, *Mutation Res.* **11**:163–173.

Gezelius, K., and Fries, N., 1952, Phage resistant mutants induced in *E. coli* by caffeine, *Hereditas* **38**:112.

Gichner, T., and Veleminsky, J., 1973, Toxic and genetic factors limiting the increase of mutation frequency after treatment by alkylating nitrosocompounds and X-ray on *Arabidopsis* seeds. *Biol. Plant* **15**:350–353.

Gichner, T., Michaelis, A., and Riger, R., 1963, Radiomimethic effects of 1-methyl-3-nitro-1-nitrosoguanidine in *Vicia faba, Biochem. Biophys. Res. Commun.* **11**:120–124.

Gierer, A., and Mundry, K. W., 1958, Production of mutants of tobacco mosaic virus by chemical alteration of its ribonucleic acid in vitro, *Nature* **182**:1457.

Gilot-Delhalle, J., Thakare, R., and Moutschen, J., 1973, Fast rejoining processes in *Nigella damascena* chromosomes revealed lag fractionated ^{60}Co γ-ray exposures, *Radiation Bot.* **13**:229–242.

Giovanni, de, R., 1960, The effects of deuterium oxide on certain microorganisms, *Ann. N.I. Acad. Sci.* **84**:644–647.

Goldacre, R. J., Loveless, A., and Ross, W. C. J., 1949, Mode of production of chromosome abnormalities by the nitrogen mustards, *Nature* **163**:667–679.

Goldfarb, D. M., and Belyaev, D. L., 1966, Mutagenesis in phage T2 under the combined action of chemical mutagens, *Genetika* **6**:113–117 (in Russian).

Goldfarb, D. M., and Chubukov, V. F., 1963, Mutations of ht–character in bacteriophage T2 induced by combined action of chemical mutagens, *Bull. Exper. Biol. and Med.* **No. 2**:61–62 (in Russian).

Goldfarb, D. M., Nesterova, G. F., and Kusnetsova, V. N., 1966, Localization of h⁺–mutations of T2 phage, induced by various mutagens, *Genetika* **2**:19–22 (in Russian).

Goldin, A., Venditti, G. M., and Kline, I., 1959, Evaluation of antileukemic agents employing advanced leukemia L 1210 in mice, *Cancer Res.* **19** (Screening Data):429–466.

Goroshkina, G. I., and Soyfer, V. N., 1971, Influence of caffeine on size of mitotic index and the length of cell cycle of HeLa cells, *Cytologia,* **13**:1031–1034 (in Russian).

Granoff, A., 1961, Induction of Newcastle disease virus mutants with nitrous acid, *Virology* **13**:402–408.

Green, D. M., and Krieg, D. R., 1962, The delayed origin and mutants induced by exposure of extracellular phage T4 to ethyl methane sulfonate, *Proc. Natl. Acad. Sci.* **47**:64–72.

Greene, M. O., and Greenberg, J., 1960, The activity of nitrosoguanidines against ascites tumours in mice, *Cancer Res.* **20**:1166–1171.

Greer, S., and Zamenhof, S., 1962, Studies on depurination by heat, *J. Mol. Biol.* **4**:123–141.

Grossman, L., 1968, Studies on mutagenesis induced in vitro, *Photochem. Photobiol.* **7**:727–735.

Gupta, A. K., and Grover, N. S., 1970, Hydrazine induced breaks in chromosomes of *Vicia faba, Mutation Res.* **10**:519–520.

Gustafsson, A., 1960, Chemical mutagenesis in higher plants, *In* H. Böhme (ed.), *Erwin-Baur-Gedächtnisvorlesungen,* 1959:14–29, Akademie Verlag, Berlin.

Hagen, U., and Wild, R., 1964, Untersuchungen über die Strahlenempfindlichkeit der Desoxyribonukleinsäure, I. Die Empfindlichkeit der Wasserstoffbrücken, *Strahlentherapie,* **124**:275–283.

Hakala, M. T., 1959, Mode of action of 5-bromodeoxyuridine on mammalian cells in culture, *J. Biol. Chem.* **234**:3072–3076.

Hall, Z. W., and Lehman, I. R., 1968, An in vitro transversion by a mutationally altered T4-induced DNA polymerase, *J. Molec. Biol.* **36**:321–333.

Hanawalt, P. C., and Hayness, R. H., 1965, Repair replication of DNA in bacteria: irrelevance of chemical nature of base defect, *Biochem. Biophys. Res. Commun.* **19**:462–467.

Harrington, H., 1966, Effect of X-irradiation on the physicochemical properties of DNA, *Arch. Biochem. Biophys.* **117**:615–622.

Hartley, M. J., 1970, The frequency of reverse mutation at the XDH loci of *Aspergillus nidulans, Mutation Res.* **10**:175–183.

Hartman, P. E., Berger, H., and Hartman, Z., 1973, Comparison of hycanthine ('Etrenol'), some hycantone analogs, myxin and 4-nitroquinoline-1-oxide as frameshift mutagens, *J. Pharmacology Exper. Therapy* **186**:390–398.

Haynes, R. H., and Inch, W., 1963, Synergistic action of nitrogen mustard and radiation in microorganisms, *Proc. Natl. Acad. Sci.* **50**:839–846.

Haynes, J. A., Reese, C. B., and Lord Todd, 1964, The methylation of nucleosides and mononucleotides with diazomethane, *J. Chem. Soc.* 1406–1412.

Helström, K. E., Helström, J., and Sjogren, H. O., 1963, Further studies on karyotypes of a variety of primary transplanted mouse polyoma tumors, *J. Natl. Cancer Inst.* **31**:1239–1253.

Hems, G., 1958, Effects of ionizing radiations on aqueous solutions of guanilic acid and guanosine, *Nature* **181**:1721–1722.

Herrington, M. B., and Takahashi, I., 1973, Mutagenesis of bacteriophage PBS 2, *Mutation Res.* **20**:275–278.

Herriott, R. M., 1966, Mutagenesis, *Cancer Res.* **26**:1971–1979.

Hess, D., 1969, Versuche zur Transformation an Höheren Pflanzen: Induktion und konstante Weitergabe der Anthocyansynthese bei *Petunia hybrida, Z. Pflanzenphysiol.* **60**:348–356.

Hewlings, M. J. E., and Brown, D. M., 1968, See Grossman, L., 1968, and see also Brown, D. M., Hewlings, M. J. E., and Schell, P., 1968, The tautomeric state of N(4)-hydroxy-and of N(4)-amino-cytosine derivatives, *J. Chem. Soc. C*:1925–1929.

Higuchi, M., Endo, K., Hoshino, E., and Araya, S., 1973, Preferential induction of rough variants in *Streptococcus* mutants by ethidium bromide, *J. Dent. Res.* **52**:1070–1075.

Hirota, Y., 1960, The effect of acridine dyes on mating type factors in *E. coli, Proc. Natl. Acad. Sci.* **46**:57–64.

Holden, H. E., Ray, V., and Wahrenbourg, M. G., 1973, Mutagenicity studies with 6-mercaptopurine. I. Cytogenetic activity in vivo, *Mutation Res.* **20**:257–263.

Holmes, A. I., and Eisenstark, A., 1968, The mutagenic effect of thymine-starvation on *Salmonella typhimurium, Mutation Res.* **5**:15–21.

Horikawa, M., Nikaido, O., and Sugahara, T., 1968, Dark reactivation of damage induced by ultraviolet light in mammalian cells *in vitro, Nature* **218**:489–491.

Horn, E. E., and Herriot, R., 1962, The mutagenic action of nitrous acid on "single-stranded" (denatured) Hemophilis transforming DNA, *Proc. Natl. Acad. Sci.* **48**:1409–1416.

Howard, B., and Tessman, I., 1964a, Identification of altered bases in mutated single-stranded DNA. II. *In vivo* mutagenesis by 5-bromodeoxyuridine and 2-aminopurine, *J. Mol. Biol.* **9**:364–371.

Howard, B. D., and Tessman, I., 1964b, Identification of the altered bases in mutated single-stranded DNA. III. Mutagenesis by ultraviolet light, *J. Mol. Biol.* **9**:372–375.

Howell-Saxton, E., Zamenhof, S., and Zamenhof, P. J., 1973, Response of *Escherichia coli* to ethylmethanesulfonate. Effect of growth phase and repair ability on survival and mutagenesis, *Mutation Res.* **20**:327–337.

Iida, S., and Hayatsu, H., 1971, The permanganate oxidation of thymine and thymidilic acid, *Biochim. Biophys. Acta* **228**:1–8.

Ino, I., and Yourno, J., 1974, X-ray mutagenesis: base-pair deletion at a frameshift hotspot in Salmonella, *J. Mol. Biol.* **85**:301–307.

Isabaeva, M. K., Valdstein, E. A., and Zhestianikov, V. D., 1967, The sensitivity of *E. coli B* strains, having different ability to repair of radiation damages, to the radiomimetics and chemical mutagens, *Cytology* **11**:1481–1488 (In Russian).

Isono, K., and Yourno, J., 1974, Chemical carcinogens as frameshift mutagens: Salmonella DNA sequence sensitive to mutagenesis by polycyclic carcinogens, *Proc. Natl. Acad. Sci.* **71**:1612–1617.

Iyer, V. N., and Szybalski, W., 1964, Mitomycin and porfiromycin: chemical mechanism of activation and cross-linking of DNA, *Science* **145**:55–58.

Iyer, V. N., and Szybalski, W., 1969, Mutagenic effect of azaserine in relation to azaserine resistance in *Escherichia coli, Science* **129**:839–840.

Jenkins, J. B., 1967, Mutagenesis at a complex locus in *Drosophila* with the monofunctional alkylating agent, ethyl methanesulfonate. *Genetics,* **57**:783–793.

Jensen, K. A., Kølmark, G., and Westergaard, M., 1949, Back-mutations in *Neurospora crassa* induced by diazomethane, *Hereditas* **35**:521–525.

Jensen, K. A., Kirk, J., Kølmark, J., and Westergaard, M., 1951, Chemically induced mutations in *Neurospora, Cold Spring Harbor Symp. Quant. Biol.* **16**:245–259.

Jha, K. K., 1970, Mutagenic action of nitrous acid on two transversions in the histidine-3 locus of *Neurospora crassa, Mutation Res.* **9**:467–472.

Johansen, I., Gulbransen, R., and Pettersen, R., 1974, Effectiveness of oxygen in promoting X-ray induced single-stranded breaks in circular phage λ DNA and killing of radiation-sensitive mutants of *Escherichia coli, Radiation Res.* **58**:384–397.

Johns, H. E., LeBlanc, Y. C., and Freeman, K. B., 1965, Reversal and deamination rates of the main ultraviolet photoproduct of cytidylic acid, *J. Mol. Biol.* **13**:849–861.

Kak, S. N., and Kaul, B. L., 1973, Role of manganese ions on the modification of the mutagenic activity of some alkylating agents, *Cytologia* **38**:577–585.

Kanazir, D., 1958, The apparent mutagenicity of thymine deficiency, *Biochim. Biophys. Acta* **30**:20–23.

Kaney, A. R., and Atwood, K. C., 1964, Radiomimetic action of polyimine chemosterilants in *Neurospora, Nature* **201**:1006–1008.

Kaplan, H. S., and Moses, L. E., 1964, Biological complexity and radiosensitivity, *Science* **145**:21–25.

Kaplan, R. W., 1949, Mutations by photodynamic action in *Bacterium prodigiosum, Nature* **163**:573–574.

Kaplan, R. W., 1950a, Auslösung von Phagenresistenzmutationen bei *Bakterium coli* durch Erythrosin mit und ohne Belichtung, *Naturwiss.* **37**:308.

Kaplan, R. W., 1950b, Photodynamische Auslösung von Mutationen in den Sporen von *Penicillium notatum, Planta* **38**:1–11.

Kaplan, R. W., 1955, Influence of the water content on the UV-sensitivity of DNA and its bearing on biological effects, *Naturwiss.* **42**:466–467.

Kaplan, R. W., 1956, Dose-effect curves of s-mutation and killing in *Serratia marcescens, Arch. Microbiol.* **24**:60–79.

Kaplan, R., Winkler, U., and Wolf-Ellmauer, H., 1960, Induction and reversion of C-mutations by irradiation of the extracellular κ-phage of *Serratia, Nature* **186**:3–30.

Katritzky, A. R., and Waring, A. J., 1962, Tautomeric azines. Part I. The Tautomerism of 1-methyluracil and 5-bromo-1-methyluracil, *J. Chem. Soc.* 1540–1544.

Kaufmann, B. P., and Gay, H., 1970, Induction by 5-bromodeoxyuridine of sex-linked lethal mutations in spermatogeneous cells of *Drosophila melanogaster, Mutation Res.* **10**:591–595.

Kaufmann, W., Gebhart, E., and Horbach, L., 1973, Determination of the threshold value of the mutagenic activity of Tenimon on human lymphocytes *in vitro, Humangenetik* **20**:1–8.

Kaul, B. L., and Zutschi, U., 1973, On the production of chromosome breakage in *Vicia faba* by caffeine, *Cytobios.* **7**:261–264.

Kehler, E., Roehborn, G., Propping, P., and Buselmaier, W., 1973, Isoniazid as a mutagen, *Pneumonologie* **148**:223–226.

Kerkis, J. Y., Sablina, O. V., Radjabli, S. I., and Bocharov, E. F., 1967, A study of chromosome abnormalities in leukocytes of the peripheral blood in patients with acute epidemic hepatitis, *Genetika* **5**:83–94 (in Russian).

Kihlman, B. A., 1956, Factors effecting the production of chromosome aberrations by chemicals, *J. Biophys. Biochem. Cytol.* **2**:543–555.

Kihlman, B. A., and Levan, A., 1949, The cytological effect of caffeine, *Hereditas* **35**:109–111.

Kimball, R. F., 1965, The induction of reparable premutational damage in *Paramecium aurelia* by the alkylating agent triethylene melamine, *Mutation Res.* **2**:413–425.

Kimball, R. F., 1970, Studies on the mutagenic action of N-methyl-N′-nitro-N-nitrosoguanidine in *Paramecium aurelia* with emphasis on repair processes, *Mutation Res.* **9**:261–271.

Kimball, R. F., and Setlow, J. K., 1974, Mutation fixation in MNNG treated *Haemophylus influenzae* as determined by transformation, *Mutation Res.* **22**:1–14.

Knaap, A., and Kramers, P., 1974, Mutagenicity of hycanthone in *Drosophila melanogaster, Mutation Res.* **22**:55–62.

Kochetkov, N. K., and Budowsky, E. I., 1969, The chemical modification of nucleic acids, *Progress in Nucleic Acids Res. and Molec. Biol.* **9**:403.

Koenig, V. L., and Perrings, J. D., 1953, Physicochemical effects of radiation. II. Effect of X-ray on thymus sodium deoxyribonucleate as revealed by sedimentation and viscosity, *Arch. Biochem. Biophys.* **44**:443–454.

Kohn, K. W., Steigbigel, N. H., and Spears, C. L., 1965, Crosslinking and repair of DNA in sensitive and resistant strains of *E. coli* treated with nitrogen mustard, *Proc. Natl. Acad. Sci.* **53**:1154–1161.

Kolb, H., and Kaudewitz, F., 1970, Action of N´-methyl-N-nitro-N-nitrosoguanidine on transforming DNA of *Bacillus subtilis, Mutation Res.* **10**:85–90.

Kondo, S., and Kato, T., 1966, Photoreactivation of mutation and killing in *Escherichia coli, In* 2nd Internat. Biophys. Congr. Vienna.

Konrad, M., 1960, The mutagenic effect of D₂O on bacteriophage T4, *Ann. N.Y. Acad. Sci.* **84**:678–684.

Krieg, D. R., 1963*a*, Ethyl methanesulfonate-induced reversion of bacteriophage T4 rII mutants, *Genetics* **48**:561–580.

Krieg, D. R., 1963*b*, Specificity of chemical mutagenesis, *Progr. Nucleic Acid Res.* **2**:125–168.

Krivisky, A. S., 1961, Hereditary variability of bacteriophages irradiated with short-wavelength rays: In: *Abstracts of Interinstitutes Conference on Experimental Genetics,* Part 1, p. 78–80 (in Russian).

Krivisky, A. S., and Solovieva, N. J., 1963, The mutagenic action of UV-rays on extracellular bacteriophage, *Microbiologia* **32**:1006 (in Russian).

Krivisky, A. S., and Tichomirova, L. P., 1970, The lethal and mutagenic effects under the combined action of UV-rays and hydroxylamine in extracellular phage, *Genetika* **6**:84–93 (in Russian).

Laterjet, R., Ekert, B., Apelgot, S., and Rebeyrotte, N., 1961, Études radiobiochimiques sur l'ADN, *J. Chim. Phys.* **58**:1046–1057.

Latarjet, R., Rebeyrotte, N., Demerseman, and Pand, P., 1958, Action de peroxydes organiques sur un facteur transformant du pneumocoque, *In* Latarjet, R. (ed.), *Organic peroxides in radiobiology,* 61–72, London-N.Y. Pergamon Press.

Lawley, P. D., 1957*a*, Relative reactivities of deoxyribonucleotides and of the bases of deoxyribonucleic acid (DNA) towards alkylating agents, *Biochim. Biophys. Acta* **26**:450–451.

Lawley, P. D., 1957*b*, The hydrolysis of methylated deoxyguanilic acid at pH 7 to yield 7-methylguanine, *Proc. Chem. Soc.* 290–296.

Lawley, P. D., 1968, Methylation of DNA by N-methyl-N´-nitrosourethane and N-methyl-N-nitroso-N´-nitrosoguanidine, *Nature* **218**:580–581.

Lawley, P. D., 1971 (Pub. 1972), Action of alkylating mutagens and carcinogens on nucleic acid. N-Methyl-N-nitroso compounds as methylating agents, *Topics Chemical Carcinogens. Proc. Int. Symp.* 1971, 237–250, Univ. Park Press, Baltimore.

Lawley, P. D., and Brookes, P., 1961, Acidic dissociation of 7:9-dialkylguanines and its possible relation to mutagenic properties of alkylating agents, *Nature* **192**:1081–1082.

Lawley, P. D., and Brookes, P., 1962, Ionization of DNA bases or base analogues as a possible explanation of mutagenesis, with especial reference to 5-bromodeoxyuridine, *J. Mol. Biol.* **4**:216–219.

Lawley, P. D., and Brookes, P., 1963, Further studies on the alkylation of nucleic acids and their constituent nucleotides, *J. Biochem.* **89**:127–138.

Lawley, P. D., and Brookes, P., 1967*a*, Interstrand cross-linking of DNA by difunctional alkylating agents, *J. Mol. Biol.* **25**:143.

Lawley, P. D., and Brookes, P., 1967*b*, Interstrand crosslinking by difunctional alkylating agents, *J. Mol. Biol.* **29**:537.

Lawley, P. D., Orr, D. J., Shan, S. A., Farmer, P. B., and Jarman, M., 1973, Reaction products from N-methyl-N-nitrosourea and deoxyribonucleic acid containing thymine residues. Synthesis and identification of a new methylation product, O⁴-methylthymidine, *Biochem. J.* **135**:193–201.

Lederberg, J., 1950, Isolation and characterization of biochemical mutants of bacteria, *In* Comre, I. H., *Methods in medical research* **3**:5, Year Book Publs. Chicago.

Ledoux, L., Huart, R., and Jacobs, M., 1971, Fate of exogenous DNA in *Arabidopsis thaliana*. Evidence for replication and preliminary results at the biological level, *In*: *Informative Molecules in Biological Systems*, 159–175 North-Holland, Amsterdam.

Ledoux, L., Huart, R., and Jacobs, M., 1974, DNA-mediated genetic correction of thiamine-less *Arabidopsis thaliana*, *Nature* **249**:17–21.

Lehmann, A. R., and Kirk-Bell, S., 1974, Effect of caffeine and theophylline on DNA synthesis in unirradiated mammalian cells, *Mutation Res.* **26**:73–82.

Leiter, J., and Schneiderman, M. A., 1959, Screening data from the cancer chemotherapy. National service screening laboratories, *Cancer Res.* **19**(Screening Data):31–279.

Lerman, L. S., 1961, Structural considerations in the interaction of DNA and acridines, *J. Mol. Biol.* **3**:18–31.

Lerman, L. S., 1963, The structure of DNA-acridine complex, *Proc. Natl. Acad. Sci.* **49**:94–102.

Lett, J. T., Alexander, P., and Parkins, G. M., 1962, Physicochemical changes produced in DNA after alkylation, *Arch. Biochem. Biophys.* **97**:80–93.

Levisohn, R., 1970, Phenotypic reversion by hydroxylamine: a new group of suppressible phage T4 rII mutants, *Genetics* **64**:1–9.

Lindström, D. M., and Drake, J. W., 1970, Mechanisms of frameshift mutagenesis in bacteriophage T4: role of chromosome tips, *Proc. Natl. Acad. Sci.* **65**:617–624.

Litman, R. M., and Ephrussi-Taylor, H., 1959, Inactivation et mutation des facteurs génétiques de l'acide deoxyribonucleique du pneumocoque par l'ultraviolet et par l'acide nitreux, *Compt. rend. Acad. Sci.* **249**:838–840.

Litman, R. M., and Pardee, A. B., 1956, Production of bacteriophage mutants by a disturbance of deoxyribonucleic acid metabolism, *Nature* **178**:529–531.

Lobashov, M. E., 1937, About the mutagenic action of chemical agents on mutational process in *Dros. melanogaster*, *Trans. Leningrad. Soc. Naturalists* **3**:346–376 (In Russian).

Löbbecke, F. A., 1963, Nitrogen mustard induced mutation in extra-cellular bacteriophage, *Genetics* **48**:691–696.

Loprieno, N., and Clarke, C. H., 1965, Reversions to methionine independence induced by mutagens in *Schizosaccharomyces pombe*, *Mutation Res.* **2**:312–319.

Lotz, W., Kaplan, R. W., and Mennigmann, H. D., 1968, Wirking von Aminoacridinen auf die intrazelluläre Vermehrung und die Auslösung von Mutationen beim Phagen κ von *Serratia marcescens*, *Mutation Res.* **6**:329–343.

Loveless, A., 1959, The influence of radiomimetic substances on deoxyribonucleic acid synthesis and function studies in *Escherichia coli* system. III. Mutation of T2 bacteriophage as a consequence of alkylation in vitro: uniqueness of ethylation, *Proc. Roy. Soc.* (B) **150**:497–508.

Loveless, A., 1966, *Genetic and allied effects of alkylating agents*, London. Butterworths.

Loveless, A., 1969, Possible relevance of 0-6-alkylation of deoxyguanosine to the mutagenicity and carcinogenicity of nitrosamines and nitrosamides, *Nature* **223**:206–207.

Luzatti, D., 1962, The action of nitrous acid on transforming deoxyribonucleic acid, *Biochem. Biophys. Res. Commun.* **9**:508–516.

Lyon, M. F., Phillips, J. S. R., and Searle, A. Y., 1962, A test for mutagenicity of caffeine in mice, *Z. Vererbungslehre* **93**:7–13.

Magee, P. N., 1972, Possible mechanisms of carcinogenesis and mutagenesis by nitrosoamines, Topics Chemical Carcinogens, *Proc. Intern. Sympos.* 1971, 259–278.

Magni, G. E., 1965, Origin and nature of spontaneous mutations in meiotic organisms, *J. Cell. Comp. Physiol.* **64** suppl. 1:165–172.

Magni, G. E., and Borstel, R. C., 1962, Different rates of spontaneous mutation during mitosis and meiosis in yeast, *Genetics* **47**:1097–1108.

Magni, G. E., and Puglisi, P. P., 1966, Mutagenesis of supersuppressors in yeast, *Cold Spring Harbor Symp. Quant. Biol.* **31**:699–704.

Magni, G. E., Borstel, R. C., and Sora, S., 1964, Mutagenic action during meiosis and antimutagenic action during mitosis by 5-aminoacridine in yeast, *Mutation Res.*, **1**:227–230.

Mahler, H. R., and Baylor, M. B., 1967, Effects of steroidal diamines on DNA duplication and mutagenesis, *Proc. Natl. Acad. Sci.* **58**:256–263.

Makino, F., and Okada, S., 1974, Comparative studies of the effects of carcinogenic and antitumor agents on the DNA replication of cultured mammalian cells, *Mutation Res.* **23**:387–394.

Malkin, M. F., and Zahalsky, A. C., 1966, Interaction of the water-soluble carcinogen, 4-nitroquinoline-N-oxide with DNA, *Science* **154**:1665–1667.

Malling, H. V., 1966a, Hydroxylamine as a mutagenic agent for *Neurospora crassa, Mutation Res.* **3**:470–476.

Malling, H. V., 1966b, Mutagenicity of two potent carcinogens, dimethylnitrosamine and diethylnitrosamine, in *Neurospora crassa, Mutation Res.* **3**:537–540.

Malling, H. V., 1967, The mutagenicity of the acridine mustard (ICR-170) and the structurally related compounds in *Neurospora, Mutation Res.* **4**:265–274.

Malling, H. V., and De Serres, F. J., 1968, Correlation between base-pair transition and complementation pattern in nitrous acid-induced ad-3-mutants of *Neurospora crassa, Mutation Res.* **5**:359–371.

Malling, H. V., and De Serres, F. J., 1971, Hydroxylamine-induced purple adenine (ad-3) mutants in *Neurospora crassa.* I. Characterization of mutants by genetic tests, *Mutation Res.* **12**:35–46.

Mandel, J., and Greenberg, J., 1960, A new chemical mutagen for bacteria: 1-methyl-3-nitro-1-nitrosoguanidine, *Biochem. Biophys. Res. Commun.* **3**:575–577.

Marmur, I., Schildkraut, C. L., and Doty, P., 1961, The reversible denaturation of DNA nucleic acid homologies and the biological relatedness of microorganisms, *J. Chem. Physiol.* 945–955.

Marquardt, H., Zimmermann, F. K., and Schwaier, R., 1964, Die Wirkung krebsauslösender Nitrosamine und Nitrosamide auf das Adenin-6-45-Rückmutationssystem von *Saccharomyces cerevisiae, Z. Vererbungslehre* **95**:82–96.

Martignoni, K. D., and Smith, K. C., 1973, The synergistic action of ultraviolet and X-radiation on mutants of *Escherichia coli* K12, *Photochem. Photobiol.* **18**:1–8.

Martin, R. G., 1967, Frameshift mutants in the histidine operon of *Salmonella typhimurium, J. Mol. Biol.* **26**:311–328.

Mathew, C., 1965, The production of recessive lethals by calf-thymus DNA in *Drosophila, Genetic. Res.* Camb. **6**:163.

Mathews, M. M., 1963, Comparative study of lethal photosensitization of *Sarcina lutea* by 8-methoxypsoralen and by toluidine blue, *J. Bacteriol.* **85**:322–328.

Matsuura, T., and Saito, J., 1967, Photosensitized oxydation of hydroxylated purines, *Chem. Commun.* 693–694.

McCalla, D. R., 1965, Chloroplast mutagenesis: effect of N-methyl-N′-nitro-N-nitrosoguanidine and some other agents on *Euglena, Science* **148**:497–499.

McCalla, D. R., and Voutsinos, D., 1974, Mutagenicity of nitrofurans, *Mutation Res.* **26**:3–16.

McKay, A. F., and Wright, C. F., 1947, Preparation and properties of N-methyl-N-nitroso-N´-nitroguanidine, *J. Amer. Chem. Soc.* **69**:3028-3030.

McLaren, A. D., and Shugar, D., 1964, *Photochemistry of proteins and nucleic acid,* Oxford, Pergamon Press.

Meadows, M. G., Quah, S. K., and von Borstel, R., 1973, Mutagenic action of hycanthone and IA-4 [8-chloro-2(diethylamonoethyl)-2H-[1] benzothiopyrano [4,3,2,-cd] indazole-5-methanol monomethanesulfonate] on yeast, *J. Pharmacol. Exp. Therap.* **187**:444-450.

Meissel, M. N., 1928, Die Wirkung des Chloroforms auf die Entwicklung der Hefe, *Wochenschrift für Brauerei No. 43,* **27** Oktober (See also *Microbiol. J.* 1928, **4**:255-259, in Russian).

Meissel, M. N., 1932, The action of cyanide salts on yeast development, *Proc. U.S.S.R. Acad. Sci.* 1337-1344 (in Russian).

Meissel, M. N., 1933, Wirkung der Cyansalze auf die Entwicklung der Hefe, *Zentralblatt für Bakteriologie, Parasitenkunde Infektionskrankheiten,* **88**, Abt. II, 449-459.

Merril, C. R., Geier, M. R., and Petricciani, J. C., 1973, Bacterial virus gene expression in human cells, *Nature* **223**:398-401.

Michaelis, A., Nikoloff, H., and Rieger, R., 1962, Influences of EDTA on the induction of chromatid aberrations by triethylenemelamine and ethyl alcohol, *Biochem. Biophys. Res. Commun.* **9**:280-284.

Miller, J. A., Cramer, J. W., and Miller, E. C., 1960, The N- and ring-hydroxylation of 2-acetylaminofluorene during cancerogenesis in the rat, *Cancer Res.* **20**:950-962.

Mittler, S., Mittler, I. E., Tonetti, A. M., and Szymczak, M. E., 1967, The effect of caffeine on chromosome loss and nondisjunction in *Drosophila, Mutation Res.* **4**:708-710.

Moh, C. C., 1970, Mutagenic effect of cycasin in beans (*Phaseolus vulgaris L.*), *Mutation Res.* **10**:251-253.

Mohn, G., 1973, 5-Methyltryptophan resistance mutations in *Escherichia coli* K12. Mutagenic activity of monofunctional alkylating agents including organophosphorus insecticides, *Mutation Res.* **20**:7-15.

Moore, A. M., 1958, Ultraviolet irradiation of pyrimidine derivatives. II. Note on the synthesis of the product of reversible photolysis of uracil, *Canad. J. Chem.* **36**:281-283.

Moroson, H., and Furlan, M., 1969, Single-strand breaks in DNA of Ehrlich ascites tumor cells produced by methyl hydrazine, *Radiation Res.* **40**:351-362.

Mortimer, R. K., 1958, Radiobiological and genetic studies on polyploid series (haploid to hexaploid) of *Saccharomyces cerevisiae, Radiation Res.* **9**:312-326.

Mulder, C., and Doty, P., 1968, Residual activity of denatured transforming DNA of *Haemophilus influenzae*: a naturally occurring cross-linked DNA, *J. Mol. Biol.* **32**:423-435.

Müller, A. J., and Gichner, T., 1964, Mutagenic activity of 1-methyl-3-nitro-1-nitrosoguanidine on *Arabidopsis, Nature* **201**:1149-1150.

Müller, H. J., 1927, Artificial transmutation of the gene, *Science* **66**:84-87.

Müller, H. J., 1941, Induced mutations in *Drosophila, Cold Spring Harbor Symp. Quant. Biol.* **9**:151.

Munson, R. J., and Bridges, B. A., 1969, Lethal and mutagenic lesions induced by ionizing radiation in *E. coli* and DNA strand breaks, *Biophysik* **6**:1-12.

Munz, P., and Leupold, U., 1970, Characterization of ICR-170 induced mutation in *Schizosaccharomyces pombe, Mutation Res.* **10**:199-212.

Murakami, A., 1973, Mutagenesis of acridine orange in mitotic cleavage nuclei of the silkworm, *Bombyx mori, Mutation Res.* **20**:67-70.

Nadson, G. A., and Philippov, G. S., 1925, Influence of X-rays on sexual process and mutation formation in low yeasts (Mucoraceae), *Proc. Rentgenology Radiobiology* **3**:305-310 (in Russian).

Nagata, C., Imamura, A., Saito, H., Fukui, K., 1963, Changes of electron distribution of deoxyribonucleic acid after alkylation and their possible relation to the biological effect, *Gann.* **54**:109–117.

Nakai, S., and Saeki, T., 1964, Induction of mutation by photodynamic action in *Escherichia coli, Genetical Res.* **5**:158.

Namdeo, K. N., and Dube, J. N., 1973, Herbicidal influence on growth sensitivity and mutagenic transformation in *Rhizobia, Indian J. Exp. Biol.* **11**:114–116.

Naono, S., and Gros, F., 1960, Syntèse par *E. coli* d'une phosphatase modifièe en prèsence d'un analogue pyrimidique, *Compt. Rend. Sean. Acad. Sci.* **250**:3889–3891.

Newell, G. W., and Maxwell, W. A., 1972, Mutagenic effect of ionol, C.P. (Butylated hydroxytoluene), *U.S. Nat. Tech. Inform. Serv.* PB rep. No 221827/9:106.

Nichols, W. W., 1963, Relationship of viruses, chromosomes and carcinogenesis, *Hereditas* **50**:53–80.

Nirenberg, M., Leder, P., Bernfield, M., Brimacombe, R., Trupin, J., Rottman, E., and O'Neal, C., 1965, RNA codewords and protein synthesis. VII. On the general nature of the RNA code, *Proc. Natl. Acad. Sci.* **53**:1161–1168.

Novick, A., 1956, Mutagens and antimutagens, *Brookhaven Symp. Biol.* **8**:201–215.

Novick, A., and Szillard, L., 1952, Anti-mutagens, *Nature* **170**:926–927.

Oehlkers, F., 1943, Die Auslösung von Chromosomenmutationen in der Meiosis durch Einwirkung von Chemikalien, *Z. Vererbl.* **81**:313–341.

Olast, M., and Bertinchamps, A. J., 1973, Electron transfer in gamma irradiated complexes between aromatic amino acids and DNA derivatives, *Int. J. of Radiat. Biol.* **26**:589–594.

Ong, T., 1970, Mutagenicity of aflatoxins in *Neurospora crassa, Mutation Res.* **9**:615–618.

Ono, J., Wilson, R. G., and Grossman, L., 1965, Effects of ultraviolet light on the template properties of polycytidilic acid, *J. Mol. Biol.* **11**:600–612.

Orgel, A., and Brenner, S., 1961, Mutagenesis in bacteriophage T4 by acridines, *J. Mol. Biol.* **3**:762–768.

Orgel, A., and Orgel, L. E., 1965, Induction of mutations in bacteriophage T4 with divalent manganese, *J. Mol. Biol.* **14**:453–457.

Orgel, L. E., 1965, The chemical basis of mutation, *Advances Enzymol.* **27**:289–346.

Ostertag, W., 1966, Chemische Mutagenese an menschlichen Zellen in Kultur, *Abhandl. Mainzer Akad. Wiss. Lit. Mainz* 1–124.

Panfilova, Z. I., Voronina, E. N., Poslovina, A. S., Goryunova, N. M., and Salganik, R. I., 1966, The study of combined action of chemical mutagens and UV-rays on appearing of back mutation in *Escherichia coli, Genetika* **2**:49–56 (in Russian).

Papirmeister, B., and Davison, C. L., 1964, Elimination of sulfur mustard induced products from DNA of *Escherichia coli, Biochem. Biophys. Res. Commun.* **17**:608–617.

Parkash, O., 1965, Induction of sex-linked recessive lethals and visible mutations by feeding X-irradiated DNA in *Drosophila melanogaster, Nature* **205**:312.

Pauling, E. C., 1968, The specificity of thymineless mutagenesis, *In* Rich, A., and Davison, N., (ed.), *Structural chemistry and molecular biology*; a volume dedicated to Linus Pauling by his students, colleagues and friends, W. H. Freeman and Co., San Francisco.

Peacocke, A. R., and Preston, B. N., 1960, The action of γ-rays on sodium deoxyribonucleate in solution. II. Degradation, *Proc. Roy. Soc. London (B)* **153**:90–110.

Pearson, M. L., Ottensmeyer, F. P., and Johns, H. E., 1965, Properties of an unusual photoproduct of UV irradiated thymidyl-thymidine, *Photochem. Photobiol.* **4**:739–747.

Phillips, J. H., and Brown, D. M., 1968, The mutagenic action of hydroxylamine. *In* Davidson, J. N. and Cohn, W. E., (eds.), *Progress in nucleic acids research*, 349–368, N.Y.-London, Acad. Press.

Pletikha-Lansky, R., 1972, Radiation effects induced by γ-radiation of pyrimidines in water solutions, *Tsitol. Genet.* **6:**400–409 (in Russian).

Plough, H. H., 1941, Spontaneous mutability in *Drosophila, Cold Spring Harbor Symp. Quant. Biol.* **9:**127–137.

Pons, F. W., 1973, Formation of pure and mixed clones of *c* mutants in serratiaphage sigma (σ) after treatment of the free virion with UV or hydroxylamine, *Mutation Res.* **20:**149–157.

Prakash, L., Stewart, J. W., and Sherman, F., 1974, Specific induction of transitions and transversions of G C base pairs by 4-nitroquinoline-1-oxide in iso-1-cytochrome *c* mutants of yeast, *J. Mol. Biol.* **85:**51–65.

Putrament, A., and Baranowska, H., 1973, Mechanism of mutagenesis in microorganisms according to new data, *Postepy Mikrobiol.* **12:**143–172.

Putrament, A., Baranowska, H., and Pachecka, J., 1973, Mutagenic action of hydroxylamine and methoxyamine. I. Hydroxylamine. II. Methoxyamine, *Molec. Gen. Genet.* **122:**61–80.

Rapoport, I. A., 1940, Influence of thymonucleic and nucleic acids, some of their components and salts on mutation formation, *Proc. Acad. Sci. U.S.S.R.* **27:**1033.

Rapoport, I. A., 1947, The hereditary changes obtaining under the influence of diethylsulfate and dimethylsulfate, *Proc. U.S.S.R. Acad. Agricult. Sciences* **12:**12–15 (in Russian).

Rapoport, I. A., 1948*a*, Action of ethylenoxyde, glycide and glycoles on genic mutations, *Proc. Acad. Sci. U.S.S.R.* **60:**469–472 (in Russian).

Rapoport, I. A., 1948*b*, The alkylation of genic molecule, *Proc. Acad. Sci. U.S.S.R.* **59:**1183–1186 (in Russian).

Rapoport, I. A., 1963, Mutagenic effects of commercial drugs and some toxic substances. Problems of general and industrial toxicology, Leningrad, p. 65–75 (in Russian).

Rapoport, I. A., (ed.), 1966, *Supermutagens, Moscow, Publ. House "Nauka"* (in Russian).

Reddy, T. P., Reddy, C. S., and Reddy, G. M., 1974, Interaction of certain base-specific chemicals and diethyl sulfate in the induction of chlorophyll mutations in *Oryza sativa, Mutation Res.* **22:**127–132.

Regan, J. D., Trosco, J. E., and Carrier, W. L., 1968, Evidence for excision of ultraviolet-induced pyrimidine dimers from DNA of human cells in vitro, *Biophys. J.* **8:**319–325.

Reiner, B., and Zamenhof, S., 1957, Studies on the chemically reactive groups of deoxyribonucleic acids, *Biol. Chem.* **228:**475–480.

Reznik, L. C., and Shapiro, N. I., 1973, Mutagenic action of O-methyl hydroxylamine on Chinese hamster cells in vitro, *Genetika* **9:**49–54 (in Russian).

Rhaese, H. J., and Freese, E., 1968, Chemical analysis of DNA alteration. Base liberation and backbone breakage of DNA and oligodeoxyadenilic acid induced by hydrogen peroxide and hydroxylamine, *Biochim. Biophys. Acta* **155:**476–490.

Rieger, R., 1965, On the experimental induction of chromosomal aberrations in plants, *In: Induction of mutations and the mutation process,* Publ. House of the Czechoslovak Acad. Sci.

Rieger, R., and Michaelis, A., 1964, On the distribution amongst cells of chemically induced chromatid aberrations in root tip meristemes of *Vicia faba, Mutation Res.* **1:**109–112.

Rieger, R., and Michaelis, A., 1965, Chemical induction of heat-reversible 'Potential lesion' in *Vicia faba* chromosomes, *Nature* **206:**741–742.

Ritchie, D. A., 1964, Mutagenesis with light and proflavin in phage T4, *Genetical Res.* **5:**168–169.

Ritchie, D. A., 1965, Mutagenesis with light and proflavin in phage T4, II. Properties of the mutants, *Genetical Res.* **6:**474–478.

Roberts, J. J., and Ward, K. N., 1973, Inhibition of post-replication repair of alkylated DNA by caffeine in Chinese hamster cells but not HeLa cells, *Chem. Biol. Interactions* **7:**241–264.

Roberts, J. J., Sturrock, J. E., and Ward, K. N., 1974, Enhancement by caffeine of alkylation-induced cell death, mutations and chromosomal aberrations in Chinese hamster cells, as result of inhibition of post-replicative repair, *Mutation Res.* **26**:129–143.

Ross, W., 1962, *Biological alkylating agents*, London, Butterworths.

Roti, J. L., Stein, G. S., and Cerutti, P. A., 1974, Reactivity of thymine to γ-rays in HeLa chromatin and nucleoprotein preparations, *Biochemistry* **13**:1900–1905.

Rozenkranz, H. S., Hyorth, R., and Carr, H. S., 1971, Studies with hydroxyurea: the biologic and metabolic properties of formamidooxide, *Biochim. Biophys. Acta* **232**:48–60.

Rudner, R., 1961, Mutation as an error in base pairing. I. The mutagenicity of base analogues and their incorporation into the DNA of *Salmonella typhimurium, Z. Vererbl.* **92**:336–379.

Rupp, W. D., and Howard-Flanders, P., 1968, Discontinuites on the DNA synthesis in an exicision-defective strain of *Escherichia coli* following ultraviolet irradiation, *J. Mol. Biol.* **31**:291–304.

Rurishauser, A., and Bollag, W., 1963, Cytological investigations with a new class of cytotoxic agents: methylhydrazine derivatives, *Experientia* **19**:131–132.

Rutman, R. J., Chu, E. H. L., and Jones, J., 1969, Observations on the mechanism of the alkylation reaction between nitrogen mustard and DNA, *Biochim. Biophys. Acta* **174**:663–673.

Saccharov, V. V., 1932, Iodine as chemical factor acting on mutational process in *Drosophila melanogaster, Biol. J.* **1,** No. 3–4:1–8 (in Russian).

Saffhill, R., 1974, The incorporation of wrong bases by DNA-polymerase I following γ-irradiation of DNA-like templates, *Biochim. Biophys. Acta* **349**:23–31.

Safonova, T. S., 1973, Chemistry of alkylating drugs, *Zhurn. Vses. Khimich. Obschestva* **18**:657–669 (in Russian).

Salganik, R. I., 1963, Application of radiation denaturation of DNA in combination with chemical mutagens to affect the mutational process. *In: Genetics Today,* 1.

Salganik, R. I., 1969, Specificity of radiation and chemical mutagenesis in microorganisms. *In Genetical basis of selection of microorganisms,* Moscow, 126–139 (in Russian).

Salganik, R. I., Drevich, V. E., and Vasyunina, E. A., 1967, Isolation of ultraviolet-denatured regions of DNA and their base composition, *J. Mol. Biol.* **30**:219–222.

Salganik, R. I., Vasyunina, E. A., Poslovina, A. S., and Andreeva, I. S., 1973, Mutagenic action of N-4-hydroxycytidine on *Escherichia coli* B cyt⁻, *Mutation Res.* **20**:1–5.

Sauerbier, W., 1964, Role of cytosine photoproducts in ultraviolet inactivation of bacteriophage T1, *J. Mol. Biol.* **10**:551–553.

Sauro, F. M., and Green, S., 1973, *In vivo* cytogenetic evaluation of chloroindazole thioxanthene IA-4 (a hycanthone analog) and niridazole in rat bone marrow, *J. Pharmacol. Exp. Therap.* **186**:399–401.

Sax, K., 1941, Types and frequences of chromosomal aberrations induced by X-rays, *Cold Spring Harb. Symp.* **9**:93–103.

Sax, K., and Sax, H. J., 1966, Radiomimetic beverages, drugs, and mutagens, *Proc. Nat. Acad. Sci.,* **55**:1431–1435.

Sax, K., and Sax, H. J., 1968, Possible mutagenic hazards of some food additives, beverages and insecticides. *Japan J. Genet.,* **43**:89–94.

Schoenthal, R., 1960, Carcinogenic action of diazomethane and nitroso-N-methylurethane, *Nature* **188**:420–421.

Scholes, G., 1963, The radiation chemistry of aqueous solutions of nucleic acids and nucleoproteins, *Progr. Biophys. Mol. Biol.* **13**:59–104.

Scholes, G., and Weiss, J. J., 1954, Chemical action of X-rays on nucleic acids and related substances in aqueous systems, *Biochem. J.* **56**:65–72.

Scholes, G., and Weiss, J. J., 1960, Organic hydroxy-hydroperoxides: A class of hydroperoxides formed under the influence of ionizing radiation, *Nature* 185:305–306.

Scholes, G., Weiss, J. J., and Wheeler, C. M., 1956, Formation of hydroperoxides from nucleic acids by irradiation with X-rays in aqueous systems, *Nature* 178:157.

Schöneich, J., 1967, The induction of chromosomal aberrations by hydrogen peroxide in strains of ascites tumors in mice, *Mutation Res.* 4:384–388.

Schöneich, J., Michaelis, A., and Rieger, R., 1970, Coffein und die chemische Induktion von Chromatidenaberrationen bei *Vicia faba* und Ascitestumoren der Maus, *Biol. Zentralbl.* 89:49–63.

Schreiber, J. P., and Danne, M. P., 1974, Fluorescence of complexes of acridine dye with synthetic polydeoxyribonucleotides: a physical model of frame-shift mutation, *J. Mol. Biol.* 83:487–501.

Schuster, H., 1960, The reactions of nitrous acid with deoxyribonucleic acid, *Biochem. Biophys. Res. Commun.* 2:320–323.

Schuster, H., 1961, The reaction of tobacco mosaic virus ribonucleic acid with hydroxylamine, *J. Mol. Biol.* 3:447–451.

Schuster, H., 1964, Photochemie von Ribonukleinsäure. I. Photoprodukte von Uridilsäure und Cytidilsäure und ihr Nachweis in UV-bestrehlten Ribonukleinsäuren, *Z. Naturforsch.* 19 b:815–830.

Schuster, H., and Wilhelm, R. C., 1963, Reaction differences between tobacco mosaic virus and its ribonucleic acid with nitrous acid, *Biochim. Biophys. Acta* 68:554–560.

Schuster, H., Gierer, A., and Mundry, K. W., 1960, Inaktivierende und mutagene Wirkung der chemischen Veränderung von Nucleotiden in Virus Nukleinsäure, *In*: Chemische Mutagenese, *Abh. Deutsch. Akad. Wiss.* Nr. 1 76–85.

Schwartz, N. M., 1963, Nature of methanesulfonate-induced reversions of Lac Mutants of *Escherichia coli, Genetics* 48:1357–1375.

Sechaud, J., Streisinger, G., Emrich, J., Newton, J., Lanford, H., Reinhold, H., and Stahl, M. M., 1965, Chromosome structure in phage T4. II. Terminal redundancy and heterozygosis, *Proc. Natl. Acad. Sci.* 54:1333–1339.

Sederoff, R. R., 1966, The mutational mechanism of ICR-170, a manyfunctional quinacrine mustard, on T4 bacteriophage, Ph.D. Dissertation, Univ. of Calif., Los Angeles.

Sesnowitz-Horn, S., and Adelberg, E. A., 1968, Proflavine treatment of *Escherichia coli*: generation of frameshift mutations, *Cold Spring Harbor Symp. Quant. Biol.* 33:393–402.

Setlow, R. B., 1966, Cyclobutane-type pyrimidine dimers in polynucleotides, *Science* 153:379–386.

Setlow, R. B., Regan, J. D., German, J., and Carrier, W. L., 1969, Evidence that xeroderma pigmentosum cells do not perform the first step in the repair of ultraviolet damage to their DNA, *Proc. Natl. Acad. Sci.* 64:1055–1061.

Shambulingappa, K. G., 1968, Radiomimetic action of N-methyl-N-nitrosoaniline, *Mutation Res.* 6:322–324.

Shankel, D., 1963, Synergistic mutational effects between radiation and chemicals, *Genetics Today* 1.

Shapiro, R., 1966, Isolation of a 2-nitropurine from the reaction of guanosine with nitrous acid, *J. Am. Chem. Soc.* 86:2948–2949.

Shapiro, R., and Klein, R. S., 1966, The deamination of cytidine and cytosine by acidic buffer solutions. Mutagenic implications, *Biochemistry* 5:2358–2362.

Sharma, R., and Glover, P., 1970, Interaction of mutational lesions induced by ethylmethanesulfonate and γ-rays in males of *Drosophila, Mutation Res.* 10:221–226.

Shein, H., and Enders, J., 1962, Transformation induced by Simian Virus 40 in human renal cell cultures. I. Morphology and growth characteristics, *Proc. Natl. Acad. Sci.* 48:1164–1172.

Shugar, D., 1962, The photochemistry of nucleic acids and their components, *In*; *Nucleic acids*. Moscow, 34–87 (in Russian).

Shugar, D., and Baranowska, J., 1960, Aggregation and staining behavior of ultraviolet-irradiated films of nucleic acid, *Nature* 185:33–34.

Sicard, A. M., 1965, Mutagenèse par la proflavine chez le pneumocoque, *C. R. Acad. Sci.* 261:4917–4918.

Silver, S., Levine, E., and Pilelman, P. M., 1968, Acridine binding by *Escherichia coli*: pH dependency and strain differences, *J. Bacteriol.* 95:333–339.

Simon, M. I., and Van Vunakis, H., 1962, The photodynamic reaction of methylene blue with deoxyribonucleic acid, *J. Mol. Biol.* 4:480–499.

Simon, M. I., and Van Vunakis, H., 1964, The dye-synthesized photooxidation of purine and pyrimidine derivatives, *Arch. Biochem. Biophys.* 105:197–206.

Singer, B., and Fraenkel-Conrat, H., 1966, Dye-catalyzed photo-inactivation of tobacco mosaic virus ribonucleic acid, *Biochemistry* 5:2446–2450.

Singer, B., and Fraenkel-Conrat, H., 1967, Chemical modification of viral RNA. VI. The action of N-methyl-N′-nitro-N-nitrosoguanidine, *Proc. Natl. Acad. Sci.* 58:234–239.

Singer, B., and Fraenkel-Conrat, H., 1969, The role of conformation of chemical mutagenesis, *Progress in Nucleic Acids Res. Molec. Biol.* 9:1–29.

Singht, U. P., 1973, Effect of acriflavine on UV-induced mutants of *Fusarium* species, *Mycopathol. Mycol. Appl.* 50:183–193.

Sinsheimer, R. L., and Hastings, R., 1949, A reversible photochemical alteration of uracil and uridine, *Science* 110:525–526.

Skinner, W. A., Gram, H. F., Greene, M. O., Greenberg, J., and Baker, B. R., 1960, Potential anticancer agents. XXXI. The relationship of chemical structure to antileucemic activity with analogues of 1-methyl-3-nitro-1-nitrosoguanidine, *J. Med. Pharm. Chem.* 2:299–333.

Sklyadneva, V. B., Kisseliova, N. P., Budovsky, E. I., and Tichonenko, T. I., 1970, The reaction of O-methylhydroxylamine with free and intraphage DNA, *Molec. Biol.* 1:110 (in Russian).

Small, G. D., and Gordon, M. P., 1968, Reaction of hydroxylamine and methoxyamine with the ultraviolet-induced hydrate of cytidine, *J. Mol. Biol.* 34:281–291.

Small, G. D., Tao, M., and Gordon, M. P., 1969, Pyrimidine hydrates and dimers in ultraviolet-irradiated tobacco mosaic virus ribonucleic acid, *J. Mol. Biol.* 38:75–87.

Smith, K. C., 1964, Photochemistry of nucleic acids, *In Photophysiology* 2:329–388. Giese, A. C. (ed.), N.Y. Acad. Press.

Smith, K. C., 1966, Physical and chemical changes induced in nucleic acids by ultraviolet light, *Radiat. Res. Suppl.* 6:54–79.

Smith, M. D., Green, R. R., Ripley, L. S., and Drake, J. W., 1973, Thymineless mutagenesis in bacteriophage, *Genetics* 74:393–403.

Sobels, F. H., 1963, Peroxides and the induction of mutations by X-rays, ultraviolet and formaldehyde, *Radiation Res.* 3:171–183.

Sobels, F. H., 1965, Organic peroxides and mutagenic effects in *Drosophila, Nature* 177:979–980.

Sobels, F. H., 1973, Some aspects of chemical mutagenesis, *Agents and Actions* 3:68–72.

Somers, C. E., and Hsu, T. C., 1962, Chromosome damage induced by hydroxylamine in mammalian cells, *Proc. Natl. Acad. Sci.* 48:937–943.

Sora, S., Panzeri, L., and Magni, G. E., 1973, Molecular specificity of 2-aminopurine in *Saccharomyces cerevisiae, Mutation Res.* 20:207–213.

Soyfer, V. N., 1965, Mutagenesis in extracellular bacteriophage T2. *In: Experimental mutagenesis in animal, plants and microorganisms* 1:137–139, Moscow (In Russian).

Soyfer, V. N., 1969, *The molecular mechanisms of mutagenesis*, Publ. House "Nauka," Moscow (in Russian), See also translation into German (Akademie Verlag, 1975).

Soyfer, V. N., 1970a, *The essays on the history of molecular genetics*, Publ. House "Nauka," Moscow (in Russian).

Soyfer, V. N., 1970b, Synergistic effect of combined action of hydroxylamine and acridine on mutation frequency in T2 phage and problem of reparability, *Microb. Gen. Bull.* **32**:12.

Soyfer, V. N., 1971a, The role of school of G. A. Nadson in studying of induced mutagenesis, *Abstr. Rep., XIII Intern. Congr. Hist. Sci. Sect.* **9**:127–128, Moscow. (See also Proc. Hist. of Sci. Congress, 1974, 165–167).

Soyfer, V. N., 1971b, Neueste Forschungsergebnisse und Perspektiven der Mutationstheorie, *Moderne Medizin,* Stuttgart **1**:134–141.

Soyfer, V. N., 1972, Participation of the intracellular enzymes in the control of the mutation process. *In: Molecular mechanisms of genetic process,* Moscow, 364–377 (in Russian). See also Soyfer, V. N., 1972, *Abstr. IV Intern. Biophys. Congr.,* V–VII sections, 320–321.

Soyfer, V. N., and Cieminis, K. K., 1974, Dark repair in higher plants, *Proc. Acad. Sci. U.S.S.R.* **15**:1261–1264.

Soyfer, V. N., and Dorohov, Y. L., 1972, The frequency of complete and mosaic mutations in *E. coli* uvr⁺ and uvr⁻ cells under the irradiation of them by high doses of UV-rays, *In: Molecular mechanisms of genetic processes,* 54–58, Moscow (in Russian).

Soyfer, V. N., and Turbin, N. V., 1974, Genetic transformation of waxy character in barley, 2-nd Europ. Meet. *Transformation and transfection,* Krakow 120–123.

Soyfer, V. N., and Yakovleva, N. I., 1972, Kinetics of excision of thymine dimers from DNA of human cells irradiated with UV-rays, *Abstr. IV Intern. Biophys. Congr. Sections I–IV* 135–136.

Soyfer, V. N., and Yakovleva, N. I., 1974, Inhibition of DNA synthesis in bacteria after treatment with hydroxylamine. *Proc. Acad. Sci. U.S.S.R.,* **217**:235–237.

Soyfer, V. N., Matussevitch, L. L., and Goroshkina, G. I., 1970, Dimerization of pyrimidine bases in DNA of HeLa cells under UV-irradiation and excision of dimers in the course of dark repair, *Radiobiologia* **10**:275–278 (in Russian).

Sparrow, A. H., 1965, Relationship between chromosome volume and radiation sensitivity in plant cells, *In: Cellular Radiation Biology,* Williams and Wilkins, Baltimore 199–222.

Sparrow, A. H., and Rosenfeld, R. M., 1946, X-ray-induced depolymerization of thymonucleohistone and of sodium thymonucleate, *Science* **104**:245–246.

Sprague, G. F., McKinney, H. H., and Greely, L., 1963, Virus as a mutagenic agent in maize, *Science* **141**:1052–1053.

Springgate, C. F., and Loeb, L. A., 1973, Mutagenic DNA polymerase in human leukemic cells, *Proc. Natl. Acad. Sci.* **70**:245–249.

Šram, R. J., and Zudova, Z., 1974, The relationship of mutagenic activity and the chemical structure of alkylaminoaziridine, *Folia Biol.* **20**:1–13.

Sree, R. K., 1970, Sensitivity and induction of mutation in Sorghum, *Mutation Res.* **10**:197–206.

Stacey, K. A., Cousens, S. F., and Alexander, P., 1958, Reaction of the radiomimetic alkylating agents with macromolecules *in vitro. Ann. N.Y. Acad. Sci.* **68**:682–701.

Stafford, R. S., and Donnellan, J. E., 1968, Photochemical evidence for conformation changes in DNA during germination of bacterial spores, *Proc. Natl. Acad. Sci.* **59**:822–828.

Stahl, F. W., Crasemann, J. M., Okun, L., Fox, E., and Laird, C., 1961, Radiation-sensitivity of bacteriophage containing 5-bromodeoxyuridine, *Virology* **13**, 98–104.

Stark, R. M., and Littlefield, J. W., 1974, Mutagenic effect of BUDR in diploid human fibroblasts, *Mutation Res.* **22**:281–286.

Steward, D. L., and Humphrey, R. M., 1966, Induction of thymine dimers in synchronized populations of Chinese hamster cells, *Nature* **212**:298–300.

Stewart, C. R., 1968, Mutagenesis by acridine yellow in *Bacillus subtilis, Genetics* **59**:23–31.

Stich, H. F., Arila, L., and John, D. S., 1968, Virus in mammalian chromosomes. IX. The capacity of UV-impaired adenovirus type 18 to induce chromosome aberrations, *Exptl. Cell. Res.* **53**:44–54.

Stollier, A., and Collman, R. D., 1965, Virus aetiology for Down's syndrome (mongolism), *Nature* **208**:903–904.

Strack, H. B., Freese, E. B., and Freese, E., 1964, Comparison of mutation and inactivation rates induced in bacteriophage and transforming DNA by various mutagens, *Mutation Res.* **1**:10–21.

Strauss, B. S., 1961, Specificity of the mutagenic action of alkylating agents, *Nature* **191**:730–731.

Strauss, B. S., 1962, Response of *E. coli* auxotrophs to heat after treatment with mutagenic alkyl methanesulfonates, *J. Bacteriol.* **83**:241–249.

Streisinger, G., Okada, Y., Embrich, J., Newton, J., Tsugita, A., Terzaghi, E., and Inouye, M., 1966, Frameshift mutations and the genetic code, *Cold Spring Harb.* **31**:77–84.

Strelzoff, E., 1961, Identification of base pairs involved in mutations induced by base analogues, *Biochem. Biophys. Res. Commun.* **5**:384–388.

Strelzoff, E., 1962, DNA synthesis in the presence of 5-bromo-uracil. II. Induction of mutations, *Z. Vererbung.* **93**:301–318.

Strigini, P., 1965, On the mechanism of spontaneous reversion and genetic recombination in bacteriophage T4, *Genetics* **52**:759–776.

Stubbe, H., 1937, *Spontane und strahleninduzierte Mutabilitat*, Leipzig.

Sussenbach, J. S., and Berends, W., 1965, Photodynamic degradation of guanine, *Biochim. Biophys. Acta* **95**:184–185.

Swartz, M. N., Trautner, T. A., and Kornberg, A., 1962, Enzymatic synthesis of deoxyribonucleic acid. XI. Further study of nearest neighbor base sequences in deoxyribonucleic acids, *J. Biol. Chem.* **237**:1961–1967.

Swietlinska, Z., and Zuk, J., 1974, Effect of caffeine on chromosome damage induced by chemical mutagens and ionizing radiation in *Vicia faba* and *Secale cereale*, *Mutation Res.* **26**:89–97.

Swinehart, J. L., Lin, W. S., and Cerutti, P. A., 1974, Gamma-ray induced damage in thymine in mononucleotide mixtures and in single- and double-stranded DNA. *Radiation Res.* **58**:166–175.

Sztumpf-Kulikowska, E., Shugar, D., and Boag, I. W., 1967, Kinetics of photodimerization of orotic acid in aqueous medium, *Photochem. Photobiol.* **6**:41–54.

Szybalska, E.-H., and Szybalski, W., 1962, Genetics of human cell lines. IV. DNA-mediated heritable transformation of a biochemical trait, *Proc. Natl. Acad. Sci.* **48**:2026–2034.

Szybalski, W., 1958, Observation on chemical mutagenesis in microorganisms, *Ann. N.Y. Acad. Sci.* **76**:475–489.

Szybalski, W., and Djordjevic, B., 1959, Radiation sensitivity of chemically modified human cells, *Genetics* **44**:540–541.

Szybalski, W. A., and Lorkiewich, Z., 1962, The principal targets of lethal and mutagenic radiation effects. *Strahleninduzierte Mutagenese, Abhandl. deutchen Akad. Wiss.* Berlin, Klass fur Medizine, N 1:63–71.

Szybalski, W., Opara-Kubinska, Z., Lorkiewicz, Z., Ephrati-Elizur, E., and Zamenhof, S., 1960, Transforming activity of deoxyribonucleic acid labeled with 5-bromouracil, *Nature* **188**:743–745.

Takemura, S., 1957, Hydrazinolysis of yeast ribonucleic acid. Formation of "ribo-apyrimidinic acid," *J. Biochem.* **44**:321–325.

Takemura, S., 1959, Hydrazinolysis of nucleic acids. I. The formation of deoxyapyrimidinic acid from herring sperm deoxyribonucleic acid, *Bull. Chem. Soc. Japan* **32**:920–926.

Tamm, C., and Hodes, M. E., and Chargaff, E., 1952, The formation of apurinic acid from the deoxyribonucleic acid of calf thymus, *J. Biol. Chem.* **195**:49–63.

Tamm, C., Shapiro, H. S., Lipschitz, R., and Chargaff, E., 1953, Distribution density of nucleotides within a deoxyribonucleic acid chain, *J. Biol. Chem.* **203**:673–688.

Tarmy, E. M., Venitt, S., and Brookes, P., 1973, Mutagenicity of the carcinogen 7-bromomethyl-benz[a]anthracene. Quantitative study in repair-deficient strains of *Escherichia coli, Mutation Res.* **19**:153–166.

Taylor, B., Greenstein, J. P., and Hollaender, A., 1948, Effects of X-radiation on sodium thymus nucleate, *Arch. Biochem.* **16**:19–31.

Tazima, Y., 1974, Naturally occuring mutagens of biological origin. A review, *Mutation Res.* **26**:225–234.

Tegtmeyer, P., Dohan, C., and Reznikoff, C., 1970, Inactivating and mutagenic effects of nitrosoguanidine on Simian virus 40, *Proc. Natl. Acad. Sci.* **66**:745–752.

Terawaki, A., and Greenberg, J., 1965, Effect of some radiomimetic agents on deoxyribonucleic acid synthesis in *Escherichia coli* and transformation in *Bacillus subtilis, Biochim. Biophys. Acta* **95**:170–173.

Terzaghi, B. E., Streisinger, G., and Stahl, F. W., 1962, The mechanism of 5-bromouracil mutagenesis in the bacteriophage T4, *Proc. Natl. Acad. Sci.* **48**:1519–1524.

Tessman, I., 1962, The induction of large deletions by nitrous acid, *J. Mol. Biol.* **5**:442–445.

Tessman, E. S., 1965, Growth and mutation of phage T1 on ultraviolet irradiated host-cells, *Virology* **2**:679–688.

Tessman, I., Poddar, R. K., and Kumar, S., 1964, Identification of the altered bases in mutated single-stranded DNA. I. *In vitro* mutagenesis by hydroxylamine, ethyl methanesulfonate and nitrous acid, *J. Mol. Biol.* **9**:352–363.

Tessman, I., Ishiwa, H., and Kumar, S., 1965, Mutagenic effects of hydroxylamine *in vivo, Science* **148**:507–508.

Theleu, T. H., and Shoffner, R., 1970, Viral-induced mutagenesis: tests in *Gallus domesticus, Mutation Res.* **9**:425–433.

Tichonenko, T. I., Kisseliova, N. P., Ulanov, B. P., Andronikova, M. L., Velikodvorskaya, G. A., and Budowsky, E. I., 1970, On the action of O-methylhydroxylamine on phage and intraphage DNA, *Probl. Virol.* No. **5**:622–627 (in Russian).

Tichonenko, T. I., Budowsky, E. I., Sklyadneva, V. B., and Khromov, I. S., 1971, The secondary structure of bacteriophage DNA in situ. III. Reactions of S_D phage with O-methylhydroxylamine, *J. Mol. Biol.* **55**:535–547.

Trautner, T. A., Swartz, M. N., and Kornberg, A., 1962, Enzymatic synthesis of deoxyribonucleic acid. X. Influence of bromouracil substitutions on replication, *Proc. Natl. Acad. Sci.* **48**:449–455.

Troll, W., Bellman, S., and Levine, E., 1963, The effect of metabolites of 2-naphtylamine and the mutagen hydroxylamine on the thermal stability of DNA and polyribonucleotides, *Cancer Res.* **23**:841–847.

Trosco, J. E., Chu, E. H. Y., and Carrier, W. L., 1965, The induction of thymine dimers in ultraviolet-irradiated mammalian cells. *Radiat. Res.,* **24**:667–672.

Trosco, J. E., Frank, P., Chu, E., and Becker, J. E., 1973, Caffeine inhibition of postreplicative repair on N-acetoxy-2-acetylaminofluorene-damaged DNA in Chinese hamster cells, *Cancer. Res.* **33**:2444–2449.

Trosco, Y. E., and Mansour, V. H., 1969, Photoreactivation of ultraviolet light-induced pyrimidine dimers in Gingko cells grown *in vitro, Mutation Res.,* **7**:120–121.

Tschernik, T. P., and Krivisky, A. S., 1965, Induction of mutations by UV-rays and nitrous acid under the treatment of extracellular phage ϕX 174, *Genetika* No. **2**:39–46 (in Russian).

Tsugita, A., 1962, The proteins of mutants of TMV: composition and structure of chemically evoked mutants of TMV RNA, *J. Mol. Biol.* **5**:284–289.

Turbin, N. V., Soyfer, V. N., Kartel, N. A., Chekalin, N. M., Dorohov, Y. L., Titov, Y. B., and Cieminis, K. K., 1973, Modifications of the waxy character in barley under the action of exogenous DNA of wild type, Moscow (in Russian).

Turbin, N. V., Soyfer, V. N., Kartel, N. A., Chekalin, N. M., Dorohov, Y. L., Titov, Y. B., and Cieminis, K. K., 1975, Genetic modification of the *waxy* character in barley under the action of exogenous DNA of the wild variety, *Mutation Res.* **27**:59–68.

Unrau, P., Wheatcroft, R., Cox, B., and Olive, T., 1973, The formation of pyrimidine dimers in the DNA of fungi and bacteria, *Biochim. Biophys. Acta* **312**:626–632.

Van der Schans, G. P., Bleichrodt, J. E., and Blok, J., 1973, Contribution of various types of damage to inactivation of a biologically-active double-stranded circular DNA by gamma-radiation, *Int. J. Radiat. Biol. Stud. Phys. Chem. Med.* **23**:133–150.

Varghese, A. R., and Wang, S. Y., 1968, Thymine-thymine adduct as a photoproduct of thymine, *Science* **160**:186–187.

Veleminsky, J., and Gichner, T., 1970, The influence of pH on the mutagenic effectiveness of nitroso compounds in *Arabidopsis, Mutation Res.* **10**:43–52.

Veleminsky, J., Osterman-Golkar, S., and Ehrenberg, L., 1970, Reaction rates and biological action of N-methyl and N-ethyl-N-nitrosourea, *Mutation Res.* **10**:169–174.

Verby, M. G., Brakier, L., and Feit, P. W., 1971, Inactivation of the T7 coliphage by diepoxybutane stereoisomers, *Biochim. Biophys. Acta* **228**:400–406.

Verly, W. G., 1974, Monofunctional alkylating agents and apurinic sites in DNA, *Biochem. Pharmacology* **23**:3–8.

Verwoerd, D. W., Kohlage, H., and Zillig, W., 1961, Special partial hydrolysis of nucleic acids in nucleotide sequence studies, *Nature* **192**:1038–1040.

Vielmetter, W., and Schuster, H., 1960, The base specificity of mutation induced by nitrous acid in phage T2, *Biochem. Biophys. Res. Commun.* **2**:324–328.

Vogt, M., and Dulbecco, R., 1963, Steps in the neoplastic transformation of hamster embryo cells by polyoma virus, *Proc. Natl. Acad. Sci.* **49**:171–179.

Wacker, A., 1965, The molecular mechanisms of mutagenic action of irradiation, *In: Nucleic Acids,* Moscow 412–444 (in Russian).

Wacker, A., Dellweg, H., and Weinblum, D., 1960a, Strahlenchemische Veränderung der Bakterien Desoxyribonukleinsäre *in vivo, Naturwiss.* **47**:477.

Wacker, A., Kirschfeld, S., and Träger, S., 1960b, Über den Einbau Purinanaloger Verbindungen in die Bakterien-Nukleinsäure, *J. Mol. Biol.* **2**:241–242.

Wacker, A., Turck, G., and Gerstenberger, A., 1963, Zum Wirkungsmechanismus photodinamischer Farbstoffe, *Naturwiss.* **50**:377.

Wagner, J. H., Nawar, M. M., Konzak, C. F., and Nilan, R. A., 1968, The influence of pH on the biological changes induced by ethyleneimine in barley, *Mutation Res.* **5**:57–64.

Wang, S. Y., 1958, Photochemistry of nucleic acids and related compounds. I. The first step in the ultraviolet irradiation of 1,3-di-ethyluracil, *J. Am. Chem. Soc.* **80**:6190–6201.

Wang, S. Y., and Vargese, A. J., 1967, Cytosine-thymine addition product from DNA irradiated with ultraviolet light, *Biochim. Biophys. Res. Commun.* **29**:543–545.

Ward, J. E., 1972, Mechanisms of radiation-induced strand break formation in DNA, *Isr. J. Chem.* **10**:1123–1138.

Ward, C. B., and Glaser, D. A., 1969, Origin and direction of DNA synthesis in *E. coli* B/r, *Proc. Natl. Acad. Sci.* **62**:881–886.

Waring, M. J., 1973, Interaction of indazole analogs of lucanthone and hycanthone with closed circular duplex deoxyribonucleic acid, *J. Pharmacol. Exp. Ther.* **186**:385–389.

Watanabe, T., and Fukasava, T., 1961, Episome-mediated transfer of drug resistance in

Enterobacteriaceae. II. Eliminated of resistance factors with acridine dyes, *J. Bacteriol.* **81**:679–683.

Watson, J. D., and Crick, F. H. C., 1953, Molecular structure of nucleic acids: A structure of desoxyribose nucleic acids, *Nature* **171**:737–738.

Webb, R. B., and Kubitschek, H. E., 1963, Mutagenic and antimutagenic effect of acridine orange in *Escherichia coli, Biochem. Biophys. Res. Commun.* **13**:90–94.

Webb, R. B., and Malling, M. M., 1970, Mutagenic effects of near ultraviolet and visible radiant energy on continuous cultures of *Escherichia coli, Photochem. Photobiol.* **12**:457–468.

Weigert, M. G., Lanka, E., and Garen, A., 1967, Base composition of nonsense codon in *Escherichia coli.* II. The N2 codon UAA, *J. Mol. Biol.* **23**:391–400.

Weinberg, R., and Latham, A. B., 1956, Apparent mutagenic effect of thymine deficiency for a thymine requiring strain of *Escherichia coli, J. Bacteriol.* **72**:570–572.

Weinblum, D., 1967, Characterization of photodimers from DNA, *Biochem. Biophys. Res. Commun.* **27**:384–390.

Weinblum, D., and Johns, H. E., 1966, Isolation and properties of isomeric thymine dimers, *Biochim. Biophys. Acta* **114**:450–459.

Weissbach, A., and Lisio, A., 1965, Alkylation of nucleic acids by mitomycin C and porfiromycin, *Biochemistry* **4**:196–200.

Westergaard, M., 1957, Chemical mutagenesis in relation to the concept of the gene, *Experientia* **13**:224–234.

Weygand, F., Wacker, A., and Dellweg, H., 1952, Stoffwecheluntersuchungen bei Mikroorganismen mit Hilfe radioaktives Isotope. II. Kompetitive und nichtkompetitive Enthemung von 5-Br^{82}-Uracil. *Z. Naturforsch* **8b**:19–25.

Wheeler, G. P., 1962, Studies related to the mechanism of action of cytotoxic alkylating agents: A review, *Cancer Res.* **22**:651–688.

Whitfield, H. G., Martin, R. G., and Ames. B. N., 1966, Classification of aminotransferase (C gene) mutants in the histidine operon, *J. Mol. Biol.* **21**:335–355.

Willets, N. S., 1967, The elimination of F-lac$^+$ from *Escherichia coli* by mutagenic agents, *Biochem. Biophys. Res. Commun.* **27**:112–117.

Williamson, J. H., 1970, Ethyl methanesulfonate-induced mutants in the Y chromosome of *Drosophila melanogaster, Mutation Res.* **10**:597–605.

Winkler, G. N., and Smislova, G. I., 1967, The uncovering of natural autoantimutagenic system, *Genetica* **No. 5**:52–56 (in Russian).

Witkin, E. M., 1947, Mutations in *Escherichia coli* induced by chemical agents, *Cold Spring Harbor Symp. Quant. Biol.* **12**:256–269.

Witkin, E. M., 1968, The role of DNA repair and recombination in mutagenesis, *In: Proc. XII Intern. Congr. Genetics.*

Wittman, H. G., 1962, Proteinuntersuchungen am Mutanten des Tabakmosaikvirus als Beitrag zum Problem des genetischen Codes, *Z. Vererb.* **93**:491–530.

Wittman, H. G., and Wittman-Liebold, B., 1966, Protein chemical studies of two RNA viruses and their mutants, *Cold Spring Harbor Symp. Quant. Biol.* **31**:163–172.

Wullf, D. L., and Fraenkel, G., 1961, On the nature of thymine photoproduct, *Biochim. Biophys. Acta* **15**:332–339.

Wyss, O., Wiese, C., and Schaiberger, G. E., 1963, Peroxydes and mutations, *Radiat. Res.* **suppl. 3**:184–191.

Yamane, T., Wyluda, B. J., and Shulman, R. G., 1967, Dihydrothymine from UV-irradiated DNA, *Proc. Natl. Acad. Sci.* **58**:439–442.

Yanofsky, C., Ito, I., and Horn, V., 1966, Amino acid replacement and the genetic code, *Cold Spring Harbor Symp. Quant. Biol.* **31**:151–162.

Yoder, J., Watson, M., and Benson, W., 1973, Lymphocyte chromosome analysis of agricultural worker during extensive occupational exposure to pesticides, *Mutation Res.* **21**:335–340.

Zadrazil, S., Veleminsky, J., Pokorny, V., and Gichner, T., 1973, Repair of DNA lesions in barley induced by monofunctional alkylating agents, *Studia Biophysika* **36/37**:271–276.

Zadrazil, S., Pokorny, V., Veleminsky, J., and Gichner, T., 1974, Changes in the DNA lesions induced by alkylating agents during the post-treatment washing and redrying of barley seeds, *Biol. Plant* **16**:7–13.

Zamenhof, S., 1967, Nucleic acids and mitability. In: *Progress in nucleic acid research and molecular biology,* J. N. Davidson & W. E. Cohn (eds.) **6**:1–38.

Zamenhof, S., and Arikawa, S., 1970, Comparative studies on alkylation of bacterial DNA *in vivo* and *in vitro, Mutation Res.* **9**:141–148.

Zamenhof, S., and Greer, S., 1958, Heat as agent producing high frequency of mutations and unstable genes in *Escherichia coli, Nature* **182**:611–613.

Zamenhof, S., and Griboff, G., 1954, *E. coli* containing 5-bromo-uracil in its deoxyribonucleic acid, *Nature* **174**:307–308.

Zamenhof, S., Alexander, H. E., and Leidy, G., 1953, Studies on the chemistry of transforming activity. I. Resistance to physical and chemical agents, *J. Exptl. Med.* **98**:373.

Zampieri, A., and Greenberg, J., 1965, Mutagenesis by acridine orange and proflavin in *Escherichia coli, Mutation Res.* **2**:552–556.

Zampieri, A., Greenberg, J., and Warren, G., 1968, Inactivating and mutagenic effects of 1-methyl-3-nitro-1-nitrosoguanidine on intracellular bacteriophage, *J. Virology* **2**:901–904.

Zeller, E. A., Gurmann, H., Hegedüs, B., Kaiser, A., Langemann, A., and Müller, M., 1963, Methylhydrazine derivatives, a new class of cytotoxic agents, *Experientia* **19**:129–130.

Zimmermann, F. K., Schweier, R., and Laer, U. V., 1965, The influence of pH on the mutagenicity on yeast of N-methyl-nitrosamides and nitrous acid, *Z. Vererbungslehre* **97**:68–71.

Zirkle, R. E., and Bloom, W., 1953, Irradiation of parts of individual cells, *Science* **117**:487–493.

4

The Classical Case of Character Displacement

P. R. GRANT

Department of Biology
McGill University
Montreal, Quebec, Canada

INTRODUCTION

If textbooks are a reliable guide, the theory of character displacement (Brown and Wilson, 1956) has been widely accepted without critical debate (e.g., Odum, 1971; Orians, 1968; Lack, 1971; Emlen, 1973, etc.) except over semantics (Mayr, 1963; Brown, 1964). The theory attempts to explain differences between sympatric populations of two systematically related species, as opposed to similarities of their allopatric populations, in terms of selection acting in sympatry against individuals of the two species which are so similar that they compete for food and/or hybridize (or tend to do so).

The most frequently quoted example concerns two species of birds (rock nuthatches), *Sitta tephronota* and *Sitta neumayer* (Vaurie 1950, 1951). Brown and Wilson (1956) used it as their prime example, anticipating that it would become "the classical case of character displacement." The two species coexist in western Iran and in small areas of Iraq and Armenia (U.S.S.R.) (Fig. 1). In this zone they differ markedly in bill size, wing length, and prominence of a black post-ocular stripe. In the zones of allopatry the differences progressively diminish away from the sympatric zone so that, as illustrated in Fig. 2, the westernmost *S. neumayer* in Yugoslavia and Greece (area 2 in Fig. 2) is almost identical to the easternmost *S. tephronota* in the Tien Shan mountains of the U.S.S.R. (area 19 in Fig. 2).

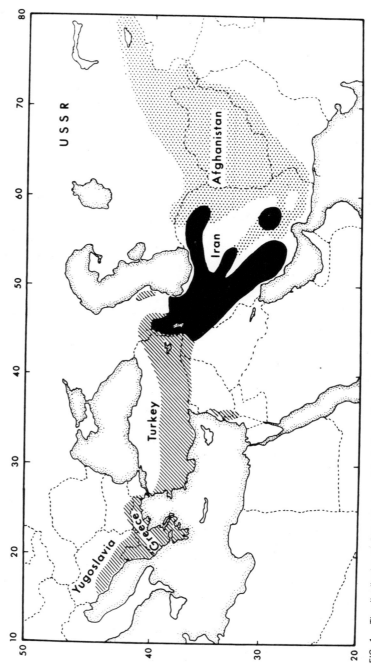

FIG. 1. The distribution of, *Sitta neumayer* (shaded) and *Sitta tephronota* (stippled); they occur together in the area shown in black. Based on information from Balát (1962), Dementiev and Gladkov (1970), Hüe and Etchecopar (1970), Paludan (1959), Streseman (1920) and Vaurie (1951).

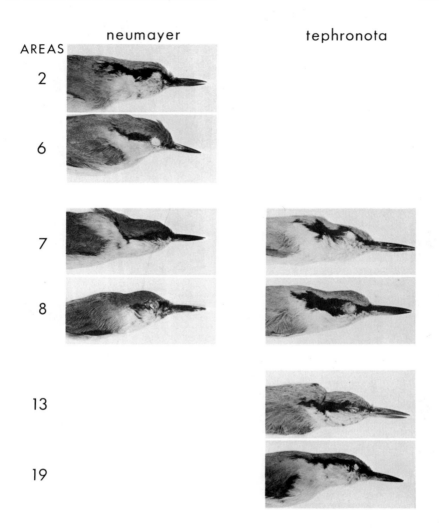

FIG. 2. **Morphological variation of *Sitta neumayer*** and *Sitta tephronota* in sympatry (areas 7 and 8) and allopatry. For location of the areas, see Fig. 4. Scale: approximately half natural size.

The application of character displacement theory to this pair of species gives rise to an interpretative model similar to that portrayed in Fig. 3 (Grant, 1972*a*). The interpretation is that mutual divergence of bill and eye stripe sizes resulted from selection in sympatry against individuals of the two species which initially had similar bill sizes, and hence competed for

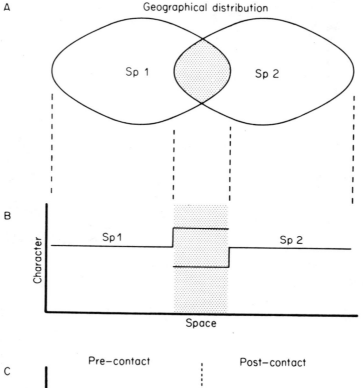

A

Geographical distribution

Sp 1

Sp 2

B

Character

Sp 1

Sp 2

Space

C

Pre-contact

Post-contact

Character

Sp 1

Sp 2

Time ——▶

FIG. 3. Character displacement inferred from character states in sympatry and allopatry. Transitions in character states at the boundaries (center of Figure) are shown as sharp "steps" for emphasis whereas in fact curves are more likely (see e.g., Crozier 1974, Hanson 1966). Reprinted from Grant (1972*a*) with kind permission of the Linnean Society (Great Britain) and Academic Press (London).

food, and which had similar eye stripe sizes and hence wasted reproductive effort in courting inappropriate potential mates (Brown and Wilson, 1956; Vaurie, 1950, 1951). It is assumed that the initial similarity of the species represents recent derivation from a common ancestor through allopatric speciation. Certainly the two species are systematically closely related (Vaurie, 1957; Voous and van Marle, 1953). Allopatric speciation is assumed because the weight of evidence and reasoning to date is heavily in favor of the allopatric model in general (Mayr, 1963; Scudder, 1974), even though sympatric speciation is a theoretical possibility (e.g., Bush, 1969; Pimentel *et al.*, 1967; Smith, 1966; Soans *et al.*, 1974).

The two species are unusual among nuthatches in living in environments largely or entirely devoid of trees, and in using trees very little when they are present (Paludan, 1940). The foraging and nesting activities of both species are more or less restricted to the ground and surfaces of cliffs and rock faces, often in steep ravines. Orians and Wilson (1964) have established a strong association between interspecific territoriality in birds and structurally simple habitats such as these; and Cody (1969, 1973), expanding on remarks by Orians and Wilson, has developed a theory that predicts selection for similarity, or convergence, coupled with interspecific territoriality, in structurally simple habitats. Thus the rock nuthatches present us with a dilemma. They occur together in environments where convergence is to be expected, yet apparently exhibit divergence.

This dilemma will be resolved by an evolutionary argument constructed with the answers to four questions. In what ways has morphological divergence or convergence occurred in sympatry which can be attributed to selection arising from the presence of the other species? Do the species differ in diet where sympatric, and if so by how much? Does the black eye stripe function as a species discrimination signal in a reproductive context? Finally, are the species interspecifically territorial with correlated similarity of song (Grant, 1966a)? The answers to the last three questions enable us to assess the significance of the morphological patterns established in answering the first question.

The first question might be thought unnecessary in view of the general acceptance of the *Sitta* evidence for character displacement. However, a recent analysis (Grant, 1972a) of Vaurie's original data showed that other interpretations of the observed patterns of variation are possible, as discussed below. Indeed a review of other purported cases of the ecological aspect of character displacement revealed further weaknesses in the interpretation of observed variation, and the conclusion was reached that "the detection of character displacement in its ecological aspect, particularly in mainland regions, will require a level of detail that has so far not been produced in a single study" (Grant, 1972a, p. 65).

This paper attempts to achieve that level of detail in a morphological, ecological, and behavioral analysis of the *Sitta* situation. It complements the review paper (Grant, 1972*a*). Inasmuch as it illustrates the problems of (a) detecting character displacement among organisms in general, and (b) of understanding past evolutionary processes from present information only, its significance extends well beyond the confines of the two species under study.

I shall use the term character displacement to mean "the process by which a morphological character state of a species changes under natural selection arising from the presence, in the same environment, of one or more species similar to it ecologically and/or reproductively" (Grant, 1972*a*, p. 44). It encompasses both divergent and convergent changes. Divergent changes are much easier to understand, but there are theoretical reasons, in addition to those connected with interspecific territorality, for expecting convergence in morphological characters with ecological (but not reproductive) functions in some circumstances (e.g., MacArthur and Levins, 1967; Schoener, 1969). Changes in behavior, ecology or physiology brought about by the same type of selection are analogues of morphological character displacement. Character displacement as defined here, is but one type of adjustment that species make under natural selection to the presence of other similar species in their environment.

THE MORPHOLOGICAL CHARACTER DISPLACEMENT PROBLEM

In both *tephronota* and *neumayer* there is a decreasing size trend from west to east. It is evident in the bill and wing length data originally provided by Vaurie (1950, 1951; see also Stepanyan, 1961), as pointed out before (Grant, 1972*a*), but it affects other dimensions as well. It can be shown by correlating various dimensions of each species separately with longitude throughout their ranges (Table I). Most correlations are significant, all are negative. Correlations are generally weaker with latitude than with longitude.

The important point of the associations with longitude is that, contrary to what is often believed, a demonstration of a size difference in dimensions between allopatric and sympatric populations of a species is, in itself, not sufficient evidence that character displacement has occurred. A knowledge of clinal variation in allopatry alone allows one to "predict" expected sizes in sympatry, on the assumption that the same factors are affecting the dimensions in a similar manner in the two zones. An allopatric cline projected

TABLE I. Correlation Coefficients for Population Means of Several Dimensions and Longitude and Latitude[a]

Dimension	S. neumayer		S. tephronota	
	Longitude	Latitude	Longitude	Latitude
Wing length	−0.650*	0.790**	−0.568*	−0.386
Tarsus length	−0.792**	0.658*	−0.808****	−0.517
Bill length	−0.759**	0.599	−0.813****	−0.688**
Bill width	−0.619*	0.464	−0.609*	−0.553*
Eye stripe length	−0.870***	0.826***	−0.301	−0.017
Eye stripe area	−0.851*	0.860*	−0.669*	−0.574

[a] One to four asterisks indicate statistical significance at the 5, 1, 0.5, and 0.1% level, respectively. Adult male and female samples have been combined. Sample sizes are 11 for *neumayer* and 13 for *tephronota*.

into sympatry constitutes the prediction. The predictions are for large size in sympatry for *tephronota* and small size in sympatry for *neumayer*. If character displacement has occurred the assumption is incorrect and observed sizes in sympatry will differ from the expected; deviation from expectation may occur as convergence towards or divergence from the other species.

There are two possible approaches to the detection of character displacement. One is the regression approach in which the relationship between a dimension and longitude (or other environmental variable) is studied in allopatry and sympatry separately, and then the two compared. It depends upon the existence of a statistically demonstrable relationship in allopatry. The other approach involves a comparison of adjacent populations and is not dependent on any regularity in allopatry. If character displacement has occurred it is detected by a sharp discontinuity in character states between populations either side of the sympatry/allopatry boundary (henceforth referred to as the zone boundary). The significance attached to a discontinuity depends upon whether there exist other such discontinuities between neighboring populations *within* allopatry and *within* sympatry.

The first approach has been used with Vaurie's wing and bill data for *tephronota* (Grant, 1972a). For each dimension, allopatric and sympatric regression equations were similar, with slopes and intercepts in sympatry not departing significantly from those in allopatry. These results mean that, for both dimensions, variation in allopatry adequately predicts variation in sympatry. The procedure failed to detect evidence of character displacement in this species.

This analysis was limited to two dimensions of the one species for which data were available. However, the result is sufficiently interesting to make it worthwhile extending the analysis to eye stripe features of *tephronota,* and to all dimensions of *neumayer.* All available museum specimens of these two species were examined and measured, resulting in sample sizes approximately twice as large as those used by Vaurie, and making it possible to apply the two approaches to the detection of character displacement in both species.

METHODS

Five hundred and fourteen specimens of rock nuthatches were measured. Forty-eight were juveniles and/or specimens of unknown sex, and these were treated separately. Measurements in millimeters were taken as follows: *Wing length:* the extent of the flattened and folded wing. *Tarsus* (*tarsometatarsus*) *length:* tibiotarsus articulation to an identified distal scute, usually the lowest undivided one. *Bill length:* exposed culmen from base of the front feathers to the tip. *Bill width:* width of the upper mandible at the anterior level of the nares. *Eye stripe length:* from the posterior edge of the eye to the furthest (posterior) point of the continuous black feathers which constitute the post-ocular stripe. *Eye stripe area:* measured from photographs (print magnification × 2) by a Quantimet 720 image analyser. This instrument records the number of points registered as black in a specified field of the image, and area is calculated from the fact that there are 64 points to one square millimeter. The sensitivity of the machine can be made to vary from low intensities of grey (non-white) to absolute black. Calibration is required to ensure that all of the eye stripe but none of the neighboring grey color is recorded as black. This control of intensity record-ing was utilized to determine the reflectance value of the black eye stripe. A specified small area was selected entirely within the black eye stripe of a few specimens. A dial was adjusted so that the entire area was only just registered as black, and the value recorded. The Quantimet was thus used as a crude densitometer. The value was then converted to a reflectance density on the Kodak Grey Scale from a previously established relationship between this scale and dial readings. The specimens themselves rather than photographs were used for all determinations. Statistical analyses were performed on the IBM 360/75 computer at McGill University.

Localities from which specimens were collected were plotted on a map and found to cluster into 20 areas of approximately similar size (Fig. 4) which were gratifyingly discrete. The data analysis is based upon the divi-

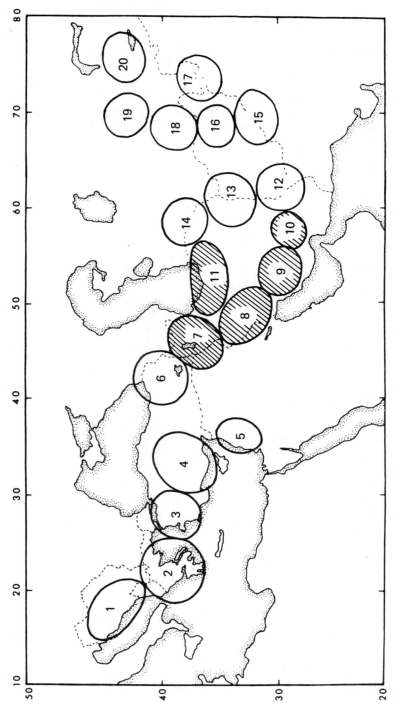

FIG. 4. The areas from which samples of *Sitta neumayer* and *Sitta tephronota* were drawn.

sion of samples of specimens into these 20 areas; for convenience the specimens are referred to as being drawn from 20 populations, even though these populations are not likely to be discrete in nature. The zone boundaries are not known precisely, but from published distribution maps (e.g., Dementiev and Gladkov, 1970; Hüe and Etchecopar, 1970; Vaurie, 1951) they appear to run consistently between the areas recognized in this study rather than transecting them, so all areas can be referred to as either in sympatry or allopatry.

Geographical Variation in Morphology

Choice of Environmental Predictor

Multiple linear regression analysis was used to identify the environmental variable or variables capable of explaining the most variation in each of the morphological dimensions of the two species. Independent variables used were longitude, latitude, average July maximum and average January minimum temperature, rainfall, relative humidity, and Emberger's quotient which is a composite of temperature and moisture often referred to as a zerothermic index. Analyses were performed with allopatric and sympatric populations of each species treated separately and together. All independent variables were incorporated when the degrees of freedom allowed, otherwise they were introduced in groups.

The result of the several analyses was that only longitude or latitude correlated significantly with any of the dimensions. Latitude correlated less well than longitude, except for wing and eye stripe area of total adults (i.e. males and females) of *neumayer,* sympatric and allopatric populations combined, and except for bill length of allopatric adult male *tephronota.* Bill length of total adult *tephronota,* sympatric and allopatric populations combined, regressed significantly on both longitude and latitude. Contributions to the coefficient of determination ($100.R^2$) were 50.21 and 28.14%, respectively. In all other regressions a dimension regressed significantly on no more than one independent variable.

Longitude was therefore chosen as the environmental predictor of morphological dimensions. Part of the reason for its superiority over latitude must lie in its greater range of values (see Fig. 1).

Bill Size

Bill Length. Since there are small but consistent differences between male and female mean bill lengths (and other dimensions) of both species, the regression analyses used data of only adult males (male samples were generally larger than female samples).

For both species the regression approach yields the result that geographic variation in allopatry apparently predicts what is actually found in sympatry. Only one regression is significant at the 5% level (Table II). Considering all regressions and disregarding the level of significance we find that there are no statistically significant differences in slope or intercept between allopatric and sympatric regressions on longitude of either species (Table II). If character displacement has occurred it has not been detected by this approach. Undoubtedly the small sample sizes make it difficult to detect differences, if such exist, but the ostensible similarity of the estimates of slopes and of intercepts argues against any major shifts in sympatry.

The more detailed approach comparing adjacent populations yields essentially the same result. The data are given in full for both sexes in Table III, and geographical variation is illustrated in Fig. 5 with just adult males.

When viewing the illustration it is necessary to bear in mind that there are deviations from a linear array of populations along a longitudinal gradient. These can be best appreciated by referring to Fig. 4. Population 5 is a southern offshoot from the longitudinal chain of *neumayer* populations, and population 11 does not lie between 10 and 12 but to the north of 10. Similarly *tephronota* populations 12, 13, and 14 are aligned more in a northerly than an easterly direction. With these deviations in mind it becomes evident that sympatric population 11 of *tephronota* should be compared with adjacent allopatric populations 13 and 14, and sympatric population 10 should be compared with allopatric population 12.

Figure 5 should be "read" from allopatry to sympatry. Mean bill length of *tephronota* shows a gradual increase in allopatry from east to west that continues apparently uninterruptedly as one passes from area 12 to 10

TABLE II. Regression of Adult Male Mean Bill Lengths on Longitude[a]

	N	F	r	$a \pm S.E.$	$b \pm S.E.$
S. neumayer					
Allopatry	6	7.277	−0.808	23.737 ± 0.747	−0.066 ± 0.024
Sympatry	4	62.906**	−0.984**	25.281 ± 0.063	−0.080 ± 0.001
S. tephronota					
Allopatry	9	6.036	−0.621	30.936 ± 3.244	−0.101 ± 0.048
Sympatry	4	1.304	−0.620	31.524 ± 4.087	−0.090 ± 0.081

[a] Symbols: N sample size; r correlation coefficient; a intercept; b slope; F variance ratio; $S.E.$ standard error; asterisks as in Table I.

TABLE III. Bill Length Characteristics of *Sitta neumayer* and *Sitta tephronota* Populations[a]

	Adult males			Adult females		
Populations	N	$\bar{x}$	$S\bar{x}$	N	$\bar{x}$	$S\bar{x}$
S. neumayer						
1	22	22.56	0.18	15	22.31	0.15
2	20	22.64	0.24	10	22.34	0.20
3	15	21.18	0.17	13	20.99	0.21
4	7	21.83	0.16	6	21.57	0.29
5	4	21.87	0.32	2	(22.4, 22.8)	—
6	4	20.58	0.36	3	22.23	(21.5, 22.3, 22.9)
7	9	21.82	0.28	5	21.42	0.20
8	21	21.44	0.18	13	21.31	0.14
9	9	21.08	0.18	2	(20.3, 21.6)	—
10	2	(21.5, 21.6)	—	4	21.06	0.05
11	12	21.16	0.23	7	21.89	0.25
S. tephronota						
7	6	27.90	0.30	4	27.51	0.57
8	19	27.35	0.25	18	26.51	0.27
9	16	25.76	0.23	13	26.08	0.28
10	6	26.90	0.25	5	26.33	0.50
11	2	(23.9, 25.2)	—	2	(24.6, 24.7)	—
12	12	25.71	0.24	4	26.34	0.75
13	19	25.12	0.25	13	24.64	0.23
14	6	23.77	0.40	5	23.57	0.42
15	4	24.44	0.63	4	23.95	0.11
16	13	24.58	0.17	9	24.23	0.27
17	12	24.43	0.34	10	24.28	0.39
18	9	23.26	0.36	5	23.12	0.50
19	12	23.52	0.36	9	22.97	0.32
20	6	22.62	0.54	6	22.53	0.56

[a] Symbols: N. sample size; $\bar{x}$ mean; $S\bar{x}$ standard error. In this and subsequent tables means are calculated for samples of three or more, and individual measurements are shown in brackets for samples of three or less. Other minimum samples for calculation are four for standard error.

into the sympatric zone. However, the poorly known population 11 appears to be actually smaller than the adjacent allopatric population 13, in both sexes (Table III). The small sample size of population 11 renders any firm conclusion unwise, but there is a possibility here of convergence toward *neumayer*. Similarly population 9 appears to have converged toward *neumayer* since it is significantly smaller (two-tailed *t* test, $P < 0.01$), not larger, than population 10, as well as being significantly smaller than neighboring population 8 ($P < 0.001$).

Population 12 is significantly smaller than 10 ($P < 0.01$), a difference expected under a divergent character displacement hypothesis. But there are also significant differences in the same direction between neighboring populations within sympatry, specified above, and allopatry—in shorthand $16 > 18$ ($P < 0.01$) and $13 > 14$ ($P < 0.05$). These significant differences are not confined to adult males, since for females $13 > 14$ ($P < 0.05$) $13 > 15$ ($P < 0.02$), $12 > 15$ ($P < 0.05$) and $12 > 13$ ($P < 0.05$). Therefore the difference between populations 10 and 12 at the zone boundary is not unusual, nor is its direction. Furthermore, the magnitude of the difference is not unusually large. The mean of 12 exceeds the mean of 10 by 4.6%, but within sympatry the mean of 8 exceeds the mean of 9 by as much as 6.2%. *S. tephronota* therefore shows no clear evidence of divergence in sympatry.

Figure 5 makes it clear that *neumayer,* if anything, has converged in bill length in sympatry toward the *tephronota* condition. Population 7 in sympatry is significantly larger than 6 in allopatry ($P < 0.05$). There are indications of this difference in the earlier literature (Dunajewski, 1934; Vaurie, 1950; see also Grant, 1972a, Fig. 5). Population 7 is also significantly larger than 11 ($P < 0.05$), but none of the sympatric populations is significantly smaller than any of the allopatric populations east of Greece (2).

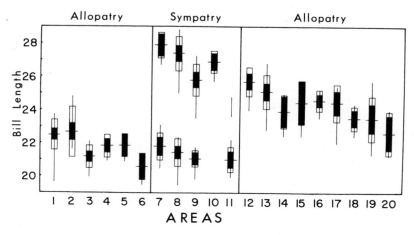

FIG. 5. Bill length characteristics of *Sitta neumayer* (lower, areas 1–11) and *S. tephronota* (upper, areas 7–20) populations, adult males only. In each bar diagram the horizontal line indicates the mean, the solid bar represents two standard errors either side of the mean, the open bar represents one standard deviation either side of the mean and the central vertical line represents the total range of measurements. Data taken from Table III, in which sample sizes are given.

Significant differences exist between neighboring populations within allopatry, as well as between the populations either side of the zone boundary. Population 6 has a smaller mean than 5 ($P < 0.05$) and 4 ($P < 0.02$), and population 3 has a smaller mean than 4 ($P < 0.02$) and 2 ($P < 0.01$). For females also $2 > 3$ ($P < 0.01$). Any interpretation of the between-zone difference in terms of character displacement is weakened by the occurrence of within-zone differences.

It is noteworthy that the largest difference between adjacent populations is between 2 and 3 which are the only two, of either species, to be separated by a water barrier. The barrier, presumably to population mixing and gene flow, could help maintain the size difference but is not essential because a comparable difference exists between *tephronota* populations 8 and 9 in geographical contiguity.

Bill Width. This is another dimension associated with diet. Geographical variation can be examined for evidence of character displacement. However the regression approach cannot be used with this dimension. First, although it is moderately well correlated with longitude in allopatric populations of *neumayer* ($r = -0.790$, $P \sim 0.06$), it is not correlated with longitude in allopatric *tephronota* ($r = -0.520$, $p > 0.1$). Second, no regression relationship could be generated for sympatric *neumayer,* with which to compare the allopatric regression, even when the F level was set artificially low at 1.00. These attempts however did reveal that allopatric *neumayer* has a regression slope (-0.012 ± 0.005) almost identical to that of allopatric *tephronota* (-0.014 ± 0.009). This similarity of the species in allopatry but difference in sympatry may be indicative of character displacement in some nonobvious way.

Comparison of adjacent populations of *tephronota* provides a stronger and clearer indication of the possible operation of character displacement (Table IV). For both males and females, population 10 in sympatry not only has a significantly larger bill width than population 12 in allopatry ($P < 0.01$ in each case), it has ostensibly the largest bill width of all sympatric populations (Fig. 6). The steep increase in bill width across this zone boundary is evidence of divergence away from *neumayer*. But an apparent decrease is manifested across the zone boundary from 13 to 11, in both sexes (Table IV), and adult males of population 9 have a significantly narrower bill ($P < 0.01$) than population 10. These facts are not consistent with a simple divergence hypothesis. There are also significant differences between the adult males of population 15 and those of neighboring allopatric populations 12 ($P < 0.002$), 13 ($P < 0.01$) and 16 ($P < 0.01$).

Variation in bill width among *neumayer* populations is similar to that in bill length. There is no evidence of divergence, rather one of convergence since for adult males $6 < 7$ ($P < 0.002$). Population 6 is also significantly

TABLE IV. Bill Width Characteristics of *Sitta neumayer* and *Sitta tephronota* Populations[a]

Populations	Adult males			Adult females		
	N	$\bar{x}$	$S\bar{x}$	N	$\bar{x}$	$S\bar{x}$
S. neumayer						
1	21	4.63	0.06	16	4.61	0.07
2	20	4.66	0.04	10	4.56	0.05
3	15	4.57	0.06	13	4.59	0.06
4	7	4.46	0.05	6	4.55	0.05
5	4	4.60	0.06	2	(4.8, 4.9)	—
6	4	4.25	0.03	3	4.57	(4.3, 4.6, 4.8)
7	10	4.49	0.05	5	4.40	0.11
8	21	4.39	0.04	14	4.26	0.06
9	9	4.40	0.15	3	4.53	(4.1, 4.4, 5.1)
10	2	(4.5, 4.5)	—	4	4.30	0.07
11	11	4.60	0.05	7	4.67	0.13
S. tephronota						
7	6	5.82	0.15	5	5.68	0.22
8	20	5.74	0.08	18	5.71	0.10
9	16	5.59	0.06	13	5.63	0.08
10	6	6.02	0.06	6	5.72	0.11
11	2	(4.9, 4.9)	—	2	(4.9, 5.1)	—
12	12	5.39	0.07	5	5.10	0.09
13	20	5.31	0.05	13	5.18	0.08
14	6	5.17	0.13	5	5.02	0.06
15	4	4.90	0.11	4	5.00	0.00
16	13	5.30	0.06	9	5.11	0.07
17	12	5.14	0.07	10	5.27	0.05
18	9	5.20	0.12	5	5.06	0.13
19	12	5.09	0.09	9	4.99	0.08
20	6	4.97	0.13	6	4.97	0.15

[a] Symbols and conventions as in Table III.

smaller than 5 ($P < 0.002$) and 4 ($P < 0.05$). In two ways bill width variation differs from bill length variation. Populations 2 and 3 are not significantly different in bill width. And population 11 has the broadest bill in sympatry (and shortest), being significantly broader than 8 in males ($P < 0.02$) and females ($P < 0.01$). The largest sympatric *neumayer* lives in the same area as the apparently smallest sympatric *tephronota*. This does not accord with a simple hypothesis of divergence.

Bill shape, measured as width/length, is approximately constant within

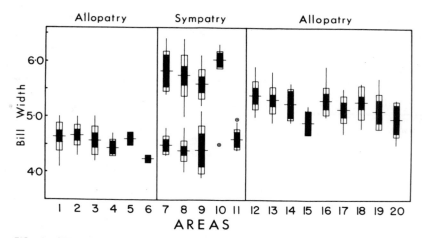

FIG. 6. Bill width characteristics of *Sitta neumayer* (lower, areas 1–11) and *Sitta tephronota* (upper, areas 7–20) populations, adult males only. Symbols as in Fig. 5. In addition a circle with point enclosed indicates two (identical) measurements. Data taken from Table IV in which sample sizes are given.

species and gives no additional indication of displacement not provided by separate analysis of the two dimensions. *S. neumayer* population 11 and *tephronota* population 10 have relatively, as well as absolutely, broad bills.

Summary. There is no clear evidence of character displacement in either bill length or width in either species. The strongest indication is of divergence in bill width of *tephronota*, but other complications make it doubtful that this can be attributed to character displacement.

Eye Stripe Size

Eye Stripe Length. This does not correlate with longitude in allopatric populations of *tephronota* ($r = 0.147$, $P > 0.1$), but does correlate strongly in allopatric populations of *neumayer* ($r = -0.943$, $P < 0.01$). However, sympatric populations of *neumayer* do not exhibit a relationship with longitude, and even with F set artificially low at 1.00 no regression equation could be generated for comparison with the allopatric one. A regression line describing the allopatric populations ($Y = 34.18 - 0.33X$) extended into the sympatric zone passes close to the means of populations 7 and 11. Populations 8, 9 and 10 are much smaller than "predicted" by this method (Fig. 7; see also Fig. 2).

Population data are given in Table V. Eye stripe length shows no systematic variation in *tephronota*, and the averages for sympatric and allopatric populations are about the same. There is a significant decrease in length from 12 to 10 ($P < 0.01$, adult males), indicating convergence toward the *neumayer* condition, whereas 11 and 13 are about the same. Within allopatry there are some significant differences between adjacent populations. For males $12 > 13$ ($P < 0.01$) and $17 < 16$, 18 ($P < 0.01$), a difference in parallel to that between 12 and 10. For females $13 < 14$ ($P < 0.02$) and $18 < 19$ ($P < 0.05$). Thus the overall pattern of variation is one of approximate constancy over the total species range.

An indicated above, the variation in *neumayer* is quite different. The

TABLE V. Post-Ocular Eye Stripe Length Characteristics of *Sitta neumayer* and *Sitta tephronota* Populations[a]

Populations	Adult males			Adult females		
	N	$\bar{x}$	$S\bar{x}$	N	$\bar{x}$	$S\bar{x}$
S. neumayer						
1	18	28.0	0.6	12	27.6	0.9
2	18	27.8	0.8	7	25.7	1.0
3	11	23.8	0.9	10	25.6	0.7
4	5	24.8	0.9	3	29.7	(28, 30, 31)
5	4	21.7	1.0	1	(29)	—
6	3	23.0	(19, 23, 27)	1	(22)	—
7	8	21.4	1.0	7	17.7	1.8
8	21	10.9	0.3	13	10.1	0.5
9	8	10.6	1.0	3	10.7	(10, 11, 11)
10	2	(10, 12)	—	3	11.0	(10, 11, 12)
11	12	20.1	1.1	3	20.0	(19, 20, 21)
S. tephronota						
7	6	32.3	1.5	5	30.8	1.6
8	18	29.4	0.6	18	30.9	0.8
9	15	30.7	0.8	13	29.8	1.2
10	6	28.3	1.0	6	29.3	1.4
11	1	(28)	—	1	(32)	—
12	11	32.5	0.8	3	27.0	(23, 26, 32)
13	16	27.1	0.8	11	26.4	0.6
14	5	28.6	1.4	4	30.5	0.9
15	1	(28)	—	1	(23)	—
16	10	28.5	1.2	5	28.4	1.2
17	12	24.1	0.8	10	25.5	1.3
18	9	28.2	0.9	4	26.5	0.5
19	9	30.2	1.6	9	30.4	1.6
20	6	31.7	1.4	7	31.8	0.7

[a] Symbols and conventions as in Table III.

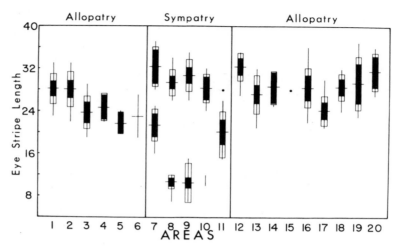

FIG. 7. Post-ocular eye stripe length characteristics of *Sitta neumayer* (lower, areas 1–11) and *Sitta tephronota* (upper, areas 7–20) populations, adult males only. Symbols as in Fig. 5. In addition a point indicates a single measurement. Data taken from Table V, in which sample sizes are given.

conspicuous element is the very short eye stripes of populations 8, 9 and 10. Populations 8 and 9 are significantly different ($P < 0.001$) in each sex from the other two sympatric populations 7 and 11. However, 7 does not differ significantly from 4, 5 or 6, so there is no evidence here of displacement. As with wing and bill length, $3 < 2$ ($P < 0.002$, adult males). The similarity of populations 4, 5, 6, 7 and 11 in eye stripe length is not a new discovery; it was commented upon, in different words, by Paludan (1940) and Vaurie (1950).

Eye Stripe Area. For this measure there are fewer data (Table VI, Fig. 8) and the regression approach cannot be used.

Variation in *tephronota* eye stripe area differs from that in length in that allopatric populations 13 to 20 have ostensibly smaller means than sympatric populations 8 to 10 (Fig. 8). But, as with eye stripe length, population 12 has a larger mean than 10 ($P < 0.05$, adult males), so there is evidence of convergence. Population 12 also has a larger mean than 13 ($P < 0.01$, adult males). The only other significant difference is $16 < 17$ ($P < 0.05$, adult females). The unusually low single value for population 11 (male) may be misleading, because a female has a value higher than the mean for neighboring population 10.

The fragmentary data for *neumayer* do not indicate a substantial difference between population 7 in sympatry and 4, 5, and 6 in allopatry, hence provide no evidence of displacement. On the other hand, populations 8 and

9 have a conspicuously small eye stripe area. Population 8 has a significantly smaller mean than 7 ($P < 0.001$, each sex) and 11 ($P < 0.01$, adult males). In addition $8 > 9$ ($P < 0.01$, adult males). Population 10 is more similar to 8 and 9 than 11, because the single female measured has an eye stripe area nearly identical to the average for population 8 females (Table VI).

In allopatry $2 > 1, 3$ ($P < 0.05$, adult males). The difference between 2 and 1 is surprising because there is no counterpart in any of the other dimensions studied.

The important element of the eye stripe functioning in communication may be the thickness or width of the stripe rather than length or area. When calculated as area/length and plotted in the same manner, width shows a pattern of variation no different from that of area for both species.

Eye Stripe Intensity.   The small samples, necessitating the lumping of data, and the absence of values from specimens from either side of the zone boundary preclude the detection of displacement in this character. However, intensity varies geographically in a way that can be related to variation in eye stripe size in some instances.

S. neumayer shows a reduction in intensity in sympatry (Table VII), which parallels the reduction in eye stripe size. Populations 7 and 11 have a lower intensity than allopatric populations 2 and 3 ($P < 0.05$). Populations 8 and 9 have an even lower average intensity, differing significantly ($P < 0.001$) from the other sympatric populations. Population 10 is probably very

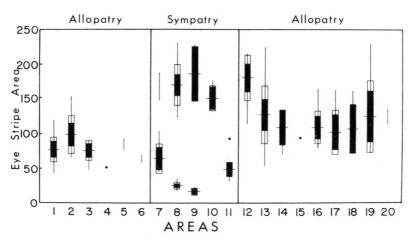

FIG. 8.   Post-ocular eye stripe area characteristics of *Sitta neumayer* (lower, areas 1–11) and *S. tephronota* (upper, areas 7–20) populations, adult males only. Symbols as in Figs. 5 and 7. Data taken from Table VI, in which sample sizes are given.

TABLE VI. Post-Ocular Eye Stripe Area Characteristics of *Sitta neumayer* and *Sitta tephronota* Populations[a]

Populations	Adult males			Adult females		
	N	$\bar{x}$	$S\bar{x}$	N	$\bar{x}$	$S\bar{x}$
S. neumayer						
1	13	77.8	5.2	10	77.2	5.5
2	12	99.2	7.9	4	73.2	5.4
3	8	76.7	4.7	6	78.8	4.1
4	1	(51)	—	—	—	—
5	2	(78, 93)	—	—	—	—
6	2	(60, 69)	—	1	(51)	—
7	6	64.2	8.8	6	52.2	4.2
8	17	25.5	1.3	10	21.8	1.7
9	7	16.9	1.6	1	(27)	—
10	—	—	—	1	(21)	—
11	4	49.7	5.1	1	(67)	—
S. tephronota						
7	2	(147, 188)	—	1	(223)	—
8	14	168.4	7.5	12	169.3	6.9
9	4	185.5	19.3	6	186.3	15.5
10	5	149.4	7.7	5	151.4	16.6
11	1	(92)	—	1	(166)	—
12	10	177.6	10.4	3	126.3	(102, 109, 168)
13	15	127.8	11.0	10	108.3	7.7
14	4	109.5	12.8	3	110.0	(99, 104, 127)
15	1	(94)	—	—	—	—
16	10	110.0	7.2	6	89.8	6.2
17	6	103.0	12.6	9	107.8	3.7
18	4	108.2	17.6	2	(95, 120)	—
19	8	125.5	18.2	5	139.0	12.4
20	2	(116, 136)	—	3	126.0	(60, 155, 163)

[a] Symbols and conventions as in Table III.

similar to 8 and 9. The differences come about through a lack or reduction of pigment in some of the feathers constituting the black eye stripe.

S. tephronota shows no significant reduction or increase in intensity in sympatry. There is a significant difference between allopatric populations 13, 16 and 17 on the one hand, and 18, 19 and 20 on the other ($P < 0.05$). This pattern is quite unrelated to variation in eye stripe area, but is more similar to variation in eye stripe length (see Fig. 7) for reasons that are not obvious. The two species are identical in eye stripe intensity at their furthest points of allopatry.

Summary. A comparison of populations either side of a zone boundary gives no evidence of divergent character displacement in eye stripe length or area for either species. In progressing from allopatry to sympatry *tephronota* shows signs of convergence in both characters. Much sharper discontinuities exist between neighboring *neumayer* populations in both characters, but the discontinuities occur within the zone of sympatry, well away from the allopatry boundary. Some sympatric *neumayer* populations have not only particularly small eye stripes, the intensity of the black is markedly low.

Body Size

S. tephronota displays weak evidence of divergence in bill width but convergence in eye stripe features; *neumayer* shows some evidence of convergence in both bill length and width. These shifts may have been produced by selection acting on these characters directly, or indirectly (in some instances) as allometric consequences of body size changes. Consider bill size shifts. Selection may have acted directly on the bill to maximize feeding efficiency. Alternatively selection may have acted primarily on body size in relation to climate, which varies geographically, in such a way as to minimize thermoregulatory stress. There is an abundance of evidence for an association of body size and body parts with climate (James, 1970; Johnston and Selander, 1964, 1971; Mayr, 1963; Selander and Johnston, 1967) in support of this suggestion. Bill size changes, as allometric consequences of body size changes, may have little significance in terms of feeding efficiency.

TABLE VII.   Reflectance Densities of the Black
Eye Stripe

Populations (sexes combined)	N	$\bar{x}$	$S\bar{x}$
S. neumayer			
2	5	1.542	0.097
3	6	1.440	0.106
5	3	1.757	(1.35, 1.89, 2.03)
7, 11	6	1.103	0.080
8, 9	7	0.571	0.071
10	1	(0.64)	
S. tephronota			
8, 9	6	1.523	0.192
13, 16, 17	7	1.101	0.087
18, 19, 20	5	1.550	0.152

I start by considering whether either species has undergone character displacement in body size.

Weight.   The most direct measure of body size (mass) is weight. Unfortunately few values of fresh (wet) weight are available (Table VIII). Nevertheless they show that allopatric *tephronota* populations 13, 16 and 17 are nearly identical in body size, and much smaller than the sympatric and nearly identical populations 8 and 9. Body weights of *neumayer* indicate more or less uniform body size in allopatry and less uniform but much smaller size in sympatry, particularly in the southern areas.

There are no data for *tephronota* populations 10, 11 and 12 or for *neumayer* population 7, so the possibility of character displacement of body size cannot be assessed by the comparison of populations either side of the zone boundary. However, the available data are consistent with the hypothesis that mutual divergence in body size has occurred in sympatry.

Wing Length.   Further inferences about body size variation can be made using indirect measures of body size. The most frequently used indirect measure is wing length. The disadvantage of being a less direct

TABLE VIII.   Fresh Weights of *Sitta neumayer* and *Sitta tephronota* in Grams, Taken from the Labels of Museum Specimens and from One Female Live Specimen of *S. neumayer* from Kavalla, Greece (area 2)[a]

S. neumayer			*S. tephronota*		
Area	Sex	Weights	Area	Sex	Weights
2	♂	31.5, 31.5, 33	17	♀	35.2, 35.4
	♀	30.5, 30.5, 35.5			
			16	♂	33, 33, 34, 34, 35, 35, 35, 38
4	♂	29.2		♀	30, 31.5, 32, 32, 33, 34
6	♂	31	13	♂	36, 36, 37
	♀	31		♀	33
8	♀	21.6, 22.1	8	♂	45.8, 46.5, 50.1, 55.0
9	♂	20.2, 21, 21.2	9	♂	47, 55
	♀	21		♀	42.7, 45
	Unknown	20.5, 20.5, 22		Unknown	44.7
10	♀	23			
11	♂	22.9, 23.1, 25.1			
	♀	23.3, 24.7, 27			

[a] Where fat condition was recorded (see Paludan, 1940), specimens had no, little, or moderate amounts of fat.

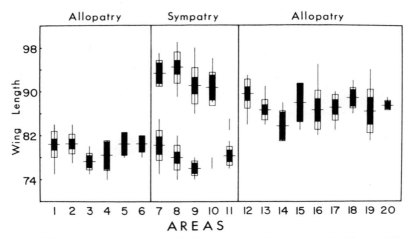

FIG. 9. Wing length characteristics of *Sitta neumayer* (lower, areas 1–11) and *Sitta tephronota* (upper, areas 7–20) populations, adult males only. Symbols as in Fig. 5. Data taken from Table IX, in which sample sizes are given.

measure than weight tends to be offset by the smaller effect of seasonal variation, no diurnal variation, larger sample sizes and samples from populations either side of the zone boundary.

The regression approach cannot be adopted with either species due to the absence of a systematic relationship between wing length and longitude (or latitude) in allopatry. In sympatry the association with longitude is stronger (*tephronota* $r = -0.833$, $P \sim 0.08$; *neumayer* $r = -0.777$, $P \sim 0.12$), and slopes of regression lines are not significantly different (*tephronota* $-0.225 + 0.106$; *neumayer* $-0.247 + 0.141$).

Data for the individual populations are given in Table IX and illustrated for adult males in Fig. 9 which shows that sympatric populations of *tephronota* are larger in wing length than allopatric populations, as was found with weight. But the difference in wing length between allopatric and sympatric populations is small compared with the difference in weight (or $\sqrt[3]{}$ weight), and there is no discontinuity in wing length at the zone boundary. Similarly, with *neumayer* the difference in wing length between allopatric and sympatric populations is small compared with the difference in weight, and there is no discontinuity at the zone boundary. Wing length, as an indirect measure of body size, gives no evidence of character displacement in either species.

Tarsus Length. This can be used as an alternative indicator of body size since it is proportional to the mass of body it supports and transports (Dilger, 1956; Fretwell, 1969; Grant, 1965, 1966*b*; Störck, 1968). No

TABLE IX. Wing Length Characteristics of *Sitta neumayer* and *Sitta tephronota* Populations[a]

Populations	Adult males			Adult females		
	N	$\bar{x}$	$S\bar{x}$	N	$\bar{x}$	$S\bar{x}$
S. neumayer						
1	22	80.2	0.5	16	79.9	0.6
2	20	80.2	0.4	9	77.7	0.9
3	15	77.3	0.4	13	76.2	0.4
4	6	79.0	1.2	6	78.3	1.2
5	4	79.5	1.0	2	(77, 80)	—
6	4	79.5	0.6	3	79.3	(78, 78, 82)
7	10	79.2	0.8	7	78.0	1.1
8	22	78.0	0.5	14	76.6	0.4
9	9	76.0	0.5	3	74.3	(73, 73, 77)
10	2	(76, 78)	—	4	75.0	0.6
11	13	78.8	0.5	7	76.9	1.0
S. tephronota						
7	6	93.3	1.0	4	91.5	1.4
8	20	94.4	0.6	17	92.9	0.5
9	16	91.1	0.8	13	89.8	0.4
10	6	90.8	1.1	6	90.2	1.1
11	2	(83, 85)	—	2	(87, 89)	—
12	12	89.7	0.7	5	88.0	1.3
13	20	86.7	0.4	12	86.2	0.6
14	6	83.3	1.0	5	84.2	0.6
15	4	88.0	1.7	4	83.8	1.0
16	13	86.7	1.0	9	83.4	1.2
17	12	87.1	0.6	8	86.3	0.5
18	9	88.3	0.8	5	84.8	0.9
19	12	86.1	1.0	9	87.7	0.8
20	6	87.5	0.3	7	87.0	0.6

[a] Symbols and conventions as in Table III.

geographical variation in the use of perches is known, or likely in the structurally simple habitats occupied by the species, that would complicate this relationship. As with wing length, the regression approach cannot be used for an assessment of character displacement.

For the comparison of *tephronota* populations, tarsus length probably gives a better picture of geographical variation in body size than does wing length, because the percent difference between sympatric populations 8 and 9 on the one hand and allopatric populations 13, 16 and 17 on the other is the same for tarsus length as for $\sqrt[3]{}$ weight (about twelve), but is only six for wing length. However, there is no discontinuity at the zone boundary (Table X, Fig. 10); allopatric population 12 is not significantly smaller than sympatric population 10 ($P > 0.1$).

Tarsus length does not appear to be as good an indicator of body size for *neumayer*. Populations 8 and 9 have a cube root of body weight about 14% less than populations 6, 4 and 2, yet the corresponding differences in tarsus and wing length are only 2% and 4%, respectively. In sympatry population 11 has the largest average weight but the smallest average tarsus. Tarsus length shows a slight and gradual decrease eastward (Fig. 10) with no significant differences between adjacent populations. Because there is no significant difference between 6 and 7, and because there is a gradual and apparently uninterrupted decrease without a marked discontinuity at the zone boundary, there is no evidence here of divergence in tarsus length or body size with which it is possibly at least partly correlated.

Summary. Weight data are consistent with an hypothesis of mutual divergence in body size but not sufficient to assess it. Wing length and tarsus length, as indirect measures of body size, provide no evidence of divergence (or convergence) by the method of comparing adjacent populations either side of the zone boundary.

Allometry. I now consider the possibility that geographical variation in bill and eye stripe sizes is related to body size variation. The simplest approach to the question of allometry is to regress log body part (e.g., bill

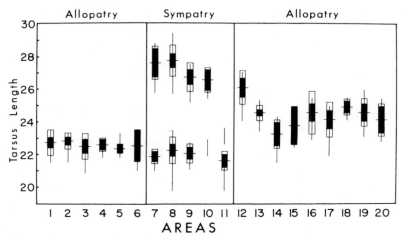

FIG. 10. Tarsus length characteristics of *Sitta neumayer* (lower, areas 1–11) and *Sitta tephronota* (upper, areas 7–20) populations, adult males only. Symbols as in Fig. 5. Data taken from Table X, in which sample sizes are given.

TABLE X.   Tarsus Length Characteristics of *Sitta neumayer* and *Sitta tephronota* Populations[a]

Populations	Adult males			Adult females		
	N	$\bar{x}$	$S\bar{x}$	N	$\bar{x}$	$S\bar{x}$
S. neumayer						
1	22	22.65	0.16	16	22.75	0.18
2	20	22.85	0.11	10	22.73	0.18
3	15	22.51	0.18	13	22.33	0.21
4	7	22.67	0.14	6	21.70	0.20
5	4	22.36	0.13	2	(22.0, 23.0)	—
6	4	22.56	0.49	3	22.67	(21.8, 22.6, 23.6)
7	9	21.86	0.15	7	21.71	0.23
8	21	22.25	0.18	14	21.87	0.11
9	9	22.05	0.19	3	21.07	(20.8, 20.9, 21.5)
10	2	(22.2, 22.4)	—	4	21.65	0.13
11	13	21.87	0.19	7	21.93	0.31
S. tephronota						
7	6	27.59	0.42	5	26.94	0.29
8	20	27.71	0.21	18	27.13	0.20
9	16	26.75	0.20	12	26.79	0.22
10	6	26.59	0.29	6	25.44	0.38
11	2	(22.9, 24.2)	—	2	(25.9, 26.7)	—
12	12	26.09	0.27	5	24.48	0.62
13	20	24.56	0.10	13	24.31	0.18
14	6	23.57	0.54	5	23.89	0.38
15	4	23.78	0.57	4	24.01	0.47
16	13	24.59	0.27	9	23.45	0.23
17	12	24.10	0.30	10	24.00	0.16
18	9	24.68	0.24	5	23.99	0.37
19	12	24.49	0.23	9	23.91	0.21
20	6	24.12	0.40	7	23.45	0.29

[a] Symbols and conventions as in Table III.

length) on log $\sqrt[3]{\text{weight}}$ for sympatric and allopatric samples separately, and to compare slopes and intercepts. If the regressions are statistically significant and null hypotheses of no difference in slopes and no difference in intercepts are accepted, we have evidence of body part size in sympatry being determined allometrically by body size in the same way as it is determined in allopatry. Rejection of the null hypotheses constitutes evidence of an altered relationship between parts and whole, in which case body size does not adequately account for the size of body parts in sympatry.

Male and female samples were combined because they are small. Variables treated were bill length and width, eye stripe length and area. Only two regressions were significant ($P < 0.05$); log bill width and log eye stripe area regressed significantly on log $\sqrt[3]{}$ weight in the allopatric *tephronota* sample. Ignoring levels of statistical significance, slopes and intercepts of all sympatric and allopatric regressions were compared. The only significant ($P < 0.05$) differences were between allopatric and sympatric *tephronota*; slopes and intercepts of eye stripe area regressions differed.

These are not satisfactory analyses. Attempts to adequately test null hypotheses are hampered by inadequate sample sizes. The results are presented as suggestive evidence that geographical variation in bill size is accounted for by geographical variation in body size, in both species, whereas geographical variation in *tephronota* eye stripe size is not. Sympatric *tephronota* have disproportionately large eye stripes.

A more superficial comparison of populations (Figs. 5–10) shows that the above suggestion is incorrect with regard to *neumayer*. Eye stripe sizes have decreased in some sympatric areas in parallel with weight, but bill, wing and tarsus have not. Therefore in this species, bill size is disproportionately large in sympatry.

Sitta europaea: a "Control" for the Geographical Variation in
Sitta neumayer

S. europaea has a geographical distribution largely coextensive with *neumayer* (e.g., see Voous, 1960). Like the latter it reaches its southeastern limit slightly to the east of the Caspian Sea. The two species rarely come into contact because they occupy different habitats. *S. europaea* is a typical nuthatch in being largely arboreal, and its activities are more or less confined to forests and woods. Only at the limited habitat interface of woods and open, rocky areas, such as at Yasuch in Fars, Iran (area 9) do the activities of the two species overlap (L. Cornwallis, *pers. comm.*).

The assumption can be made that *tephronota,* in open habitat, has no or negligible influence on *europaea* in forest habitat. In contrast *tephronota* may influence *neumayer* which is in the same habitat. The pattern of geographical variation of *europaea* can be used as a base-line or control with which to compare the pattern of geographical variation of *neumayer*. If character displacement has occurred in the latter (but not in *europaea*) the variation of the two species is not expected to be the same in the sympatric zone. If character displacement has not occurred, the variation of the two species could be the same. If the two species vary in like manner, within the sympatric zone and in passing from allopatry to sympatry, the hypothesis of character displacement is not supported.

The data for *europaea* are given in Table XI. There are enough data from eight populations to compare directly with *neumayer*. Correlation analysis (Table XII) shows a significant positive association between average bill widths of the two species and between average eye stripe areas throughout the coextensive range of the two species.

The pattern of variation of four morphological features of the two species is illustrated in Fig. 11. In passing from one area to the next there is good correspondence between the species. Differences between the species in magnitude of change are greater when the change is in the same direction than when opposite. Thus population 3 is characterized by small size in *europaea,* as it is in *neumayer,* but to a more pronounced extent. The most conspicuous discrepancy between the species, not shown in the illustration, is in bill width in area 6, where it is particularly large in *europaea,* but small in *neumayer.* The *europaea* sample size is small however, and does not permit a confidence estimate. Within sympatry the two species vary similarly in this dimension.

The two most important points which emerge from the illustration are that the two species vary in like manner in the sympatric zone, and that the differences between allopatric population 6 and sympatric population 8 are in the same direction in the two species. With regard to the first point, note that for each species eye stripe and wing length vary in parallel in allopatry, and that in sympatry (areas 8 and 9) eye stripe length becomes disproportionately short. Unfortunately there are no *europaea* data from area 7, so the comparison of populations either side of the zone boundary cannot be

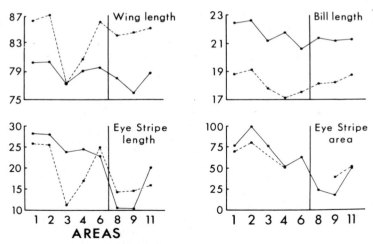

FIG. 11. Covariation in *Sitta neumayer* (solid lines) and *Sitta europaea* (broken lines). Data taken from Tables III, V, VI, IX, and XI.

TABLE XI. Morphological Characteristics of *Sitta europaea* Populations in the Region where *Sitta neumayer* Occurs[a]

Population	Adult males			Adult females		
	N	$\bar{x}$	$S\bar{x}$	N	$\bar{x}$	$S\bar{x}$
Bill length						
1	5	18.86	0.38	—	—	—
2	6	19.01	0.30	7	18.11	0.17
2a	8	18.37	0.38	—	—	—
3	10	17.86	0.24	8	17.14	0.22
4	10	17.16	0.26	9	17.27	0.14
6	1	(17.8)	—	—	—	—
8	12	18.16	0.17	4	17.69	0.21
9	8	18.25	0.19	—	—	—
11	11	18.87	0.23	4	18.19	0.55
11a	4	20.66	0.20	—	—	—
Bill width						
1	6	4.47	0.03	—	—	—
2	8	4.54	0.09	6	4.42	0.09
2a	9	4.10	0.06	—	—	—
3	10	3.87	0.04	8	3.93	0.06
4	10	3.88	0.04	9	3.91	0.05
6	1	(4.1)	—	—	—	—
8	12	3.76	0.05	4	3.63	0.06
9	8	3.85	0.07	—	—	—
11	11	4.55	0.08	4	4.25	0.12
11a	4	4.70	0.15	—	—	—
Wing length						
1	6	86.2	0.7	—	—	—
2	8	87.1	0.8	7	85.0	1.0
2a	9	85.0	0.5	—	—	—
3	10	77.3	1.5	8	72.5	1.2
4	10	81.0	1.6	11	76.1	1.3
6	1	(87)	—	—	—	—
8	12	83.9	0.5	4	81.3	1.0
9	8	84.5	0.7	—	—	—
11	11	85.4	0.4	4	82.5	1.3
11a	4	85.3	1.1	—	—	—

[a] Populations 2a and 11a are from areas where *Sitta neumayer* has not been collected. Area 2a is in north western Turkey, in both European and Asiatic parts, and therefore lies between areas 2 and 3 (see Fig. 3); area 11a is north of area 11, at the south east corner of the Caspian Sea. Symbols and conventions as in Table III.

TABLE XI.   (Continued)

Population	Adult males			Adult females		
	N	$\bar{x}$	$S\bar{x}$	N	$\bar{x}$	$S\bar{x}$
Tarsus length						
1	6	19.47	0.26	—	—	—
2	8	19.63	0.23	6	19.42	0.32
2a	8	19.65	0.15	—	—	—
3	10	17.93	0.33	7	16.89	0.22
4	10	18.50	0.27	11	17.41	0.23
6	1	(19.7)	—	—	—	—
8	12	18.67	0.15	4	18.65	0.32
9	8	18.65	0.09	—	—	—
11	10	21.16	0.37	4	21.21	1.06
11a	4	21.01	0.18	—	—	—
Eye stripe length						
1	6	26.1	1.6	—	—	—
2	8	25.9	1.4	6	26.0	1.4
2a	9	18.7	1.1	—	—	—
3	5	11.6	2.3	4	10.5	1.6
4	7	17.0	0.5	6	11.2	1.5
6	1	(17)	—	—	—	—
8	12	14.1	0.7	4	12.8	0.9
9	8	14.3	0.6	—	—	—
11	11	16.0	1.1	4	18.0	2.3
11a	3	16.7	(15, 15, 20)	—	—	—
Eye stripe area						
1	2	(53, 87)	—	—	—	—
2	3	79.7	(57, 74, 108)	5	81.6	7.4
2a	1	(53)	—	—	—	—
3	—	—	—	1	(55)	—
4	3	51.3	(41, 46, 67)	2	(46, 48)	—
6	—	—	—	—	—	—
8	—	—	—	—	—	—
9	3	39.0	(28, 43, 46)	—	—	—
11	4	50.8	5.4	—	—	—
11a	—	—	—	—	—	—

made with this species. But significant differences between other neighboring populations parallel, in some instances, significant differences between populations of *neumayer* in the same areas. For instance both species exhibit a significant difference ($P < 0.02$, adult males) in bill width, and in the same direction, between populations 8 and 11. Further evidence of

TABLE XII. Correlation Coefficients and Associated Probability Values Calculated from the Means of *Sitta neumayer* and *Sitta europaea* Populations 1, 2, 3, 4, 8, 9 and 11 (Adult Males Only) [a]

Dimension	r	P
Wing length	0.473	>0.1
Tarsus length	-0.363	>0.1
Bill length	0.373	>0.1
Bill width	0.864	<0.02
Eye stripe length	0.652	>0.1
Eye stripe area	0.988	<0.005

[a] For the eye stripe area computation, the means of populations 1, 2, 4, 9 and 11 only were used.

covariation of the two species is the ostensible similarities of regressions of morphological dimensions on longitude (Table XIII).

In summary the similarity in geographical variation of the two species, particularly in the sympatric zone, casts doubt upon the likelihood of character displacement in *neumayer*. No other species of *Sitta* is coextensive with *tephronota*, so the same method of assessment cannot be applied to this species.

Sexual Dimorphism and Population Variation

There are other ways, not previously investigated with the rock nuthatches, in which the two species might show evidence of a response to selection in sympatry minimizing the chances of competition for food and hybridization. Population variation in bill and eye stripe characteristics of one species might be constrained in the presence of the other species within narrower limits than in its absence. This would be manifested in one or both of two ways. First, within each sex, population (sample) variation would be smaller in sympatry than in allopatry. Second, the population average dimensions for males and females would be closer together in sympatry than in allopatry, giving rise to a smaller spread of population values if the within-sex variation was constant.

The degree of sexual dimorphism in bill and eye stripe characters is shown in Table XIV as the difference between male and female means

TABLE XIII. Regression of Bill Length, Bill Width and Eye Stripe Length (Allopatry Only) and $\sqrt[3]{}$ Body Weight (Allopatry and Sympatry Combined) on Longitude for *Sitta europaea* and *Sitta neumayer* Adult Males

Dimension	N	F	r	$a \pm S.E.$	$b \pm S.E.$
Bill length					
S. europaea	5	4.912	−0.788	21.091 ± 0.590	−0.114 ± 0.051
S. neumayer	6	7.277	−0.803	23.695 ± 0.535	−0.065 ± 0.024
Bill width					
S. europaea	5	8.910	−0.865	5.234 ± 0.166	−0.043 ± 0.014
S. neumayer	6	6.662	−0.790	4.894 ± 0.108	−0.012 ± 0.005
Eye stripe length					
S. europaea	5	4.055	−0.758	40.218 ± 4.619	−0.809 ± 0.402
S. neumayer	6	17.735[a]	−0.925[a]	34.177 ± 1.210	−0.333 ± 0.079
$\sqrt[3]{}$ Weight					
S. europaea	5	2.747	−0.691	2.900 ± 0.073	−0.005 ± 0.003
S. neumayer	5	4.861	−0.786	3.428 ± 0.126	−0.010 ± 0.005

[a] Significant at 5% level.

expressed as a percentage of the smaller mean. These values are generally small, as they are for wing and tarsus length, with male averages usually exceeding female averages by only a few percent. Sexual dimorphism tends to be greater in *tephronota* than in the smaller *neumayer*, but there are no clear differences within each species between sympatric and allopatric populations. Perhaps the eye stripe area character of *tephronota* provides an exception to these statements. Females have the larger average eye strip area in five of the seven populations represented by samples of four or more individuals of the same sex, and dimorphism is consistently greater in allopatry than in sympatry. Possibly the presence of *neumayer* does act as a constraint upon intersex variation in the eye stripe area of *tephronota*, although the difference between the species in sympatric areas 8, 9, and 10 is so large that a constraint does not appear likely.

A cursory examination of Figs. 5–8 shows no systematic trend toward less variation in the adult male samples in sympatry than in allopatry. A more detailed examination of the data for both sexes in Tables III–VI confirms there is no such systematic trend. Squared coefficients of variation have been calculated from the standard errors and means in the tables. There are few significant differences in variation between sympatric and allopatric populations. For bill width, adult females of *tephronota* vary

TABLE XIV. Sexual Dimorphism in *Sitta neumayer* and *Sitta tephronota*[a]

Populations	Bill length	Bill width	Wing length	Tarsus length	Eye stripe length	Eye stripe area
S. neumayer						
1	1.1	0.4	0.4	*0.4*	1.4	0.8
2	1.3	2.2	3.2	0.5	8.2	35.5
3	0.9	*0.4*	1.4	0.8	*7.6*	*2.7*
4	1.2	*2.0*	0.9	4.5	—	—
5	—	—	—	—	—	—
6	—	—	—	—	—	—
7	1.9	*1.5*	1.5	0.7	20.9	23.0
8	0.6	3.0	1.8	1.7	7.9	17.0
9	—	—	—	—	—	—
10	—	—	—	—	—	—
11	*3.4*	*2.6*	2.5	*0.3*	—	—
S. tephronota						
7	1.4	2.5	2.9	2.4	4.9	—
8	3.2	0.5	1.6	2.1	5.1	*0.5*
9	*1.2*	*0.7*	1.4	0.1	3.0	*0.4*
10	2.2	5.2	0.7	4.5	*3.5*	*1.3*
11	—	—	—	—	—	—
12	*2.4*	5.7	1.9	6.6	—	—
13	1.9	2.5	0.6	1.0	2.6	18.0
14	0.8	3.0	*1.1*	*1.4*	*6.6*	—
15	2.0	*2.0*	5.0	*1.0*	—	—
16	1.4	3.7	4.0	4.9	0.3	22.5
17	0.6	*2.5*	0.9	0.4	*5.8*	*4.7*
18	0.6	2.8	4.1	2.9	6.4	—
19	2.4	2.0	*1.9*	2.4	*0.7*	*10.8*
20	0.4	0.0	0.6	2.9	*0.3*	—

[a] The difference between population means is expressed as a percentage of the smaller mean, and this is in italics when the smaller mean is the male's.

more in area 8 than in area 17 ($F = 5.25$, two-tailed test, $P < 0.025$). This is the only significant difference in bill variation and it is in the opposite direction to that expected.

Adult female *tephronota* vary more in eye stripe length in sympatric area 9 than in allopatric area 13 ($F = 3.93$, $P < 0.025$), and adult female *neumayer* vary more in sympatric area 8 than in allopatric area 3 ($F = 4.27$, $P < 0.05$). But the single significant difference in eye stripe area variation is in the expected direction. For adult male *tephronota* $8 < 13$ ($F = 3.93$, $P <$

0.025). *S. neumayer* shows similar, but not significant, differences in the male samples (Table VI), although not in the female samples. Possibly, therefore, variation in eye stripe area in sympatric populations of both species is constrained by selection arising from the presence of the other species, but the evidence is not clear.

Alternative Hypotheses of Character Displacement

The merit of the Brown and Wilson (1956) hypothesis is its simplicity. All populations of a species within sympatry are assumed to be similar, all populations within allopatry are assumed to be similar, and the two groups of populations are compared. However, by the methods of analysis applied in this paper, no clear evidence of character displacement has been found. This may be because the initial assumptions are wrong.

In agreement with the Brown and Wilson approach, the simple assumption has been made so far that the two species occur together throughout the sympatric range. This is reasonable since there are no statements in the literature to make one think otherwise. Certainly the two species are syntopic in the central areas of 9 (*pers. obs.*,) 8 (D.A. Scott, *pers. comm.*, Érard and Etchecopar, 1970), and 10 (L. Cornwallis, *pers. comm.*). But it is far less certain that the species are synoptic in areas 7 and 11. Reasons for thinking that they might be allotopic in these areas are given in Appendix 1. If we accept as a working hypothesis that complete synotopy is achieved in areas 8, 9 and 10 only, we may re-examine the data for further evidence of character displacement. *S. neumayer* population 8 (syntopic) will be compared with neighboring populations 7 and 11 (allotopic). *S. tephronota* population 8 (syntopic) will be compared with 11 (allotopic), but since the sample size of population 11 is so small any conclusions reached from the comparison must be tentative.

The Partial Syntopy Hypothesis

From this altered perspective the evidence for character displacement of the eye stripe is much stronger, particularly in *neumayer*. In Table XV the percent mean differences between populations 8 and 11 are shown. Differences in eye stripe length and area are about the same and clearly greater than any other feature, including the $\sqrt[3]{weight}$ and $(\sqrt[3]{weight})^2$. Thus small differences in bill dimensions between the populations are accounted for by small differences in body size, but the large differences in eye stripe features are not. Differences between populations 8 and 7 are slightly greater, (see. Fig. 2) but are not given in the Table because of the lack of weight estimates for population 7.

TABLE XV. The Percent Differences between Mean Values of Allotopic (11) and Syntopic (8) Populations of *Sitta neumayer* and *Sitta europaea*

Dimension	*S. neumayer*	*S. europaea*
$\sqrt{\ }$ Weight	(4.4)	—
Wing	1.0	1.8
Tarsus	1.7	11.8
Bill length	1.3	3.8
Bill Width	4.6	17.4
Eye stripe length	45.8	11.9
Eye stripe area	48.7	—
$(\sqrt{\ }$ Weight$)^2$	(8.6)	—

a Calculations based upon adult female means are shown in brackets, all others are from adult male means.

Interpretation of the difference in eye stripe features between allotropic and syntopic *neumayer* populations is complicated by a similar difference between corresponding *europaea* populations (Fig. 11); it will be recalled that *europaea* is not likely to have undergone character displacement. The case for character displacement in *neumayer* rests on the greater magnitude of the difference, and the fact that is it out of proportion to body size differences. Table XV shows that the difference between *europaea* populations 8 and 11 is almost the same in body size, as indicated by tarsus length (11.8%), as it is in eye stripe length (11.9%). This is much less than the nearly 50% difference in eye stripe length (and area) between the *neumayer* populations. Therefore the *neumayer* differences can be attributed at least partly to character displacement, on the assumption, that syntopic population 8 (and 9 and 10) was derived from allotopic population 11 (or 7).

The situation with *tephronota* is more confusing. Allotropic population 11 has smaller mean measurements in all dimensions than syntopic population 8, indicating the possibility of character displacement in area 8. This is a weak indication because the sample of population 11 is so small. Furthermore, population 8 may not have been derived from 11.

The morphological similarity of adjacent allotropic populations of *neumayer* in areas 7 and 11, together with the absence of topographical barriers between these areas, suggests that interchange of individuals between areas is possible. Yet the *tephronota* populations in these areas are quite different. This suggests that *tephronota* population 7 was not derived from population 11 but from population 8 (see Fig. 2). If population 7 was not derived from 11, it is likely that population 8, which is morphologically very different from 11, was also not derived from 11. Population 8 is similar to 9, which is similar to 10, and therefore a southern route of colonization of western Iran is indicated rather than a northern one. If this interpretation is correct, syntopic and allotopic populations of *tephronota* should not be compared for an assessment of character displacement. Instead sympatric (10) and allopatric (12) populations should be compared, as was done earlier without yielding clear indications of character displacement.

An Hypothesis of Character Displacement Pattern Modified by Gene Flow

The morphological features of *tephronota* population 7 suggested that this allotopic population was derived from syntopic population 8. If the same type of process occurred at the eastern zone boundary as well, the result would undermine any attempts to detect character displacement by the comparison of adjacent populations at the boundary. Specifically the characteristics of population 12 might be influenced by gene flow from population 10. Or the population in area 12 may have become extinct in the past, followed by a re-invasion from area 10 and little or no subsequent morphological change. Or *neumayer* may have occurred in area 12 in the past, *tephronota* may have undergone character displacement and remained in that state even after the demise of the *neumayer* population. Any one of these complications would result in adjacent allopatric and sympatric populations being similar. There is no way of objectively testing these hypotheses because we have no basis for expecting gene flow effects to be manifested up to a certain distance from the zone boundary. But we can examine the pattern of geographical variation in *tephronota* to see if it is consistent with these hypotheses.

Significant differences occur between populations 12 and 13 in wing length, tarsus length, eye stripe length and area, but not in the two bill dimensions. These contrast with the absence of significant differences, except for eye stripe length, between populations 12 and 10. The pattern of geographical variation in wing length, tarsus length and eye stripe area can be summarized as follows. Sympatric populations and population 12 are consistently larger in these features than are the remaining allopatric populations; there is relatively little variation among populations in these two groups, yet there is a discontinuity between populations 12 and 13, i.e., there is a step in the pattern of variation.

Thus in body size and eye stripe size *tephronota* variation is consistent with an hypothesis of character displacement in sympatry followed by gene flow outward or geographical range contraction of *neumayer*.

Conclusions

This section has been concerned with the detection of a process in the past by an assessment of the evidence from present day morphology. There are obvious limitations to the success that can be achieved. These will be discussed briefly before final conclusions are drawn.

Character displacement may have occurred, but the results may have become obscured by subsequent countervailing processes which have rendered the unequivocal detection of character displacement impossible. For instance, it can be argued that gene flow between sympatry and allopatry has smoothed out the morphological discontinuities that once existed between the zones (Grant, 1972a) and which this study has sought and failed, with few exceptions, to find. According to this view character displacement has occurred but it is now no longer possible to detect it.

In principle this is plausible; in application to the *Sitta* situation it is less convincing. First the effects of gene flow are not likely to extend much beyond the zone boundary because the distances in the two zones, particularly the allopatric ones, are very great. Dispersal characteristics of the two species are not known. They are not irruptive species as the related *europaea* is at latitudes much further north (Eriksson, 1970; Gatter, 1974). Limited banding studies near Shiraz, area 9, indicate that both species are fairly sedentary (L. Cornwallis, *pers. comm.*). Hence gene flow is not likely to be extensive. Second the smoothing effects of gene flow have probably been overestimated until recently (Ehrlich and Raven, 1969). It is now known that morphological differences can be maintained by a relatively small selection differential even in the face of substantial amounts of gene flow (Endler, 1973; Thoday and Gibson, 1962). A possible example is provided by *neumayer* populations 8 and 7 (and 11). Populations are contiguous and are not known to be separated by a topographical barrier so presumably gene exchange takes place, yet the eye stripe differences between populations remain substantial (Fig. 2). Another possible example is provided by *tephronota* populations 7 and 11, which are likewise contiguous but morphologically quite different. Nevertheless, the general principle remains that the results of character displacement may have become masked by other adaptive changes.

There are also limitations arising from inadequate data. With relatively few data points, the regression approach used in this study may not be sufficiently sensitive to detect a small but meaningful shift in position (intercept) of the regression line in sympatry. The comparison of populations either side of a zone boundary is limited by the small size of samples in a similar way, particularly in the case of *tephronota* population 11. It is a pity that data limitations obtrude in a study which has used most museum material in existence, and twice as much as in the previous study (Vaurie, 1951; about 125 of the 150 specimens in Russian Museums—Stepanyan, 1961—were not available for examination in the present study). These limitations can be overcome however by collecting new material from the under-represented areas that this study has served to identify.

These considerations make it advisable to be cautious in rejecting character displacement interpretations. What can be accepted? If we accept that populations in areas 7 and 11 are allotopic then it appears that character displacement has occurred in *neumayer*, for populations syntopic with *tephronota* have markedly smaller eye stripes and body weights than do allotopic and allopatric populations. There is an obvious need for further fieldwork to establish (or deny) that populations are indeed allotopic in areas 7 and 11. *S. tephronota* shows some indications of having undergone character displacement in eye stripe and bill width, but there are enough complications to render such a judgment suspect.

The case for character displacement in *tephronota* is stronger when we change from asking "does the evidence point to character displacement" to "*could* the data be interpreted that way?" The pattern of geographical variation in body size and eye stripe size (but not bill width) is consistent with an hypothesis of character displacement in sympatry followed by gene flow outward to area 12 or geographical range contraction of *neumayer* from this area. This is not arbitrary. A major discontinuity in geographical variation exists between *tephronota* populations 12 and 13, just as it exists between syntopic and allotopic *neumayer* populations.

A further reason for considering that character displacement has occurred in both species emerges from a comparison of allopatric populations close to the zone boundary. *S. tephronota* population 13 is chosen for reasons given above, and compared with *neumayer* population 6. In body weight the two species differ by about 13% (of the smaller weight). In eye stripe area they differ by about 115%. In contrast the differences are much greater in syntopic sympatry (area 9) in body weight (125%) and eye stripe area (∼1000%) Both species have undergone shifts in sympatry and, as a result, the interspecific ratios are in the order of two to one in body weight and ten to one in eye stripe size. Both species vary very little in these two morphological features throughout their allopatric ranges. So we can say that if the allopatric populations were brought together, whether from close to the zone boundaries or from far, they would be much more similar in these two morphological features than syntopic, sympatric populations actually are. The corresponding comparisons with bill features yields a quite different result. The same allopatric populations differ in bill length and bill width by 22 and 25%, respectively, which are almost identical to the inter-specific differences in syntopic sympatry (area 9) of 22% and 27%, respectively. It is also worth noting that *neumayer* has failed to decrease in bill size in sympatry to any appreciable extent, despite a substantial decrease in body size.

I conclude that body size and eye stripe of the two species have undergone mutually divergent character displacement. Divergence in eye stripe size has been out of proportion to body size in *tephronota*, possibly also in *neumayer*, so that the displacement cannot be viewed as being brought about simply by body size changes, although these no doubt contributed. Bill length and bill width apparently have not undergone displacement. Of course, it is always possible that some displacement did occur but that gene flow smoothed out discontinuities between neighboring populations. However, if that did occur we would have to explain why the body size and eye stripe size discontinuities were not smoothed out as well.

The answer to this might be that bill characteristics have a more complex polygenic determination than body and eye stripe features, and that selection can maintain discontinuities more easily with genetically simple morphological systems than with complex ones. This is conjectural, and even if correct there still remains the problem that the two species do not coexist in regions where the difference between them in bill size is substantial. Thus in allotopic area 7, males of the two species differ in bill length and width by 28% and 30%, respectively, and in neighboring syntopic area 8 the differences are almost identical, 28% and 31%, respectively (see also Fig. 2). So if *neumayer* and *tephronota* can coexist without severe competition in area 8, for the sake of argument as a result of divergent character displacement in bill size, there is no obvious ecological reason why they cannot do so in area 7, as a result of gene flow maintaining the divergent states. Gene flow probably does occur, but it is a failure of the ecological character displacement hypothesis that it does not account for the allotopic condition. A simpler explanation is that the two species do not coexist for reproductive, not ecological, reasons. For instance, eye stripe areas differ by a factor of more than six in area 8 but by a factor of less than three in area 7, and the latter may not be sufficient for the avoidance of ambiguity.

The Significance of Bill Size: Foraging Behavior and Food

Even though neither species appears to have undergone character displacement in bill dimensions, the difference between them in sympatry may allow coexistence through differential exploitation of potential food items. To investigate this, foraging behavior and diet were studied. Systematic, quantitative observations of foraging behavior were made in 1972 on the two species in the vicinity of Shiraz in Iran (area 9), March 3–27, and on allopatric *neumayer* near Kavalla, May 7–30 and June 16, 23–24; Nauplion, June 5; Mycaenae, June 7; Delphi, June 13–14 (all of these in Greece, area 2); and near Kotor, Yugoslavia (area 1), June 29–30. Nest building was taking place in Iran in March. Young were in the nest in Greece in May, emerging in early June. Breeding was probably later than usual in both regions (L. Cornwallis, pers. comm.). Habitats at Shiraz and Kavalla are illustrated in Fig. 12.

These observations on foraging, accumulated in approximately 40 hours in sympatry and 50 hours in allopatry, revealed no conspicuous difference in foraging techniques used by the two species in sympatry, or by allopatric and sympatric *neumayer*. At this time of year at least, the species spend more than 90% of foraging time hopping over horizontal rock surfaces or climbing vertical ones, and pecking arthropods and seeds from surfaces or cracks. Of the remaining techniques the most important were digging (with the bill) and fly-catching or hovering at a flower head to take insects. Perching in trees was rarely observed, it was nearly always accompanied by posturing and singing, and birds were never seen to feed there or even to launch out on flycatching sallies from an arboreal perch.

Digging was exhibited by *tephronota* at Shiraz (area 9) and *neumayer* at Delphi (area 2). At Shiraz *tephronota* probed with the bill at the base of tufts of an unidentified grass and a *Luzula* (wood rush) species, perhaps for coleopterous larvae. Measurement of the holes made by the probing bill showed that a maximum depth of 35 mm was reached ($N = 19$, $\bar{x} = 23.4$ mm, $S\bar{x} = 1.4$), which is a little larger than the bill itself (see Table III). Presumably this is a greater depth than that which *neumayer* can reach. It may allow *tephronota* to exploit a food not available to *neumayer*, but comparable measurements of *neumayer* digging holes were not obtained for checking this.

Both species were observed to carry seeds to flat but pitted rocks. Here the seed was placed in a pit (Fig. 13), sometimes wedged tightly in, and hammered with the bill until it cracked open, thereby allowing the kernel to be extracted and consumed. This fortunate circumstance permitted the collection of large numbers of seed cases for the purpose of comparing part of the diets of the species. In a similar way the use of a rock as an anvil by European thrushes (*Turdus ericetorum*) to break the shells of snails (*Cepaea hortensis* and *C. nemoralis*) has conveniently allowed estimates of predation on these two species and on different morphs within each species (e.g., Clarke, 1962; Ford, 1971). *S. europaea* (Löhrl, 1958) and *S. pusilla* (Morse, 1967) also wedge seeds into cracks of trees and break them open.

Figure 14 illustrates the cracked seed cases found most commonly on the *Sitta* anvils. Apparently rodents also use the anvils in some instances,

FIG. 12.   Habitats of the rock nuthatches in sympatry, above (Shiraz, area 9), and *Sitta neumayer* allopatry, below (Kavalla, area 2). The broom-like shrub in the upper illustration is *Prunus (Amygdalus) scoparia* Spach.

FIG. 13. Seed case fragments of a *Prunus* species, probably *P. (Amygdalus) reuteri* Boiss et Bh., wedged in the pits of a *Sitta tephronota* anvil near Shiraz, Iran (area 9).

perhaps exploiting seeds carried there by nuthatches but not cracked by them. The tooth grooves in the broken surfaces and the nature of the crack serve to identify the ones gnawed by rodents (Fig. 14).

Measurements of the seed cases are given in Table XVI and data for the longest dimension of the cases are illustrated in Fig. 15. Despite a poor sample of seed cases cracked by *neumayer* in sympatry, these data show a clear difference between the species. *S. tephronota* exploits larger seeds than does *neumayer* in sympatry or allopatry. This confirms the speculation of Brown and Wilson (1956), Érard and Etchecopar (1970) and Lack (1971). The difficulty of getting a large sample for *neumayer* may be partly due to a more insectivorous diet at this time of year, in which case the difference in diet between the species is even greater than is shown by the data.

The seeds exploited by *tephronota* are not only larger but harder than those exploited by *neumayer*. Measurements of the hardness of four types of seeds are summarized in Table XVII. Hardness here is measured by applying a gradually increasing force to the seed until it cracks. This is not strictly appropriate since a bird cracks a seed by delivering a blow to it with its bill, but it is probably sufficient for comparative purposes. The first two types of seeds listed in the Table, *Prunus* A and *Pinus* sp, are known to be exploited by allopatric *neumayer*. They crack at an applied force of less than 10 kgf. The remaining three *Prunus* species are common in the vicinity of Shiraz (see Fig. 12). *Prunus reuteri* is almost certainly *Prunus* C in Table

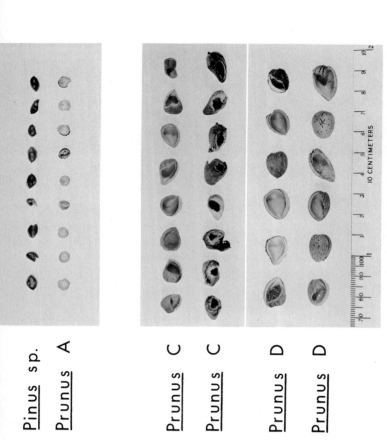

FIG. 14. Seed case fragments taken from *Sitta neumayer* and *Sitta tephronota* anvils. *Pinus* fragments were collected at Delphi (area 2), *Prunus* A fragments were collected from Kotor (area 1), both in *neumayer* allopatry, and the remainder were collected from *tephronota* anvils near Shiraz (area 9) in sympatry. The top row of *Prunus* C fragments was cracked by *tephronota*; the lower row were gnawed by rodents. *Prunus* C is almost certainly P. (*Amygdalus*) *eburnea* Spach. *Prunus* D is believed to be P. (*Amygdalus*) *scoparia* Spach.

TABLE XVI. Characteristics of Seed Cases Cracked by *Sitta neumayer* and *Sitta tephronota*[a]

	N	$\bar{x}$	$S\bar{x}$	s	*Range*
ALLOPATRY					
S. neumayer: (1) Kotor					
Prunus A					
Length	70	5.9	0.1	0.5	5–7
Width	70	5.6	0.1	0.5	5–6
Depth	76	4.4	0.1	0.7	4–6
S. neumayer: (2) Delphi					
Pinus sp.					
Length	27	7.4	0.1	0.7	6–9
Width	27	4.6	0.1	0.6	4–6
Depth	29	3.9	0.1	0.3	3–4
SYMPATRY					
S. neumayer: (1) Shiraz					
Type B					
Length	1	(9)			
Width	1	(7)			
Prunus B					
Length	5	6.6	0.40	0.89	5–7
Width	5	5.0	0.0	0.0	5
S. tephronota: (1) Shiraz					
Type A					
Length	16	9.9	0.24	0.96	9–12
Width	23	9.4	0.30	1.44	6–12
Depth	1	(5)			
Prunus C					
Length	58	11.8	0.2	1.3	9–14
Width	56	9.3	0.2	1.2	6–11
Depth	44	7.8	0.1	0.6	6–10
Type C					
Length	3	7.0	(7, 7, 7)		7
Width	3	6.0	(6, 6, 6)		6
Depth	3	6.0	(6, 6, 6)		6

[a] Original measurements in millimeters. Where the cases have been cracked in two, total depth has been calculated by multiplying measured depth by two. Kotor (Yugoslavia) is in area 1, Delphi (Greece) in area 2 and Shiraz (Iran) in area 9. Symbols and conventions are as in Table III.

TABLE XVI.   (Continued)

	N	$\bar{x}$	$S\bar{x}$	s	Range
SYMPATRY (cont'd)					
Prunus D					
Length	18	14.9	0.3	1.5	13–18
Width	18	10.9	0.3	1.1	8–13
Depth	14	8.0	0.3	1.2	7–10
Large Plum					
Length	3	19.3	(17, 20, 21)		17–21
Width	3	12.7	(12, 13, 13)		12–13
Depth	3	7.0	(6, 7, 8)		6–8
Type D					
Length	2	(9, 9)			
Width	2	(9, 8)			
Depth	2	(6, 8)			

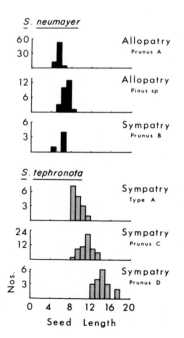

FIG. 15.   Frequency distributions of seed case fragment lengths (mm) taken from *Sitta neumayer* and *Sitta tephronota* anvils. Data taken from Table XVI.

TABLE XVII.   Seed Hardnesses, Measured as the Force Required
to Crack a Seed[a]

Seed	N	$\bar{x}$	$S\bar{x}$	s	Range
Prunus A from Kotor (1)	1	(8.0)			
Pinus sp. from Delphi (2)	20	7.03	0.17	0.77	5.7–8.1
Prunus reuteri from Shiraz (9)	20	37.71	0.84	3.75	31.0–43.9
Prunus scoparia from Shiraz (9)	8	51.24	2.37	6.69	41.8–61.8
Prunus eburnea from Shiraz (9)	5	39.52	1.67	3.74	33.2–42.7

[a] For seeds <45 kgf a pliers device (McGill seed cracker) was used (see Abbott *et al.*, 1975). For seeds >45 kgf the McGill University Faculty of Engineering's instron was used. Area numbers are given in brackets after the localities.

XVI and hence is exploited by *tephronota*. These crack at an applied force several times greater than that necessary to crack the two types exploited by allopatric *neumayer*. Since sympatric and allopatric *neumayer* exploit seeds of similar linear dimensions (Fig. 13), the hardnesses of their seeds may be similar too, in which case the two species in sympatry exploit seeds of substantially different hardnesses.

The data are not sufficient to explore the possibility of a difference in diet between sympatric and allopatric *neumayer*. The linear dimensions of the seeds exploited by *neumayer* in sympatry and allopatry appear to be the same (Fig. 13). A difference in diet, if one exists, would be difficult to detect because it is likely to be small since the difference in bill length is only about 8% (see Table 3). Much larger samples of seed case measurements are required. In contrast the difference in bill length between the two species in sympatric area 9 is about 25%.

S. tephronota took about 10 seconds to crack seed *Prunus* C once it was firmly positioned in a pit. *S. neumayer* at Delphi (area 2) took a shorter time to crack pine seeds, but this species was not observed cracking seeds at Shiraz (area 9). An attempt was made to investigate the seed cracking abilities of the two species experimentally. Pistachio nuts, known to be cracked and eaten by *tephronota* (L. Cornwallis *pers. comm.*), were placed on a rock known to be frequented by a pair of birds in the Shiraz study area. It was hoped to measure cracking and (kernel) extracting times, and to do the same for *neumayer* later. However, in a subsequent 3 hour, nearly continuous, observation period the birds did not visit the nuts. The experiment with free-ranging birds is probably not feasible with species like this, which occupy large territories. This particular pair ranged over an area of at least 200 × 150 yards. It would be better to use caged birds.

In summary, both *neumayer* and *tephronota* feed on arthropods and seeds in a similar manner. In sympatry they exploit seeds of different sizes, and probably hardnesses, which correlates well with the substantial difference in bill dimensions. The assumption that the two species in sympatry exploit different foods as a result of possessing bills of different size is fully justified by the data available. In contrast, sympatric and allopatric populations of *neumayer* appear to exploit seeds of similar sizes.

The Significance of Eye Stripe Size and Body Size: Mate Selection and Species Recognition

Aim

To assess the significance of eye stripe size as a visual cue facilitating mate and species recognition, discrimination experiments were performed in the field at Shiraz in Iran (sympatric area 9) and Kavalla in Greece (allopatric area 2). Following precedents set by Noble and Vogt (1935), Noble (1936), Lack (1939, 1947), Lanyon (1957, 1963), Gill and Lanyon (1964), Kroodsma (1974), Lewis (1972), Löhrl (1958), and Venables (1938), museum specimens were presented to birds in the vicinity of the nest, and playback of rock nuthatch song from a tape recorder was used to attract the birds to the nest vicinity when necessary. Two specimens were presented at a time. In essence the experiments asked the birds to express a preference, if possessed, for one specimen or another in terms of a differential in attention directed to them; which one would you prefer as mate, which one would you prefer to hate?! (The results of the experiments give an answer to one or both questions and the distinction between them will be discussed later). As detailed below, specimens were presented in pairs to maximize or minimize contrast in the eye stripe feature. The other morphological character in which the species differ conspicuously and which could act as a visual cue is body size. Other experiments were conducted in which contrast in body size (and associated bill size) was maximized or minimized.

Experimental Design

In sympatry the main questions are, can each species distinguish between models of the two species by appearance alone, and if so what are the cues used—eye stripe size, body size or both?

The first question was asked by simultaneous presentation of specimens of *tephronota* and *neumayer*. The second question was asked by using specimens with the eye stripes painted to resemble those of the other species. This manipulatory technique was first used by Noble (1936) in mate recognition experiments with the sexual dimorphic flickers, *Colaptes auratus*. The eye stripes of a sympatric *neumayer* specimen were painted large with a water soluble, nonodorous black paint to resemble those of sympatric *tephronota* (Fig. 16). The eye stripes of a sympatric *tephronota* specimen were painted with grey above and white below to reduce the area of black feathers to the size of sympatric *neumayer* eye stripes. Table XVIII gives the combination of specimens used and the nature of the discrimination tests.

Time limitations prevented each series of experiments from being replicated adequately for statistical analyses. Consequently the decision was taken to perform at least one experiment per series rather than, say, a minimum of five experiments in one series before starting on a second series. Therefore the coverage is extensive rather than intensive, and it is hoped that the results are reliable. Even so time did not permit the performance of a series D experiment with *neumayer* which would have completed the minimum requirements of the program.

FIG. 16.   Control (above) and experimental (below) modes of sympatric *Sitta neumayer*. Scale: approximately natural size.

TABLE XVIII. Six Series of Discrimination Experiments with *S. tephronota* and *S. neumayer* at Shiraz, Iran, in Sympatry (area 9)[a]

| Series | Number of experiments | Models | In relation to responding bird | | Nature of the discrimination test |
			Body size of model	Eye stripe size of model	
S. tephronota					
A	3	*tephronota* control	Same	Same	Between species
		neumayer control	Smaller	Smaller	
B	8	*tephronota* control	Same	Same	Between eye stripe sizes
		tephronota exptl	Same	Smaller	
C	2	*tephronota* control	Same	Same	Between body sizes
		neumayer exptl	Smaller	Same	
D	2	*tephronota* exptl	Same	Smaller	Between body and eye stripe sizes separately
		neumayer exptl	Smaller	Same	
S. neumayer					
A	3	*tephronota* control	Larger	Larger	Between species
		neumayer control	Same	Same	
E	2	*neumayer* control	Same	Same	Between eye stripe sizes
		neumayer exptl	Same	Larger	
F	1	*neumayer* control	Same	Same	Between body sizes
		tephronota exptl	Larger	Same	

[a] The *tephronota* experimental model has eye stripes characteristic of sympatric *S. neumayer*, and the *neumayer* experimental model has eye stripes characteristic of sympatric *S. tephronota*. These were confirmed by Quantimet determinations.

In the *neumayer* allopatric zone (Greece) the main question is, what would happen if sympatric *tephronota* entered the area? Would *neumayer* be able to distinguish between another (allopatric) *neumayer* and a *tephronota*, and if so what are the morphological cues serving as the basis of the discrimination. This hypothetical meeting of the two species presumably occurred in the past, although not in Greece, and if present day forms just outside the sympatric zone are a reliable guide the two species had eye stripes and body sizes not very different from those of the specimens used in these experiments (see Figs. 2 and 8, and Table XIX). The results of these

TABLE XIX.   Eight Series of Discrimination Experiments with *neumayer* at Kavalla, Greece, in Allopatry (area 2)[a]

| Series | Number of experiments | Models | In relation to responding bird | | Nature of the discrimination test |
			Body size of model	Eye stripe size of model	
A	2	*neumayer* S	Smaller	Smaller	Maximum contrast
		tephronota S	Larger	Larger	
B	2	*neumayer* A	Same	Same	Between subspecies
		neumayer S	Smaller	Smaller	
C	6	*neumayer* A	Same	Same	Between species
		tephronota S	Larger	Larger	
D	8	*neumayer* A *a*	Same	Same	Double control
		neumayer A *b*	Same	Same	
E	8	*neumayer* A	Same	Same	Between body sizes
		tephronota S experimental	Larger	Same	
F	2	*tephronota* S	Larger	Larger	Between body sizes
		neumayer A experimental	Same	Larger	
G	6	*neumayer* A	Same	Same	Between eye stripe sizes
		neumayer A experimental	Same	Larger	
H	3	*tephronota* S	Larger	Larger	Between eye stripe sizes
		tephronota S experimental	Larger	Same	

[a] S and A refer to sympatric and allopatric respectively, *a* and *b* to two identical specimens of allopatric *neumayer* (specimen *a* used throughout where not specified). The *neumayer* A experimental model has eye stripes characteristic of sympatric *tephronota* and the *tephronota* S experimental model has eye stripes characteristic of allopatric *neumayer* (i.e., local), confirmed by Quantimet determinations.

experiments should provide insight into what happened in the past, and what might happen now at their westernmost point of contact.

As in the Shiraz experiments, specimens with painted eye stripes were used to assess the importance of visual cues. Table XIX gives the details of eight series of experiments performed with control and experimental (painted) models.

The *tephronota* specimen painted with sympatric *neumayer* eye stripes in the Shiraz (area 9) experiments was painted with allopatric *neumayer* eye stripes in the Kavalla (area 2) experiments. Eye stripes were painted on an allopatric *neumayer* specimen to resemble those of sympatric *tephronota*. The specimens used as controls in the Shiraz experiments were also used as controls in the Kavalla experiments. In addition two allopatric *neumayer* specimens with the same appearance as the (living) birds being tested were used as controls (Table XIX). No discrimination is expected between these near identical specimens when presented together. Series A experiments were performed in both places for direct comparison of the discrimination abilities of allopatric and sympatric *neumayer*.

Procedure

Experiments were carried out about 10 km south west of Shiraz and about 10 km south east of Shiraz near the village of Shapurjan, 8–27 March, 1972; and about 10 km west of Kavalla, 10–27 May, 1972.

First an active nest was found, then the two specimens were attached to a rock by adhesive tape around the stick that supports the specimen and projects from the belly region to the tip of the tail (Fig. 17). The specimens were positioned no closer than two meters to each other, equidistant and no closer than five meters from and below the nest, and with similar orientation and angle of attachment to the rock. The distance from the nest was important, as was learned in the first few experiments. Too far from the nest and the birds did not respond (observed); too close to the nest and there was a risk that the birds would attack anything and not express a preference that might exist (not observed, but inferred). A loudspeaker was placed between the specimens, but further from the nest, and connected by a 25-m cable to an amplifier, two 6-V batteries and a Uher 4000-L tape recorder.

With the equipment in place a loop of recorded nuthatch song was broadcast, amplified, through the loud speaker; in most cases this was an allopatric *neumayer* trill (Type A, Fig. 19). The song was broadcast briefly and in burst if the birds were nearby, longer to attract them if far away. The idea was to shut off the song just before a bird arrived so that it associated the song with the models it perceived as it arrived. This worked in all but

FIG. 17. Experimental model of sympatric *Sitta neumayer* in position.

one experiment (see Results). Playback was never delivered when a bird was in the immediate vicinity of the nest and facing the specimens and speaker. Behavior of the birds was recorded verbally on the tape recorder when transfer of blank tape was possible, otherwise it was written in a notebook. As detailed a description of the behavior as possible was made. Approaches to each specimen and displays were recorded. Attacks were defined as one or more pecks delivered to a specimen, almost always to the head and shoulders; an attack was deemed to end when the bird hopped away. Mounting was defined as climbing or flying onto a model. In some instances this was followed by an attack before the bird flew down. The experiment was considered to have ended when the bird or birds flew away, or remained in the vicinity of the nest for 5 minutes without an approach to either model, or the experimenter's patience ran out while waiting for a bird to arrive.

A maximum of four experiments, never the same ones, were carried out at any one nest. On two occasions only two experiments were conducted in succession at a nest on the same day. Sequence of testing was neither random nor regular; the possible effect of sequence upon results is discussed with the results. With one or two exceptions, specified in the Results, the exact position of the specimens was changed from one experiment to another, and when the same specimen was used at a nest in successive experiments its position was reversed (from right to left, or *vice versa*). These protocols were adopted to eliminate a possible "position effect" upon the results.

The experimental results are given first, then the behavior of the responding birds is described so that the full significance of the results can be assessed.

Results

Sympatry. At Shiraz fifteen experiments were performed at eight *tephronota* nests and six were performed at two *neumayer* nests (Table XVIII). Of these, eight *tephronota* experiments and three *neumayer* experiments yielded no orientation to the models, either because the birds were aware of my presence or because the models were placed too far from an active nest. Responses can be classified as strong when a model was attacked and/or mounted and weak when a model was approached and displayed to, but not touched. Six experiments yielded strong responses, four yielded weak responses. The results of three of the experiments yielding weak responses have been discarded because only one model was approached, hence there is no indication that the responding birds were aware of both models. The results of the remaining seven successful experiments are set out in full in Table XX.

Data are tested by the Sign test. One-tailed tests have been performed on the results of series A experiments (double control), where the expectation is for the responding bird to direct most attacks to the conspecific model. No strong expectation can be entertained for the remainder, because each involved an experimental model, so two-tailed nests have been applied to their results.

The result of experiment 19 (series A) gives an answer to the first question posed above. *S. tephronota* can discriminate between the species on the basis of morphological cues alone. Both models were approached, but attacks were directed to a disproportionate extent toward the *tephronota* model (one-tailed Sign test, $P = 0.035$).

Similarly, *neumayer* can apparently discriminate between the species solely by morphological cues as evidenced by the one-sided nature of displays to the models in experiment 18 (series A). The number of approaches is less informative than the amount of time spent in the vicinity of each model. However, this was a weak response, and it is unfortunate that the experiment could not be replicated in the time available.

The second question is what visual cues are used in the discrimination?

TABLE XX. Results of Discrimination Experiments at Shiraz, Iran (area 9)[a]

Experiment	Nest	Date	Time	Models	Approaches	Attacks	Mountings	Remarks
S. tephronota								
(1) Series A								
19	7	March 26	0855–0955	1. *tephronota* control	7	7	2	Male response only.
				2. *neumayer* control	4	1	1	Female present, approached each model once. Model 1 fell over at attack No. two. Each mounting involved an attack.
(2) Series B								
8	1	March 13	0900–1041	1. *tephronota* control	5	5	2	Male responses shown.
				2. *tephronota* exptl.	2	0	0	Sequence: mount, attack from ground, mount, 4 attacks from ground. Approached model 2 after first mounting. Female approached each model; most time spent displaying to 1; chased away from each by male.
(3) Series C								
10	6	March 26	1008–1030	1. *tephronota* control	3	3	0	Female response only.
				2. *neumayer* exptl.	3	0	0	Alternated approaches to model 2 with attacks on model 1.

(4) Series D

				Model			
20	7	March 26	1010–1045	1. *tephronota* exptl.	18	0	0
				2. *neumayer* exptl.	13	4	0

Male response shown. Model 2 fell over at second attack. Female present, approached each model six times, attacked 2 once, after it had fallen, chased away from each model by male.

S. neumayer

(5) Series A

				Model			
18	1	March 25	1029–1100	1. *neumayer* control	3	0	0
				2. *tephronota* control	2	0	0

Female response only. Male absent. 52 secs. display to model 1, 5 secs. to model 2.

(6) Series E

				Model			
13	1	March 19	1045–1126	1. *neumayer* control	23	4	22
				2. *neumayer* exptl.	5	1	0

Male response shown. First approach to model 2 after seventh approach to model 1. Many of the mountings included copulatory behavior, four involved attacks. Female approached each model once, chased away by male.

(7) Series F

				Model			
14	1	March 20	1008–1048	1. *neumayer* control	12	0	12
				2. *tephronota* exptl.	3	0	2

Male response only. Copulatory behavior in each mounting. First approach to model 2 after five approaches to model 1. Female present, briefly.

[a] Experiments yielding no orientation to either model are omitted.

The results of the remaining experiments suggest that both eye stripe and body size (and associated bill size) are used by both species.

Large body size and eye stripes elicited significantly more attacks from *tephronota* than did small body size and eye stripes (series A). When body size was constant (and large) large eye stripes elicited attacks from *tephronota* whereas small eye stripes did not (series B; $P = 0.062$). When eye stripe was constant (and large) large body size elicited attacks whereas small body size did not (series C; data insufficient for testing). Thus both eye stripe and body size cues are used, both separately and together. Of the two, eye stripe size is indicated to be the most important by the result of experiment 20 (series D). Although both models were approached approximately equally, the model with large eye stripe and small body size was attacked, but the other with small eye stripe and large body size was not (data insufficient for testing).

S. tephronota attacked models with large eye stripe and body size, like itself; similarly *neumayer* attacked models with small eye stripe and body size, in turn like itself. In experiment 18 (series A) *neumayer* displayed longer to a model with small eye stripe and body size together than to another with large eye stripe and body size. When body size was constant (and small) small eye stripes elicited more attacks (but $P = 0.376$) and significantly more approaches ($P < 0.002$) and mountings ($P < 0.002$) than did large eye stripes (series E). When eye stripe was constant (and small), small body size elicited significantly more approaches ($P = 0.036$) and mountings ($P = 0.012$) than did large body size (series F). Experimental data are lacking on the question of which is the more important cue.

Time spent in proximity to each model (within ~60 cm) was recorded precisely in some experiments, estimated approximately in others. In all seven experiments in which orientation was shown to both models time spent in proximity was far from equally shared between the models. Models receiving most attacks, mountings and/or displays received most time to a highly disproportionate extent. Therefore, in none of the experiments was there a lack of discrimination. Experiment 19 was typical. The pair of *tephronota* arrived at 0915 and immediately approached the *tephronota* control model. The male mounted, pecked at the head and shoulders and hopped down. At the second mounting and pecking the model fell over. After 32 seconds of attention to this model the bird hopped over to the *neumayer* control model and spent four seconds there before returning to the fallen *tephronota* model. A third, fourth, fifth, and sixth attack was delivered to the fallen model. In between successive attacks, visits of 3, 8, and 6 seconds were made to the *neumayer* model. During the last of these the *neumayer* model was mounted, but not pecked, for 1 second only. Finally a seventh attack was delivered to the *tephronota* model. Altogether a total of

21 seconds was spent in the vicinity of the *neumayer* model and 207 seconds were spent in the vicinity of the *tephronota* model.

Thus these experiments, although unfortunately not replicated, show that the two species are capable of discrimination using eye stripe and body size cues. Eye stripe size may be more important than body size in the discrimination exhibited by *tephronota*.

Allopatry. At Kavalla 37 experiments were conducted at 11 nests (Tables XIX, XXI). As many as four experiments were carried out on different days at the same nest. The possible bearing of sequence of testing upon the results will be dealt with later.

Eight experiments did not work; the birds failed to respond to playback in experiments 7, 14, and 15, they failed to orient to both models in experiment 32, they appeared to be aware of my presence in experiment 22, and of the loudspeaker in experiment 17, in experiment 31 the area around the models was invaded by goats two minutes after playback was given and in experiment 35 the mistake was made of putting the models in the identical positions to those used in the preceeding experiments at that nest. The results of these eight experiments are therefore discarded and the results of the remainder are set out in full in Table XXII. The higher success frequency than in the Shiraz experiments is probably due to the stronger fidelity to the nests, all of which, except number 4, had young up to the last week of May. For the same reason experiments were much shorter, eight to 48 minutes in duration, with an average of 21 minutes. Most experiments yielded strong responses, with a nonsignificant tendency ($P > 0.1$) for the model first attacked to receive the most attacks. Two-tailed Sign tests have been applied to the results.

Experiments 1 and 4 (series A) reveal a clear preference for large eye stripe and body size, in the sense that the model with these features was attacked more frequently than the smaller model with small eye stripes (expt. 1, $P \sim 0.006$; expt. 4, $P \sim 0.124$). This is exactly the opposite to that shown by the smaller form of *neumayer* in sympatry at Shiraz. The result of experiment 5 (series B) is consistent with this. Both allopatric and sympatric *neumayer* models were approached but only the larger *allopatric neumayer* model was attacked ($P = 0.032$). In the replicate experiment number 2 the differential in attacks and approaches was in the same direction but smaller in magnitude ($P > 0.1$ in each case).

The results of series C experiments give a clear indication of a preference for large eye stripe and body size. This is as unexpected as it is consistent. Two results are statistically significant; experiment 9, $P = 0.002$, experiment 29, $P = 0.012$. It is a surprising result because it is a preference for an eye stripe and body size distinctly larger (a "super normal" stimulus) than those of the responding bird itself. It is a discrimination between like

TABLE XXI. The Sequence of Discrimination Experiments Performed with *S.neumayer* at Kavalla, Greece (area 2)[a]

Experiment	Nest	Date	Time	Series	Models	
1	1	May 10	0626–0650	A	*neumayer* S	*tephronota* S
2	2		0945–1011	B	*neumayer* A	*neumayer* S
3	1	May 11	0618–0632	C	*neumayer* A	*tephronota* S
4	2		0942–1018	A	*neumayer* S	*tephronota* S
5	3		1120–1135	B	*neumayer* A	*neumayer* S
6	1	May 13	0618–0640	E	*neumayer* A	*tephronota* S experimental
7	4		1005–1035	C	*neumayer* A	*tephronota* S
8	3		1136–1159	H	*tephronota* S	*tephronota* S experimental
9	2	May 14	0848–0904	C	*neumayer* A	*tephronota* S
10	5		1004–1023	G	*neumayer* A	*neumayer* A experimental
11	1	May 15	0624–0652	F	*tephronota* S	*neumayer* A experimental
12	3		1007–1030	E	*neumayer* A	*tephronota* S experimental
13	5		1214–1223	D	*neumayer* A*a*	*neumayer* A*b*
14	4	May 17	0920–0935	E	*neumayer* A	*tephronota* S experimental
15	3		1128–1152	D	*neumayer* A*a*	*neumayer* A*b*
16	5	May 18	0813–0823	E	*neumayer* A	*tephronota* S experimental
17	2		0915–0939	D	*neumayer* A*a*	*neumayer* A*b*
18	5	May 19	0814–0835	C	*neumayer* A	*tephronota* S
19	7	May 20	0830–0842	E	*neumayer* A	*tephronota* S experimental
20	8		0935–0948	D	*neumayer* A*a*	*neumayer* A*b*
21	9		1130–1218	D	*neumayer* A*a*	*neumayer* A*b*

[a] Symbols are same as in Table XIX.

TABLE XXI. (Continued)

Experiment	Nest	Date	Time	Series	Models	
22	10	May 21	0837–0904	H	*tephronota* S	*tephronota* S experimental
23	11		1026–1037	F	*tephronota* S	*neumayer* A experimental
24	7	May 22	0822–0842	G	*neumayer* A	*neumayer* A experimental
25	8		0932–0948	E	*neumayer* A	*tephronota* S experimental
26	9		1057–1118	G	*neumayer* A	*neumayer* A experimental
27	10	May 23	0900–0925	H	*tephronota* S	*tephronota* S experimental
28	11		1042–1059	G	*neumayer* A	*neumayer* A experimental
29	7	May 24	0812–0821	C	*neumayer* A	*tephronota* S
30	8		0905–0927	C	*neumayer* A	*tephronota* S
31	9		1044–1046	E	*neumayer* A	*tephronota* S experimental
32	10	May 25	0905–0935	G	*neumayer* A	*neumayer* A experimental
33	11		1049–1112	D	*neumayer* A*a*	*neumayer* A*b*
34	12		1203–1222	G	*neumayer* A	*neumayer* A experimental
35	7	May 26	0820–0849	D	*neumayer* A*a*	*neumayer* A*b*
36	9		0956–1021	E	*neumayer* A	*tephronota* S experimental
37	12	May 27	0914–0930	D	*neumayer* A*a*	*neumayer* A*b*

and unlike, with a preference for the latter. Yet it is consistent with the results of series A and B experiments in which a preference for large eye stripe and body size was manifested under different conditions.

Combining the results of all experiments in series A to C we find that in all nine cases the model with the larger body and eye stripe size elicited the most attacks ($P = 0.004$, two-tailed test). The combining of results like

TABLE XXII.　Results of Discrimination Experiments at Kavalla, Greece (area 2)

Experiment	Nest	Models	Male Approaches	Male Attacks	Female Approaches	Female Present	Remarks
(1) Series A							
1	1	1. *neumayer S.*	1	0	0	Yes	Attacks included 6 mountings.
		2. *tephronota S.*	>20	11	1		
4	2	1. *neumayer S.*	1	1	1	Yes	Attack sequence: 2, 2, 2, 2, 1, 2, 2.
		2. *tephronota S.*	4	0	0		
(2) Series B							
2	2	1. *neumayer A.*	8	3	2	Yes	Attack sequence: 1, 1, 2, 1, 2. Time spent within 60 cm of 1 and 2 was 25 and 10 secs respectively.
		2. *neumayer S.*	3	2	1		
5	3	1. *neumayer A.*	20	6	0	Yes	
		2. *neumayer S.*	4	0	0		
(3) Series C							
3	1	1. *neumayer A.*	3	2	0	Yes	Attack sequence: 2, 1, 2, 2, 2, 1, 2, 2. Attacks on 1 brief. Time spent within 60 cm of models in same proportion as attacks.
		2. *tephronota S.*	11	7	0		
9	2	1. *neumayer A.*	2	0	0	Yes	
		2. *tephronota S.*	10	10	0		
18	5	1. *neumayer A.*	14	2	3	Yes	Attack sequence: 2, 2, 2, 2, 2, 1, 2, 2, 2, 1, 2. Last two attacks on 2 involved mounting. Male chased female away from both models, then mounted 2.
		2. *tephronota S.*	22	9	3		

29	7	1. *neumayer A.* 2. *tephronota S.*	6 16	4 16	0 0	No	Attack sequence: 2, 2, 2, 2, 2, 1, 2, 2, 2, 2, 2, 1, 2, 2, 1, 1, 2, 2, 2. Model 2 mounted at attacks nine, ten, eleven; model 1 mounted at attack thirteen. Model 2 fell over at attack eleven, and to a lower ledge at attack nineteen, where, upside down, it was attacked once more.
30	8	1. *neumayer A.* 2. *tephronota S.*	2 4	0 3	0 0	No	Male asleep at start of setting up the experiment.
(4) Series D							
13	5	1. *neumayer A.a* 2. *neumayer A.b*	8 9	7 6	0 0	No	
20	8	1. *neumayer A.a* 2. *neumayer A.b*	5 2	5 2	0 0	No	Attack sequence: 1, 2, 1, 2, 1, 1, 1. Repeated pecks at head of 1 from side. Model 2 projected from a rock making such attack difficult.
21	9	1. *neumayer A.a* 2. *neumayer A.b*	10 6	6 4	3 3	Yes	Attack sequence: 2, 1, 1, 1, 2, 2, 2, 1, 1, 1. Last four attacks involved mountings. Model 1 fell over and attack continued. Female chased away from models repeatedly. Male *Oenanthe hispanica* made one attack on 2 while male was at 1, before attack number seven.
33	11	1. *neumayer A.a* 2. *neumayer A.b*	10 11	9 8	2 4	Yes	Attack sequence: 2, 2, 1, 2, 1, 1, 2, 1, 1, 2, 1, 2, 1, 2, 1, 1, 2. Female chased away from models repeatedly.

TABLE XXII. (Continued)

Experiment	Nest	Models	Male Approaches	Male Attacks	Female Approaches	Female Present	Remarks
(4) Series D continued.							
37	12	1. *neumayer* A.a	7	7	1	Yes	Attack sequence: 1, 1, 1, 2, 1, 1, 1, 2, 2, 1, 2. Female chased away from each model by male.
		2. *neumayer* A.b	4	4	1		
(5) Series E							
6	1	1. *neumayer* A	7	2	0	No	Attack sequence: 2, 1, 2, 2, 1, 2.
		2. *tephronota* S exptl.	10	4	0	No	
12	3	1. *neumayer* A	6	0	0		
		2. *tephronota* S exptl.	4	0	0		
16	5	1. *neumayer* A	12	3	0	No	Attack sequence: 2, 1, 2, 2, 2, 1, 2, 1, 2.
		2. *tephronota* S exptl.	10	6	0		
19	7	1. *neumayer* A	5	3	2	Yes	Attack sequence: 2, 2, 1, 2, 1, 2.
		2. *tephronota* S exptl.	11	4	2		
25	8	1. *neumayer* A	8	6	1	Yes	Attack sequence: 2, 1, 2, 1, 1, 2, 1, 1, 1. Female chased away from each model by male.
		2. *tephronota* S exptl.	4	3	1		
36	9	1. *neumayer* A	5	3	0	No	Attack sequence: 1, 2, 2, 2, 2, 1, 2, 2, 1. Model 2 mounted at attack number seven. After attack number six the male chased a male *Oenanthe hispanica* away from model 2. Later the *Oenanthe* dived at the male near model 2, without obvious response.
		2. *tephronota* S exptl.	8	6	0		

(6) Series F

11	1	1. *tephronota* S.	8	3	0	No	Attack sequence: 2, 1, 2, 1, 1.
		2. *neumayer* A. exptl.	5	2	0		
23	11	1. *tephronota* S.	15	13	0	Yes	Attack sequence: 2, 2, 1, 1, 1, 1, 2, 1, 1, 2, 2, 1, 1, 2, 1, 1, 1, 2, 1, 2, 1, 1. Attacks 4, 5, 11, 12, 15, 16, 18, and 21 involved mounting of Model 1.
		2. *neumayer* A. exptl.	10	8	0		

(7) Series G

10	5	1. *neumayer* A	10	6	0	No	Attack sequence: 2, 1, 2, 1, 1, 2, 1, 1, 2, 1.
		2. *neumayer* A exptl.	8	4	0		
24	7	1. *neumayer* A	6	6	0	Yes	Attack sequence: 1, 2, 1, 1, 2, 2, 1, 1, 1.
		2. *neumayer* A exptl.	4	3	0		
26	9	1. *neumayer* A	4	3	0	Yes	Attack sequence: 2, 1, 1, 2, 2, 2, 2, 1, 2, 2. Model 2 mounted at attack number six. Female chased away by male.
		2. *neumayer* A exptl.	8	7	0		
28	11	1. *neumayer* A	6	4	0	No	Attack sequence: 2, 2, 1, 2, 1, 2, 1, 1, 2. Model 1 mounted at attack number 3, model 2 mounted at attack number 8.
		2. *neumayer* A exptl.	6	5	0		
34	12	1. *neumayer* A	7	6	3	Yes	Attack sequence: 1, 1, 1, 2, 2, 1, 1, 2, 2. Female chased by male repeatedly.
		2. *neumayer* A exptl.	5	4	2		

(8) Series H

8	3	1. *tephronota* S	11	7	0	No	Attack sequence: 1, 2, 2, 2, 2, 1, 1, 2, 2, 1, 1, 2, 1, 1.
		2. *tephronota* S exptl.	12	7	0		
27	10	1. *tephronota* S	24	24	0	No	First ten approaches and attacks on 1, very vigourous. Rested after 7 minutes. Appeared to see model 2 for first time then and alternated attacks on models thereafter. Fifteen mountings of model 1, 7 of model 2.
		2. *tephronota* S exptl.	11	11	0		

this is strictly incorrect because three birds were tested in two of the three series, but statistical significance is still reached ($N = 6$, $P = 0.032$) when each bird is scored only once.

In series D (double control) none of the experiments individually yielded statistically significant results ($P > 0.1$ in all cases). But the five experiments differed from all five experiments in series C by yielding a smaller attack differential between the models. For series C the minimum was 3.5 (i.e., 3.5:1), for series D the maximum was 2.5. A Mann-Whitney U test shows the difference between the results of the two series to be statistically significant ($U = 0$, $n_1 = n_2 = 5$, $P = 0.008$).

In each of the five series D experiments model a received more attacks than model b, although to a small extent. This is not expected as the models were initially nearly identical.

It is possibly due to a small amount of damage suffered by model b in the first experiment (number 13) in which it was used. Some feathers were lost from one side of the face around the eye in an attack, and others were disturbed so that they did not lie along the contour of the head. It was not possible to repair this damage satisfactorily. If there was a difference in the appearance of models a and b in succeeding series D experiments, its effect would be to minimize, rather than maximize, the difference between series C and D results. Therefore, it is not considered to be an important bias.

The remaining experiments were designed to test body size and eye stripe cues separately by using an experimental model. In none of these (series E to H) was a clear preference shown for either experimental or control model in terms of approaches or attacks, with the single exception of experiment 27 in series H. In this experiment the model with the larger eye stripe elicited the most attacks ($P = 0.028$). With this exception, the results of the four series of experiments are statistically indistinguishable from those of control series D. Furthermore, the maximum attack differential was similar to the 2.5 found in series D. In series E and F, which tested for discrimination on the basis of body size alone, it was 2.0, and in series G and H, which tested for discrimination on the basis of eye stripe alone, it was 2.3. Therefore the discrimination exhibited in series A to C experiments was presumably based on both body size and eye stripe size together rather than on one alone.

However of the two, body size might be the more important. Combining series E and F we find that in six out of seven experiments the larger model elicited more attacks than the smaller. Combining series G and H we find that in three out of six experiments the model with the larger eye stripe elicited the most attacks (experiments yielding no attacks, number 12, and equal attacks to the two models, number 8, have been ignored). The difference between these frequencies is marked, but not significant by Fisher's

Exact Test (P = 0.196). The difference between the body size alone (E and F) and body and eye stripe size (A to C) results is not significant by the same test (P = 0.440). But the difference between body and eye stripe results (A to C) and eye stripe alone results (G and H) is significant (P = 0.044). This indicates a greater importance of body size. This conclusion is weakened by the fact that in all experiments in series E to H the only one yielding a significant discrimination (number 27) tested for discrimination on the basis of eye stripe size alone; and by the fact that some birds were tested twice in series A to C (but not in series G and H). However the models in experiment 27 were larger than the responding bird, which reinforces the main conclusion that discrimination is based on both body size and eye stripe cues.

Before accepting this conclusion we must consider a possible bias arising from the sequence of testing. The difference between the results of series A–C and D–H is one of magnitude that could have been influenced by the level of responsiveness of the birds. Responsiveness may wane in repeated testing. If this happened it might partly account for differences in results, because series A and B (and H) were conducted in first and second tests only whereas series C to G were conducted mainly in second and third tests.

In fact there is little indication of waning responsiveness. The differences between mean number of attacks per experiment in first, second, third, and fourth experiments are not significant (Mann-Whitney U tests, P > 0.1). Birds tested three times showed a nonsignificant tendency ($P > 0.1$) to attack more frequently in the second test than in the first and third. Thus it does not seem likely that the clear difference between the results of series C and D, for example, is due to a sequence effect. Moreover, the average number of attacks per experiment was around nine or ten, sufficiently high for an attack differential of 3 or more to be manifested in all tests. These results differ somewhat from those of Gill and Lanyon (1964) who found with *Vermivora pinus* that probability of attack increased with the amount of previous experimental experience.

With this possible bias ruled out an answer can be given to the original question, What would happen if sympatric *tephronota* (Iran) and allopatric *neumayer* (Greece) encountered each other in the wild? The answer is that *neumayer* would probably respond in a positive way by orienting toward *tephronota,* a response not exhibited by sympatric *neumayer* (earlier in the breeding season at least); the Kavalla experiments reveal the potential for confusion by *neumayer* over the specific identity of *tephronota* individuals. Discussion of the consequences of this confusion is deferred to later. Qualification must be made however, that the behavior of *tephronota* in reaction to allopatric *neumayer* is not known, and may be sufficiently distinctive to broadcast unambiguously the specific identity of *tephronota* to *neumayer.*

The experiments do not reveal if *tephronota* would be similarly confused. Similar experiments with allopatric *tephronota* in Afghanistan or the U.S.S.R. are needed to answer this.

Behavior.   The experiments show that birds can discriminate among models with different eye stripe and body sizes. A detailed study of the behavior of the birds in reaction to the models and to each other in the presence of the models shows how the eye stripe and other physical attributes function. Since there are no reports in the literature with which to compare these observations the behavior of the responding birds is described in detail. The descriptions are based upon approximately 8 hours of observations at Shiraz and 12 at Kavalla, all in the vicinity of nests.

No large differences in kind or frequency of behavior were noted between species or between sympatric and allopatric *neumayer*, with the exception that copulatory behavior was exhibited by *neumayer* in experiments 13 and 14 at Shiraz but in none of the experiments at Kavalla. What follows therefore is a description applicable to all three forms. It should be borne in mind that the sexes are virtually identical in plumage characteristics, including the eye stripe (Table V and VI). The behavior of the sexes differs conspicuously. One member of a pair is typically aggressive, vocal, and has a characteristic display (see below). It is referred to as a male throughout, despite the absence of confirmation of sex identification from examination of the gonads. Its behavior is similar to the male of the related *europaea*, which has been studied so much more thoroughly (Löhrl, 1958; Venables, 1938) and to the male of North American species *canadensis* and *carolinensis* (Kilham, 1972, 1973).

The typical response of the male to playback was an arrival by flight and an approach to a model by hopping. During the flight and subsequently, there was much vocalizing, which is described in the next section. In between hops the male pumped his head up and down (called bobbing) repeatedly, with bill approximately horizontal (Fig. 18, 1). During singing the head was held more vertically. Within about 30 cm of a model the male lifted its head and adopted a still more vertical posture, with the bill nearly vertical. This is termed a vertical display (Fig. 18, 2). The other characteristics were drooping wings and erect tail. Initially this display was given facing the front of the model. Subsequently the male hopped around to the side and back of the model, stopping frequently to give the vertical display towards, parallel with, or turned away from the model, with or without song.

The approach, with bobbing giving way to vertical displays, sometimes culminated in an attack, delivered from the ground (rock) or model. In the latter, the male hopped or flew onto the model, usually from the side or back, balanced on the shoulders and pecked at the face, head and shoulders.

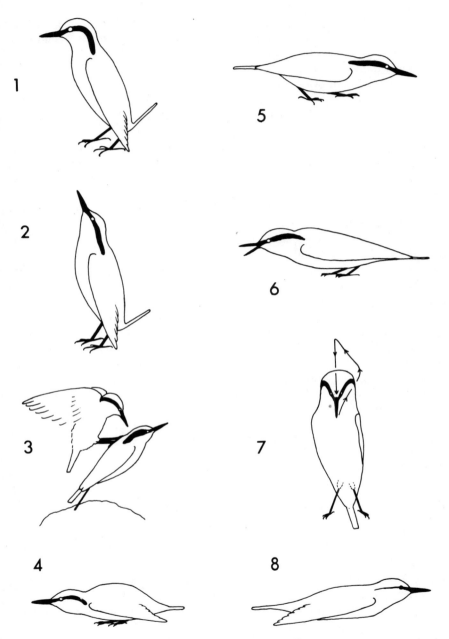

FIG. 18. The displays of *Sitta neumayer* in allopatry. 1. Bobbing. 2. Vertical. 3. Attack with mounting. 4. Hunched. 5. Horizontal. 6. Gaping. 7. Head-curving. 8. Resting/sleeping.

Rarely the approach was from the front with the bill of the model used as the initial perch. There was much wing fluttering, perhaps facilitating balance, associated with an attack while perched on the model (Fig. 18, 3). The male then hopped down and frequently gave a vertical display, facing away, within 50 cm of the model, and then sang rapidly. From this point on another attack routine sometimes ensued; at other times the male hopped around the model giving further vertical displays and at other times it flew to a prominent rock above the nest, sang and continued bobbing repeatedly but no longer gave the vertical display. A variant of the approach-attack sequence was seen at Kavalla only. The male flew straight toward a model, alighting about 50 cm away, hopped the remaining distance, jumped onto the head of the model from a distance of about four cm, gave one or two pecks to the head and jumped down. The impression was one of movement throughout the whole sequence. It was observed at two nests.

Away from the nest the behavior of the female was much the same as the males, hopping, bobbing, and calling; indeed the sexes were indistinguishable by posture. But in the vicinity of the nest and models the posture of the female was different from the male and she was invariably quiet; a very quiet trill was heard in only four experiments at Kavalla. The typical response of the female was to hop from one model to another and to give approximately equal attention to each (Table XXII). During this movement, the posture of the bird was more horizontal, and reached a most exaggerated hunched posture (Fig. 18, 4) if she was being chased by the male. The usual approach to a model was a series of hops or a run up to the model (occasionally a fluttering flight to it), stopping and "freezing" suddenly at a distance from the model of 15–25 cm; the posture of the female was nearly perfectly horizontal, (Fig. 18, 5), if anything with a slightly downward inclination. This is called a horizontal display. The orientation was either forward and facing the model from the front or at an angle of ~40–60 degrees to this direction (diagonal-horizontal). The horizontal posture was maintained for 5–10 seconds. Then the orientation of the female was rapidly switched from diagonal left (or right) to diagonal right (or left) and back again, with a pause of a few seconds in between. This is termed switching. The whole body and head were moved, not just the head. The change in direction sometimes approached as much as 180°. After the horizontal display followed by switching, the female hopped and bobbed around the specimen, repeating this pattern at the sides (head end) of the model, and then hopped or flew to the other model and did the same.

The usual response of the male to the arriving female was to ignore her if she went straight to the nest or remained about 5 m or more from the models, but to give the vertical display if she alighted near him. On most occasions when the female approached a model the male vigorously and repeatedly chased her away (Tables XX and XXII).

The usual response of the female to the male was to hop away to the other model, in a somewhat horizontal or hunched posture. Occasionally the female gave a horizontal display with switching directed toward the male, the two facing each other. The male responded by a vertical display, and the female hopped away. At Kavalla a male was once seen to peck the tail of a female it was chasing. Three times at Kavalla a female was seen to turn, face the male and give a horizontal display with bill open (Fig. 18, 6). This is called gaping. On one occasion the male responded with an identical display, on another stood still and on the third hopped away. Twice the gaping display by a female was observed directed toward a model, following a horizontal display and in substitution for switching.

The interaction between the pair sometimes led them away from the nest and models, and then their behavior changed from contrasting to mutual. During one experiment at Kavalla the male chased the female away from each model once. The second time this happened it led to mutual bobbing and vertical displays, wings drooped, tails erect, and the pair hopped away about 30 cm apart and out of sight. At Delphi, also, mutual vertical displays were observed; the pair were about 1 m apart, facing in roughly opposite directions and calling loudly.

The preceding description applied to behavior observed several times, and shows a clear distinction between male and female behavior. But once a male *tephronota* was observed to move from side to side in front of a model, somewhat reminiscent of the female switching behavior but carried out with the body held in a more vertical posture. This was in experiment 8 (Shiraz). Once in experiment 20 a male *tephronota* was observed to give a diagonal-horizontal display to a model, without switching, immediately after the female had done so repeatedly. In experiment 10 only the female *tephronota* was present at the models while the male called from about 30 cm away. Initially she gave vertical displays, and then changed to horizontal displays and continued with these for the rest of the experiment. These observations indicate the possibility that the distinction between male and female behavior is not so sharp in this species as in *neumayer*.

A female display observed rarely at Kavalla shows the connectedness of horizontal and vertical displays. The context is important. In experiment 21 the female flew to the rock above the nest and alighted near the male. She then gave the head-curving display (Fig. 18, 7). The body was upright but the bill pointed to the feet initially. Then the head was raised slowly sideways and upward and then, continuing upward, was brought back into the plane of the body. At the end the posture was identical to that of the vertical display. The bird paused for 2 or 3 seconds in this position. Then the head was lowered to the starting point and the upward movement repeated but on the other side of the body. The total behavior comprised two head curving movements and two stationary vertical postures. This behavior was also witnessed once in experiment 25 in a similar context—not close to the models.

Two miscellaneous features of behavior deserve brief mention. On June 14 at Delphi a strangely colored *neumayer* was observed in a loose group of five. The back was normally colored and the eye stripes were usual but the ventral surface was a dark chocolate brown except for a thin dull white line beneath the eye stripe. It behaved as a young bird. It first solicited attention from another two birds, possibly a pair, which it followed, hopping, from rock to rock, but only succeeded in eliciting aggressive behaviour from one of them—loud calling ("tui" and "cheu", see p. 311), vertical displays, and "rushes." After about two minutes the solicited bird turned, the two birds engaged in mutual vertical displays facing in nearly opposite directions, and then the solicited bird chased the brown one from rock to rock on 20 to 30 m flights, accompanied at a short distance by the other member of a possible pair. After about 4 minutes they were lost from sight. It is tempting to deduce from these observations that a bird with aberrant plumage is at a disadvantage in social interactions which lead to pairing and mating, since extensive chases of this sort were never observed among birds of normal plumage.

The second miscellaneous observation is that the displays of *tephronota* and *neumayer* resemble those of the related but arboreal *europaea* (Löhrl, 1958). In particular the display of the eye stripe in *europaea* is a consequence of the oblique locomotion around a tree trunk necessitated by the bird's hind limb and body structure (Löhrl, 1958; Norris, 1958; Richardson, 1942). Such displays are retained by *tephronota* and *neumayer* in a terrestrial situation.

The Interpretation of Behavior and Significance of the Experimental Results. The fact that displays commonly given to the models are also given to the other member of the pair shows that they are unlikely to be artifacts of the experimental situation. An interpretation of their significance is attempted as follows:

The displays given to the models are initiation or invitation displays, jointly communicating information and soliciting a response. The visual properties of an intruder, together with its behavioral response (which the model is unable to give) serve to identify it as a member of the same or different species, and when the same species, whether male or female. The observations show that the two rock nuthatch species differ in body size and eye stripe but not in behavior. Males and females of each species differ in behavior but not in body size and eye stripe. Therefore it is reasonable to infer that appearance communicates species identity, behavior communicates sex identity. The communication functions are interrelated however. Although body size and eye stripe size can be appreciated by a pair of interacting birds by virtue of their close proximity, certain features of their displays probably enhance the perception of these morphological characteristics. The vertical display communicates information about body size, in particular body length. Two birds of different size giving vertical displays facing each other should be able to perceive the difference in body size. In one of the interspecific interactions at Shiraz (see below) mutual vertical displays were observed briefly, before the *neumayer* fled.

The horizontal display and switching behavior of the female simultaneously communicates information about the eye stripe which is "presented" to the other bird, and allows the female to examine the face, and eye stripe in particular, of the other bird. Bill size and head size information transfer may occur at this time as well. A further indication of the importance of the eye stripe is given by the variation in its appearance. Feather posture is under the bird's control, and according to whether the feathers are raised or lowered the eye stripe appears conspicuous or not. This is difficult to assess, but is clear when comparing the eye stripes of a bird in their most extreme appearances. While interacting with another bird or model the eye stripe is at its most conspicuous, while the bird is at rest or sleeping the eye stripe of allopatric *neumayer* is most inconspicuous (Fig. 13, 8).

The lack of a behavioral response from the model is probably important in determining subsequent behavior, including attacks. Except when the male is absent the female rarely presses an attack, but gives horizontal and switching displays only. The other female characteristic is quietness. This can be thought of as behavior soliciting reproductive and (male) sex declaring behavior from the model. The female loses "interest" in the absence of a response from the model and hops over to the other model. The male characteristics of vertical display and vocalization contrast with the female characteristics of horizontal display and silence. The absence of a response from the model induces the male to attack. In contrast, the female responds to threat from the male by avoidance or occasionally by threat (gaping), so that attacks of one member of the pair upon the other are rare. Presumably a display from an intruding male would also elicit attack from the resident male; this was never observed.

The orientation of the models is probably also important in inducing attacks. In most experiments the angle it subtended from the horizontal was about 45°, occasionally not much more than 30°. It was not realized at the time the experiments were begun that orientation might communicate information about sex identity. A lower angle would have more closely matched the typical female horizontal display and a higher angle would have resembled the male posture. With hindsight it is now clear that the experiments would have simulated a reproductive context better if the models had been placed in horizontal display positions. However, even after the significance of the orientation had been appreciated the intermediate angle was adopted for the sake of consistency. It might have been ambiguous but it probably resembled the male posture more than the female's, and this is perhaps why attack behavior was the usual response and copulatory behavior was the exceptional response. The only clear copulatory behavior observed was in the *neumayer* experiments at Shiraz. The angle of the model so treated may have been particularly low. Also at this early stage in the breeding season, the motivation to copulate may have been higher than during the Kavalla experiments, and an incomplete set of stimuli from the model may have been sufficient to elicit the response. Even so there was enough ambiguity in the situation (orientation and lack of response from the model) for the male to intersperse pecks with copulatory attempts (Table XX).

Thus the attacks themselves and the level of aggression shown were to some extent artificial products of the experimental situation induced by the absence of response from the models. Presumably the threat displays of the male to the female are also an artifact, a redirected attack induced by the models. Nevertheless the choice of which model to attack provides information about the stimuli eliciting attention to the models. The crucial question is, what do the results tell us about the likely outcome of encounters between sexually mature, unmated, members of opposite sex of the same and other species?

The experimental situation does not faithfully simulate those encounters, (cf. Gill and Lanyon, 1964). They have never been reported upon, but perhaps as with *europaea*, they occur after the breeding season (Löhrl, 1958). The similarity of eye stripes of young and adults fits this interpretation. However, the experiments give us results which enable us to infer what would happen. The only difference between the models was in morphology (species identification) not in orientation or behavior (sex identification). Therefore a difference in attacks given to them indicates a species discrimination based upon the sole morphological differences of size and eye stripe. This was not just defense of nest against all comers because (a) the models were probably not placed within the critical distance and (b) some models were not attacked, even though they were approached yet did not fly away. In any event, such defense should not be directed unequally to equidistant models. The copulatory behavior of *neumayer* in the Shiraz experiments suggests that this same species discrimination is expressed in a reproductive context.

It is concluded that the experimental results, demonstrating a discrimination based upon body size and eye stripe size and indicating potential confusion over species identity in allopatry, are relevant to events which occur during courtship and mating. Body size and eye stripe cues, and responses to them, can therefore function as reproductive isolating mechanisms. Bill size was not controlled independent of body size in the experiments. It could also serve in reproductive isolation and may have contributed to the experimental results. However its role in reproductive isolation is likely to be much less inasmuch as (a) geographical variation is smaller in this character than in the other two, and (b) the other two characters are more conspicuous.

Interspecific Territoriality and Identity of Song

The expectation of interspecific territoriality and identity of song in sympatry is not supported by observations at Shiraz.

Territoriality

Pairs of the same species have nests which are spaced out. No two active nests of *tephronota* were found closer than 200 m to each other. The observations on *neumayer* nests at Shiraz are insufficient for comparison, but Érard and Etchecopar (1970) commented on finding two nests of *neumayer* (in area 8) 100 m apart. At Kavalla no two active nests were found closer than 150 m.

In neither sympatry nor allopatry was a bird of either species observed to forage closer to another nest of a conspecific than to its own. Conspecific spacing out of activity appears to be the rule. Whether these areas should be considered territories or home ranges is an unresolved point as antagonistic behavior between pairs that might be considered indicative of territorial defense was never observed, despite the large number of hours of observation. The closest to an encounter between pairs occurred during experiment 16 on March 24 at Shiraz. A neighboring pair of *tephronota* entered the activity area of the residents who were out of sight when playback of song near their nest was begun. The entry of the neighbors may have been induced by the playback. The entering pair flew through the area within about 20 m of the models at one point, and out into their own, followed about 100 m behind by the pair of residents, for a short distance. The two pairs were calling (type D Fig. 19), so their positions, and possibly their intentions, were communicated.

Since the activity areas are large, up to 300 m × 300 m, they might not be defended in the same way that woodland birds defend territories against conspecifics. Perhaps they are best regarded as home ranges.

In contrast, pairs of the two species have largely overlapping, even near identical, home ranges, as was observed at Shiraz. L. Cornwallis (*pers. comm.*) has observed pairs of the two species feeding within a few meters of each other, without strife, in other parts of the province of Fars (area 9) in Iran. Therefore interspecific overlap of home ranges is probably a general phenomenon.

However the experiments at Shiraz did reveal interspecific aggressive interaction close to the nests of both species. *S. neumayer* nest 1 was 15 m from a *tephronota* nest being built. Playback in experiments 13, 14 and 18 at nest 1 resulted initially in the arrival of both *neumayer* and *tephronota* pairs. *S. tephronota* never approached the models, but interacted with *neumayer* in two of the experiments. During experiment 13 on March 19 the male *tephronota* made four successful supplanting attacks on a *neumayer*. The *neumayer* moved three meters or more. These took place in an oak tree about 10 m above the *neumayer* nest used by both species for song perches. It was about 12 m from the *tephronota* nest. But the male *neumayer* supplanted a *tephronota* twice at another tree perch just 2 m from the *neumayer* nest. In experiment 18 on March 25 the performance was repeated. *S. tephronota* supplanted both members of the *neumayer* pair at least twice each at the 10 m perch, but at the 2-m perch the male *neumayer* vigorously chased away a *tephronota* twice.

These observations indicate that a volume of space around the nest is defended interspecifically, as with other hole-nesting birds (e.g., Hartmann, 1956, 1957). It is not defended solely against the other *Sitta* species. At Shiraz *neumayer* was observed to chase away from the nest vicinity a Radde's Accentor (*Prunella ocularis*) and an unidentified finch. At Kavalla *neumayer* was observed to chase away a House sparrow (*Passer domesticus*) twice, a Red Rumped swallow (*Hirundo daurica*) and a Black-eared Wheatear (*Oenanthe hispanica*). Thus the response to an intruder is a general one, and not specifically directed toward another *Sitta* species, although it may be restricted to potential users of the nest as in *europaea* (Venables, 1938). It might happen more frequently between the congeners than between heterogeneric species because the two *Sitta* species build their nests close together in similar positions on vertical rock faces.

Therefore the two species do not space themselves apart by interspecific territorial behavior facilitated by similarity in appearance, behavior, etc. At least, they do not do so in area 9.

Song

Samples of the songs of both species were collected in allopatry and sympatry, and analyzed spectrographically. The areas from which they came are 1, 2, and 3 (allopatric *neumayer*), 9 (sympatry) and 13 (allopatric *tephronota*). Twenty-nine different song types were recognized by syllable structure, and are illustrated in Figs. 19 and 20. Resemblance of some to *Sitta europaea* songs is noteworthy (cf. Thielcke, 1966).

Songs consist of one, occasionally two, and rarely three syllables repeated a number of times. Syllable structure is simple. Most songs consist of ascending (A, B, E) or descending (H, I) syllables, or syllables that do both (C, F, G, J, K, N, P, R, T. Y, BB, and CC). When frequency modulation predominates in one direction, they can be roughly characterized by the words "tui" (ascending) or "cheu" (descending); these terms were found useful when describing the songs produced during an experiment. Switching between songs was usually abrupt, although some smooth transitions were heard and recorded, particularly from the "trill," type A, to the "tinkle," type B (see Fig. 19). With eggs or young in the nest, female *neumayer* in allopatry rarely sang, but in sympatry earlier in the breeding season both sexes of each species sang, often simultaneously but different songs. It was not possible to determine if any song was sung exclusively by one sex.

For purposes of classification and comparison the number of syllables produced per second (or interval between syllables) and song duration were of no use because they varied so much within individuals singing the same song type. For example, playback of song in the experiments invariably elicited an initial trilling response from the male with rapid production of syllables, followed by a slower trill (or other song). The final song at the end of an experiment in allopatry, often D, E, G, or H was usually distinctly slower than earlier songs of the same type. To a lesser extent syllable shape and frequency (pitch) also varied in association with rate of syllable production and hence with presumed motivational state of the bird.

Two types of displacement of song in sympatry can occur. The repertoire of individuals or a population can change through addition or elimination of the congener's songs, or the structural characteristics of the songs can change. Differences in frequency characteristics between allopatry and sympatry might be expected to correlate with the established differences in body size, in view of the correlation found with *Myiarchus tyrannulus* by Lanyon (1960).

Similarity of repertoires of the two species is more striking than differences, and in the case of *neumayer*, the better studied species, there is little indication of song acquisition in sympatry. All but one of the 13 songs sung by *neumayer* in sympatry are also sung in allopatry (Figs. 19 and 20).

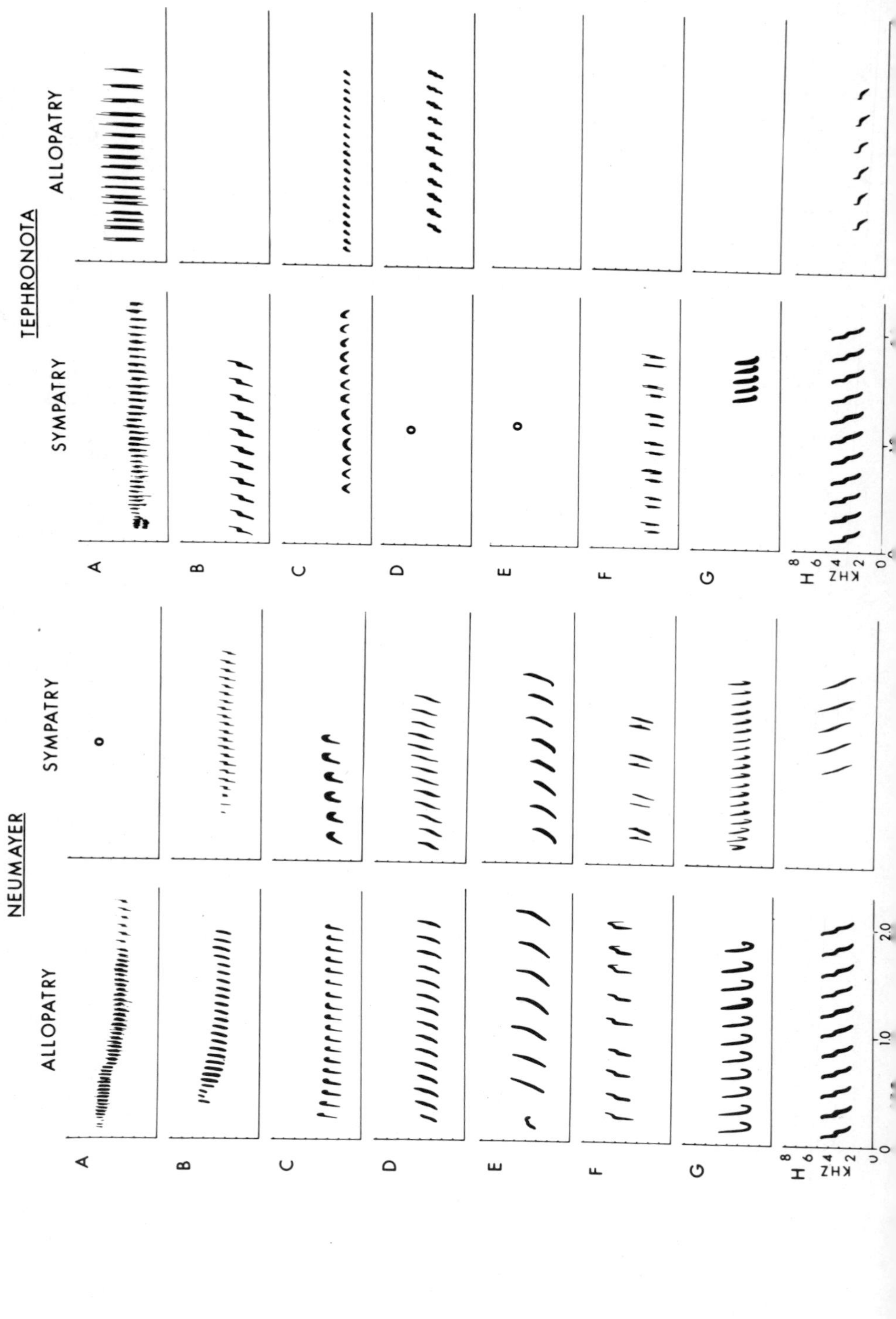
TEPHRONOTA
SYMPATRY
ALLOPATRY
NEUMAYER
SYMPATRY
ALLOPATRY
A
B
C
D
E
F
G
H
KHZ
8
6
4
2
0
1.0
2.0

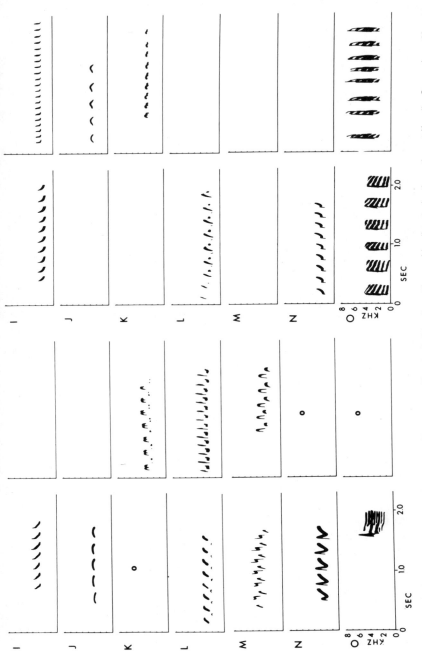

FIG. 19.   Songs of the two species in sympatry (Shiraz, area 9), recorded by the author, and in allopatry by the author, at Kavalla, Greece (area 2) and by others as follows: *Sitta neumayer*, type O, by E. D. H. Johnson at Göreme, Turkey (area 4); *Sitta tephronota*, all songs in allopatry recorded by H. Löhrl and G. Thielcke at Kabul, Afghanistan (area 16). A circle in a sonagram indicates that the song was heard but not recorded.

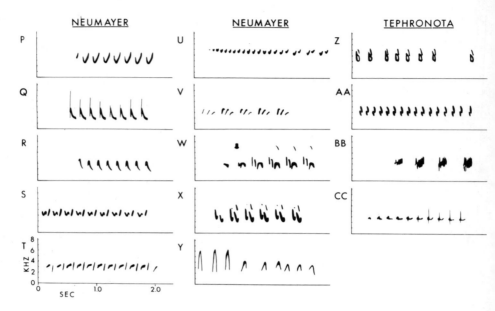

FIG. 20. Songs of the two species recorded in allopatry but not in sympatry by the author and by others as follows: *Sitta neumayer*, types T, U, W, and X, by the author at Kavalla, Greece (area 2) types V and Y by the author at Kotor, Yugoslavia (area 1) and types P, Q, R and S by H. Löhrl and G. Thielcke at Omiš, Yugoslavia (area 1); *Sitta tephronota*, all songs recorded by H. Löhrl and G. Thielcke at Kabul, Afghanistan (area 16).

The one exception does not appear to have been acquired from *tephronota* since *tephronota* was not heard to sing it in sympatry. However, *tephronota* does sing this song in allopatry (Fig. 20). Fewer songs were recorded in sympatry then in allopatry, for both species, but this does not necessarily indicate the elimination of songs from the repertoire in sympatry, as sampling effort was greater in allopatry than in sympatry, particularly for *neumayer*.

Five of the 11 songs sung by *tephronota* in sympatry were not recorded in allopatry. This may represent an elaboration of total population repertoire in sympatry through incorporation of *neumayer* songs, since all five were sung by sympatric *neumayer* also. Alternatively it may reflect the restricted nature of the collection of allopatric *tephronota* songs; 12 *tephronota* songs were recorded from one locality in allopatry, in contrast to 23 *neumayer* songs collected from four allopatric localities. It should also be mentioned that all of the 14 songs recorded in sympatry were recorded in one or both allopatric zones, and as many as 10 were sung by both species.

Having found no clear evidence of convergence in repertoire, we now consider the possibility of song modification in sympatry. This is assessed by analyzing frequency characteristics. It is made difficult by imperfections and attenuations in the sonagrams. However difficulties should apply equally to the songs of the two species in sympatry since recording conditions were identical for the two species. A parabolic reflector was used in the recording of all songs in areas 3 and 13, and some songs in area 1, but not in areas 2 an 9. Some bias may therefore enter when comparing songs recorded differently in sympatry and allopatry. The initial material for analysis are the sonagrams used for making the illustrations (Figs. 19 and 20). Strictly speaking these are not random samples of song, but nor were they chosen systematically with frequency characteristics in mind. Minimum and maximum frequencies during 2.0 seconds of song were measured, and from these were calculated the frequency mode (midpoint) and range.

Differences between the species, already noted by Érard and Etchecopar (1970) and Hüe and Etchecopar (1970), contrast with lack of differences between allopatric and sympatric populations of the same species. Frequency characteristics are given in Table XXIII for all songs except type O which is conspicuously difficult to record in its entirety. Values in the table indicate a trend toward lower frequency characteristics in sympatry than in the allopatry, for both species, and lower frequencies of *tephronota* songs than *neumayer* songs. Considering all these songs, the two species do not differ significantly in any frequency characteristic in allopatry, where variation is large, but do differ in sympatry in minimum frequency (two-tailed Mann-Whitney U test, $U = 11$, $P < 0.02$), maximum frequency ($U = 9$, $P < 0.02$) and modal frequency ($U = 6$, $P < 0.002$). Furthermore, considering only the five songs sung by both species in sympatry, the larger *tephronota* consistently has the lower maximum, modal and range of frequencies (Sign test, $P = 0.062$ in each case). This is in agreement with

TABLE XXIII. Mean and Standard Errors of Frequency (KH$_z$) Characteristics of the Songs Illustrated in Figures 19–20, but Excluding Type O

Frequency	S. neumayer		S. tephronota	
	Allopatry ($N = 23$)	Sympatry ($N = 10$)	Sympatry ($N = 8$)	Allopatry ($N = 11$)
Minimum	2.19 ± 0.13	1.90 ± 0.08	1.43 ± 0.11	2.06 ± 0.13
Maximum	4.78 ± 0.21	4.46 ± 0.17	3.45 ± 0.21	3.98 ± 0.36
Mode	3.49 ± 0.13	3.18 ± 0.09	2.44 ± 0.14	3.01 ± 0.18
Range	2.61 ± 0.23	2.56 ± 0.16	2.03 ± 0.20	1.92 ± 0.38

Lanyon's (1960) results with *Myiarchus tyrannulus* subspecies. Thus a basis exists for species discrimination by song in sympatry. Among those acoustical cues potentially available, only a small number are likely to be critical for species recognition (Emlen, 1972).

According to this result the species have not converged in sympatry but, if anything, have diverged. However, there is no evidence of a shift in song characteristics in *neumayer* between allopatry and sympatry. In *tephronota* two shifts are apparent; minimum frequencies ($U = 11$, $P < 0.02$) and modal frequencies ($U = 13$, $P < 0.02$) are lower in sympatry than in allopatry. Since all allopatric songs of *tephronota* were recorded with a reflector and all sympatric songs were recorded without, this difference might be due to recording differences rather than to production differences.

A more meaningful but restricted comparison is between songs of the same type sung by the same species in allopatry and sympatry. Unfortunately, only three of the *tephronota* songs were recorded in both zones; no trend is apparent in them. Nine songs of *neumayer* were recorded in both allopatry and sympatry, all were recorded in both zones without a reflector and therefore they may be compared without qualification. None of the four frequency characteristics differs between zones (Wilcoxon matched pairs signed rank test, $P > 0.1$ in each case). Further tests were conducted on other randomly selected sonagrams of the same songs sung by other individuals, and again no significant differences were found (for a given song type, interindividual variation in frequency characteristics is small). This result is consistent with Thielcke's (1969) failure to find evidence of "contrast reinforcement" of song following secondary contact, in several pairs of species.

A more refined frequency analysis is possible. But at the level of analysis performed, it is clear that the two species in sympatry sing similar songs but with different frequency characteristics. The differences are meaningful because both species sing from rocks or from the ground where the optimal frequencies for sound transmission are low (Jilke and Leisler, 1974). Their songs are not conspicuously more similar in sympatry than in allopatry. Another difference, not shown on the sonagrams, is in loudness; the songs of *tephronota* are much louder to the human ear than are those of *neumayer*. Interestingly, loudness is also a characteristic of terrestrial woodpeckers (Short, 1971). Despite these differences there is some overlap in the characteristics of the same song type sung by the two species. The song used in the experiments was a trill (song A) recorded from allopatric *neumayer*. It elicited responses from both species in sympatry. This might be partly because it was broadcast from the vicinity of a nest, but also due to the fact that it had frequency characteristics shared by the two species in sympatry.

CONCLUSIONS

A dilemma was outlined in the Introduction; why have the two species of rock nuthatches apparently diverged when a plausible theory predicts convergence? This will now be resolved.

Upon detailed examination it turns out that the species have not diverged as much as is generally supposed. In bill length and bill width the two species show no clear evidence of having undergone divergent character displacement. The species differ in bill characteristics to a large extent where they occur together, and this is associated with a substantial difference in diet. Presumably this means that differences between the species in bill dimensions and diet were already large before the species made contact, and the differences were sufficient to facilitate coexistence when they met. There would not be selection for convergence under these conditions. Not surprisingly, therefore, territories are not held interspecifically, and the songs of the species in sympatry are different and have not converged. Thus although the environmental conditions of structural simplicity of habitat were right for selection for convergence, the initial dissimilarities of the species were the wrong conditions. The dilemma is resolved.

Bulmer (1974) modeled ecological character displacement mathematically. He found that supposed character displacements (e.g., bill length of rock nuthatches) were too large to be accounted for by the model under conditions of optimizing selection due entirely to an external, environmentally determined, optimal value. To reconcile theory with observations he suggested (a) the character (e.g., bill length) could also be involved in mate selection, and (b) most of the optimizing selection could be determined by the genetic environment. Although these suggestions are plausible, the second more than the first in my opinion (see, e.g., Grant, 1967), they are not necessary. The large difference in bill lengths between the sympatric rock nuthatches should not be attributed to character displacement.

With regard to the eye stripes of the nuthatches, in the hot, bright environment of Iran both species might be expected to have pronounced dark patches around the eyes to reduce glare (Charman, 1972; Ficken et al., 1971). Instead they have apparently undergone mutually divergent character displacement in eye stripe features with only one, tephronota, having the expected large patch, but then to a disproportionate extent. The evidence from experiments with models and observations of behavior suggests that the eye stripe communicates species identity information. The experiments at Shiraz indicate that differences in eye stripes between the species in sympatry might be sufficient to prevent confusion as to species identity. But the experiments at Kavalla show that the potential for confusion exists in the allopatric range of *neumayer*, should *tephronota* ever invade it. The potential for confusion possibly exists also in the sympatric areas 7 and 11; and it may also exist in the allopatric range of *tephronota*.

The divergence in eye stripe area, particularly marked in *neumayer* populations 8, 9, and 10, can be understood as a response to selection acting against individuals of one species that resembled too closely those of the other species. They would be courted by heterospecifics as well as conspecifics. And if their responses were not species specific they might have hybridized, at a selective cost, as appears to be happening now in Kazakhstan between an eastern and western species of shrike (*Lanius collurio* and *L. phoenicuroides*) which have similar courtship behavior (Panov, 1972). Evidence of current hybridization between the two *Sitta* species might be sought in the Elburz mountain system (areas 7 and 11). In short, when the two species established contact they were fully isolated ecologically but not fully isolated reproductively, and the reproductive isolating mechanisms were only later "reinforced" (Mayr, 1974) in sympatric syntopy.

The role played by body size adjustments is not clear, due partly to inadequate data. Possibly mutually divergent character displacement in body size (weight) occurred. Shifts in body size appear to account largely but not entirely for the degree of divergence in eye stripe features. Experiments and observations showed that body size functioned in species discrimination, and some of the Kavalla experiments showed that a moderate difference in eye stripe alone was insufficient to elicit discrimination. These results suggest that selection, in minimizing the risk of hybridization, acted on body size as well as eye stripe size. If eye stripe size is more important than body size we might expect large differences in body size between the species in allotopic sympatry (areas 7 and 11), in contrast to the known relatively small differences in eye stripe sizes there. This could easily be checked.*

Body size may have significance in ecological as well as reproductive contexts. Some have suggested that the nature of the diet and feeding efficiency vary directly with body size to some extent independently of trophic structures (e.g., Ashmole, 1968; Hespenheide, 1971; Huxley, 1942; Lack, 1944). However, given the presumed large differences in bill sizes and hence diets between the species at the time of contact, it is doubtful if feeding characteristics contributed much to selection for divergent body sizes. In conclusion, the evidence points to mutually divergent character displacement in features associated with reproductive isolation, but to no character displacement in either species in features associated with ecological isolation.

* See Appendix II.

A RECONSTRUCTION OF EVOLUTIONARY EVENTS

The model presented in Fig. 3 is clearly not entirely applicable to the two *Sitta* species. The following is a detailed verbal alternative.

The two rock nuthatches were derived from a single ancestral form related to *europaea* (Vaurie, 1957; Voous and van Marle, 1953). The distribution of the ancestral species was similar to that of one or both present-day rock nuthatches. In the Pleistocene, due to climatic changes, a habitat barrier to dispersal developed within the distributional range and divided the population into an eastern and a western one. The western one experienced a mediterranean type of climate, the eastern one experienced a continental climate. Some adaptive change took place in isolation under a regime of climatic selection, with the western individuals becoming smaller and the eastern ones larger. Subsequently the climate changed and the barrier disappeared. As a result of range extensions of each species, secondary contact was established in Iran. In the contact zone individuals of the two populations tended to select different habitats, they fed on different foods but were only partially reproductively isolated. Selection favored divergent individuals of the two populations with respect to body size, eye stripe size, and behavioral responses to them. Full reproductive isolation was achieved thereby, and the distinction between the preferred habitats became blurred. A second climatic change occurred, resulting in another barrier from the Caspian extending west of the Zagros. This barrier isolated *tephronota* and *neumayer* to the east from western *neumayer*. The spatial isolation was of shorter duration than the first, and it ended with a climatic reversal. Eastern (Zagros) and western populations of *neumayer* made contact in the northern part of the Zagros and were reproductively compatible (the possibility of partial reproductive isolation and hence incipient speciation of these two populations needs to be examined in the field). The western *neumayer* extended eastward and *tephronota* extended westward through the Elburz where the two species met and are now at a stage similar to that at the initial contact of the two species; they have different habitat affinities and are not fully reproductively isolated. During these late stages *neumayer* population 5 and 10 became isolated, and *tephronota* may have invaded Baluchistan and Pakistan (area 12) from Kirman (area 10).

The important point of the model is that climatic change has determined evolutionary events. There is much circumstantial evidence in support of this suggestion. Climatic vicissitudes and shifts in species distributions occurred widely in the North Temperate region during the Pleistocene (Mengel, 1964; Moreau, 1954, 1955). The pattern of discontinuities between races of *europaea* can be understood as arising during this time (Voous and van Marle, 1953). Periods of restriction within eastern and western glacial refuges alternated with periods of postglacial expansions and rejoinings. Other species of birds show remarkably congruent patterns (Voous and van Marle, 1953). Therefore similar processes were probably at work with *neumayer* and *tephronota*. The morphological differences between *neumayer* populations 7 and 11 on the one hand and 8, 9, and 10 on the other can be simply interpreted as being conditioned by separate (in time) postglacial events.

The known distribution of major habitats at various stages in the Pleistocene (Frenzel, 1968) supports the view that early distributions of the ancestral and derived rock nuthatches species were similar to present-day ones, with the long axis longitudinal rather than latitudinal. Precise positioning of the distribution cannot be attempted because of the lack of sufficient palynological information from the Turko–Iranian region (B. Frenzel, *pers. comm.*). Nevertheless, it is fairly certain that the open-country steppe habitat now occupied by the rock nuthatches did not undergo large latitudinal shifts in relation to climatic vicissitudes during the Pleistocene. It is possible that a relatively small latitudinal shift of this habitat occurred around the Mediterranean littoral into North Africa, as happened with typically Mediterranean vegetation (Büdel, 1951; Moreau, 1954, 1955). If so, *neumayer* may have been isolated for a time around the littoral.

Traditionally, barriers and land bridges have been erected and dismantled by biogeographers to suit their convenience, but in the present case there is palynological evidence of a barrier in the northwest of Iran during the Pleistocene. For example during the Eemian interglacial a band of hygrophilous and thermophilous deciduous and coniferous forest extended southwestward from the Caucasus toward the Persian Gulf (Frenzel, 1968). Earlier, during the Würm glaciation, forest existed from the eastern shores of the Mediterranean to the Caspian, separating extensive areas of steppe-desert in Iran and further east from small areas of such habitat in Anatolia (Butzer, 1958). It is not known that these forests were continuous, but if they were, they must have acted as powerful barriers to dispersal of the non-forest dwelling rock nuthatches. Given the repeated fluctuations of climate in this region throughout the Pleistocene, it is possible that a barrier existed several times and affected rock nuthatches several times. If speciation and subsequent events in the rock nuthatches followed a similar course to subspeciation in *europaea*, the total time span is likely to be measured in terms of 10^5 years (Voous and van Marle, 1953), not just the last few thousand years for which palynological information is most complete (Wright, 1968; Wright *et al.*, 1967). Climate, vegetation, and species distributions have remained unchanged for the last 6000 to 11,000 years (Wright, 1968); conceivably Man, through extensive forest clearance in the Mediterranean region during the last 10,000 years, has facilitated some small range extension of *neumayer*.

Tectonics during the Pleistocene produced an uplift of 200–300 m. Frenzel (1968) has summarized the evidence for the climatological consequences in the Eurasian region under study. The principal consequence was that the climatic difference between the central (continental) and peripheral (Mediterranean) areas was enhanced (see also Dumitrashkov and Kamanin, 1946). Selection for large body size under continental climate and for small body size under Mediterranean climate would possibly have proceeded as the climates changed, particularly if the populations were isolated in relatively restricted areas and individuals were unable to move elsewhere.

The model attributes speciation and the particular size difference between the species to direct and indirect influences of climate. If climate was important in the past it should also be important in the present. It should determine the present limits to the distribution of the species. In fact this seems to be so.

The western limit to the distribution of allopatric *neumayer* is set by climatic and edaphic factors, jointly acting to determine habitat characteristics. The boundary of *neumayer* coincides with the beginning of continuous forest vegetation and the end of karstic soils (Le Grand, 1973). But for small deviations, the entire distribution of *neumayer* in allopatry coincides with the distribution of typically Mediterranean climate in which only 15% or less of the annual rainfall falls in the summer months (Trewartha, 1961). The 15% isoline nicely circumscribes the western boundary of *neumayer*.

This type of climatic regime, and associated vegetation, occurs at higher altitudes in the hotter regions to the east. In Iran, *neumayer* reaches its eastern limit in two places and in one of them, Kirman (area 10) the altitude-climate-vegetation complex appears to set the limit to the distribution of *neumayer*, because further to the east the areas of Baluchistan are low lying, hot, and dry (Vaurie, 1950, made the same point). Altitude does not play a part in determining the northeastern boundary of *neumayer*. Here the boundary coincides with, and is presumably determined by, the southeastern limit of the Caspian influence on climate. Other species are similarly affected here (see Paludan, 1940).

The entire distribution of *tephronota* corresponds approximately with the distribution of one vegetation type. This is an *Astragalus-Artemesia*-Rosaceae (including almonds and pistachios) association referred to by Zohary (1963) as Irano–Turanian. The Irano–Turanian region extends from Iran and Iraqui Kurdistan through Afghanistan and Pakistan to the west Tien-Shan mountains of the U.S.S.R. Small outliers are also found in inner Anatolia (Turkey), perhaps too isolated to be reached by *tephronota*.

Within the sympatric zone of Iran the distribution of the two species is controlled by climatic factors. D. A. Scott (*pers. comm.*) has pointed out that *tephronota* rarely occurs in areas where the July mean temperature is $\leq 20°C$. and the mean annual precipitation is ≥ 750 mm; whereas *neumayer* rarely occurs where the July mean temperature is $\geq 25°C$ and the mean annual precipitation is ≤ 500 mm—a relationship with climatic factors similar to that shown in allopatry (e.g., see data in Butzer, 1958). A knowledge of climatic characteristics in both allopatric zones would lead to a prediction of an interspecific difference in sympatry in this direction. Available climatic data are summarized in Table XXIV. They do not necessarily show the nature of the climate where the birds actually occur, nor do they take into account climatic variations with altitude. Nevertheless, they are sufficient to document the fact that the climate in the allopatric range of *neumayer* is wetter and milder in temperature than is the climate in the allopatric zone of *tephronota*. In fact the climate (Qt) in *neumayer* allopatry is wetter and milder than in all areas occupied by *tephronota*, with the possible exception of area 20.

Observations have established that *neumayer* occupies the higher altitudes in the Elburz mountains, areas 7 and 11, in sympatry (Appendix). If climate and associated vegetation are important to these species, a similar but less conspicuous difference between the species should be manifested in the Zagros mountain system, areas 8, 9, and 10, as well. This is tested with data on the altitudinal distribution of the two species in Fars province alone (area 9) kindly supplied by L. Cornwallis (*pers. comm.*). The data are plotted as cumulative frequency distributions in Fig. 21, tested with a Kolmogorov–Smirnov two-sample test (one-tailed) and found to be significantly different in the direction predicted ($x_2^2 = 17.92$, $P < 0.005$). Note also that the distribution of *tephronota* over much of its altitudinal range shows a close correspondence with the altitudinal distribution of *Prunus* (*Amygdalus*) *scoparia,* a strongly suspected food plant. It is one of the most xerophytic species of almonds, and has a geographical distribution largely coextensive with *tephronota* from northwest Iran to Turkmenistan, Russia (Sokolov, 1954).

The data in Table XXIV are not good enough to allow us to predict sympatric distributions in relation to climate from the allopatric data alone. If this were possible it might be found that predictions were not entirely matched by observations, indicating that in sympatry the two species have undergone small changes in their responses to climatic and vegetational factors. This remains for future studies to determine. In the absence of this information there is only one major inconsistency in the relationships between distribution and climate. The postulated relationships do not explain the apparently nonoverlapping altitudinal distributions of the two species in the

TABLE XXIV.   Climatic Characteristics of the 20 Areas. Qt^a is Emberger's Pluvio-thermic Quotient (Emberger, 1930, 1952), Calculated from the Data[b]

Areas	Number of localities	Average annual rainfall (inches)	Average January daily minimum temperature (°F)	Average July daily maximum temperature (°F)	Temperature difference	Qt [a]
1	4	44.18	38.0	84.7	46.7	0.317
2	7	26.36	39.1	89.3	50.2	0.190
3	1	25.5	39	92	53	0.120
4	1	41.7	43	93	50	0.190
5	2	29.15	42.5	89.5	47.0	0.185
6	4	17.65	16.7	84.7	68.0	0.115
7	1	—	24	91	67	—
8	3	11.90	27.3	102.3	75.0	0.031
9	1	13.7	32	99	68	0.020
10	1	5.4	27	101	74	0.006
11	1	9.7	27	99	72	0.020
12	2	2.35	30.0	107.0	77.0	0.001
13	3	10.20	24.0	92.0	68.0	0.021
14	1	8.9	25	97	72	0.040
15	3	8.80	26.0	98.7	72.7	0.013
16	10	13.20	16.9	90.6	73.7	0.034
17	1	21.6	22	96	74	0.080
18	1	30.3	26	95	69	0.090
19	1	14.7	21	92	71	0.075
20	1	23.5	34	90	56	0.190

[a] $Qt = \dfrac{P \times N/365 \times 100}{M - m\,(M + m)/2}$, where P is the annual precipitation, N is the number of days of precipitation, M is the average daily July maximum temperature, m is the average daily January minimum temperature.

[b] Sources for the data are World Weather Records (1967a, b) and the Royal Afghan Air Authority.

Elburz. It is possible that the climate gradient with altitude is steeper in the Elburz than in the Zagros, yet there should still be a zone of climatic conditions tolerable to both species as there is in the Zagros. A climatic hypothesis fails to account for the altitudinal "replacement" in the Elburz and an interactive hypothesis is called for instead. The altitudinal replacement is quite different from those cases where ecological incompatibility is inferred from near-identity of bill sizes and presumed diets (Diamond, 1970,

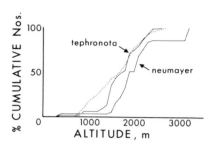

FIG. 21.   Cumulative frequency distributions of observations of *Sitta neumayer* (*N* = 37) and *Sitta tephronota* (*N* = 366) at different altitudes in Fars province (area 9), Iran. For comparison, the broken line shows the distribution of observations of *Prunus* (*Amygdalus*) *scoparia* Spach. (*N* = 72), a probable food plant of *Sitta tephronota*.

1973; Nørrevang, 1959; Watson, 1962). Therefore, an hypothesis of reproductive incompatibility was adopted (p. 275).

"Climatic selection" has also been invoked to account for the direction of the difference in body size between the species, although not the magnitude of the difference. *S. tephronota* was subjected to selection under continental climates which are more extreme in temperatures within a year than are Mediterranean ones. The most important component of the total climate conditions may have been January minimum temperatures. An examination of modern climates shows that the *tephronota* allopatric zone is colder in winter than the *neumayer* allopatric zone to a greater extent than it is hotter in summer. Large body size is selected for under cold temperature conditions, presumably for the heat conserving advantages it confers.

Does climatic selection also account for the current geographical trends in body size and exposed parts within allopatric and sympatric zones? To a small extent it might, but the answer is tentative because the climatic data are of limited usefulness. Unfortunately the climatic data are not good enough to permit a confident explanation of morphological differences between neighboring populations, or even within zones, in terms of adaptations to climate.

Linear dimensions of both species were shown to correlate with longitude (Table I). A climatic basis for this is indicated by rank correlations with Emberger's pluviothermic quotient (Qt) in four instances ($P < 0.05$ in each); *neumayer* (N = 10) wing r_s = 0.81, tarsus r_s = 0.88, and eye stripe length r_s = 0.90; *tephronota* (N = 13) bill length r_s = −0.75. However, the puzzling thing is that the sign of the correlations of the two species is different. If continental climates, with low Qt values, have selected for large size of body and parts in *tephronota* in the past, the modern negative correlation of *tephronota* bill size with Qt is understandable. It is the positive correlations of *neumayer* that need to be explained in a different way. The clinal variation in *neumayer* allopatry is noticeably shallow. I suggest that the principal reason for clinal variation in this species, and possibly also in *tephronota,* is that selection has favored different body sizes in sympatry and allopatry for reproductive reasons; gene flow between neighboring populations would be a contributing influence. This could be shown to be false if it was demonstrated that clinal variation within allopatry occurs in the absence of gene flow. The difficulties of measuring gene flow, like its importance, should not be underestimated (Hastings and Rohlf, 1974). But, apart from this, there is plenty of scope for tests of hypotheses that explain the distribution and morphological variation in the rock nuthatches in terms of reproductive isolation, ecological isolation, and climate.

In summary, present distributions can be largely explained in terms of unspecified adaptations to particular vegetations and climates, as can the fact that *tephronota* is larger than *neumayer.* Intraspecific variation in morphological characters is largely accounted for by selection for reproductive isolation; selection for ecological isolation appears to have played a minor role at most.

GENERAL DISCUSSION

Among organisms in general, the evidence for the reproductive aspect of character displacement appears much stronger than the evidence for the ecological aspect (e.g., see Blair, 1974; Levin, 1970; Mayr, 1963; Selander, 1971) although a critical review of the former is needed (see Selander, 1971; Thielcke, 1969; Walker, 1974). The conclusions of the present study conform to this pattern. Ecological character displacement has remained difficult to establish despite some good recent efforts with modern (Huey and Pianka, 1974; Huey *et al.,* 1974) and paleontological (Eldredge, 1974; Gingerich, 1974) material. Both in its extent and frequency, the ecological aspect of character displacement may have been overemphasized as an evolutionary process. This was the conclusion reached after reviewing the most prominent cases (Grant, 1972*a*).

Ecological character displacement has been invoked to explain how complex communities develop. Mutual adjustments of invading and resident species of similar trophic type allow for "close packing" of many competitor species (MacArthur, 1972; MacArthur and Levins, 1964). In view of the lack of strong evidence for ecological character displacement, it is perhaps more correct to regard the process as a relatively rare event, or at least small in magnitude. Thus complex communities are built up by the accumulation of species which have undergone adaptive changes in allopatry. Certain combinations are ecologically compatible, others are not (Diamond, 1975; Grant, 1966*c*, 1969). Adaptations already possessed by the species at the time of meeting are the principal determinants of coexistence. These adaptations include some behavioral flexibility with regard to resource utilization (food, habitat). Ecological character displacement is a secondary determinant of coexistence. Genetic shifts in the behavior involved in habitat and food selection may also occur in response to selection. These are ecobehavioral analogues of (morphological) character displacement and may be more likely to occur than morphological character displacement itself but they still, according to this view, constitute secondary determinants of coexistence. Thus evolutionary processes in sympatry represent "fine-tuning" in the structuring and functioning of trophic components of communities.

The outcome of entry of a new species into a community is presumably dependent upon how different it is from the members of that community in ecologically meaningful morphological characters, like bill size in birds, and behavior. It is also dependent upon what degree of difference is necessary to give it a chance of coexisting with potential competitors. Character displacement is viewed in this perspective in Fig. 22. The outcomes of invasion are local extinction of the invader or of a previous resident, or coexistence. The probabilities of these two outcomes are related to the degree of difference at initial contact between invader and the resident most similar to it. For simplicity I ignore other residents, even though they may have a combined competitive effect upon the invader (e.g., Pianka, 1974).

When the invader and resident are nearly identical ($< a$) extinction is nearly certain. When the invader is quite different from any resident ($> b$) coexistence is a near certainty. In the range of intermediate values of difference between invader and resident (ab) the outcome is in doubt. Here divergent character displacement is likely to contribute to that outcome by enhancing the likelihood of coexistence (the exact shape of the curves is unimportant). Note that the range of initial conditions over which divergent character displacement leads to coexistence may be quite small.

Selection for divergence may occur over the whole range of initial differences from O to b. It may be strongest when the two species are nearly identical, yet probably coexistence will fail to result (Fig. 22). This is because of insufficient time. There must be enough time to allow sufficient character displacement if competitive exclusion is to be avoided. The tendency for the invader to be excluded will be offset by immigration of additional individuals from outside the contact zone. However, this will tend to retard the displacement process to an extent dependent upon the immigration rate, so a complex balance will be struck between selection for displacement and extinction through competition.

The model is presented as an organizing and illustrative device, not as a predictive one. But the curves are not likely to be the same for all organisms

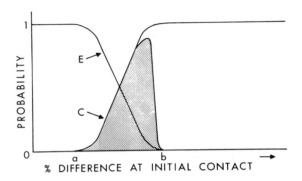

FIG. 22.   A probability model of divergent character displacement. The outcomes of invasion of a community by a species are shown as alternative probabilities of coexistence, *C*, and extinction (of invader or a resident) *E*, in relation to the degree of morphological difference between invader and a resident. The area under the coexistence curve with diagonal lines indicates the probability that divergent character displacement leading to coexistence with occur; it is a subset of total coexistence probabilities.

in all environments, and making use of this fact some predictions can be made. For example the extinction and coexistence probability curves should be displaced to the right the more potential competitors there are in the community and the lower the variety, abundance and predictability of resources (cf. May, 1972; Pianka, 1974; Schoener, 1965). If the diet of feeding generalists is less strictly determined by morphology than is the diet of specialists (i.e., there is a strong behavioral component to food choice), then generalists should have coexistence curves closer to the origin than specialists. If response to selection is a function of reproductive rate, species with rapid rates (large clutches, short generation time, etc.) should have coexistence curves less sharply ascending and from a point closer to the origin than species with slow reproductive rates. I speculate that for most organisms the genetic control of reproductive characters is simpler than that of ecological (feeding) characters, and reason from this that the displacement probability distribution is closer to the origin for reproductive than for ecological characters.

Restricting attention to the ecological aspect of character displacement, the model shows that over a certain range of initial differences coexistence is virtually impossible. This is relevant to the discussion of the limiting similarity concept of MacArthur and Levins (1967). The term limiting similarity is applied to the minimum degree of difference necessary for sustained coexistence. If coexistence probabilities pass rapidly through a transition zone (Fig. 22), the choice of a point of limiting similarity must be somewhat arbitrary. Attempts have been made to establish what that value is, approximately, for its importance lies in setting a theoretical upper limit to the number of trophically similar species in a community (MacArthur, 1972; May, 1972). May (1972) deduced that the minimum difference between the means of the resource utilization distributions of two species separated along a single niche dimension (e.g., food size) is approximately equal to one standard deviation from one of them. When ecological separation is achieved along more than one major niche dimension the limiting similarity along one or more may be smaller (cf. Grant, 1966c; May, 1972; Schoener, 1974). Resource utilization distributions are not well known (generally log normal?), and limiting similarities have not been investigated yet in any detail.

However, there are reasons for believing that this minimum difference between means along a single niche dimension, when translated into morphological terms, is equivalent to the sum of two standard deviations of each of the means. The evidence for this statement comes from birds. The most closely related sympatric species are double invasion species pairs; these are pairs of congeneric species of birds on islands which have been derived sequentially from a common ancestor. With few exceptions members of a pair differ only in bill size not in shape. Sympatric species so derived have a minimum difference in bill length or other, single, linear dimension of about 15% (Grant, 1968, 1972b). Taking 3.5% of the mean to be a reasonable estimate of the magnitude of one standard deviation when extraneous sources of variation are removed (e.g., Grant, 1967), a 15% difference in means is equivalent to the sum of two standard deviations of each. When the standard deviations are larger than 3.5% of the mean, the difference between the means is larger too by the appropriate amount, and the four standard deviation difference is maintained. For example, sympatric species of ground finches on the Galápagos, *Geospiza magnirostris*, *G. fortis* and *G. fuliginosa* differ in overall bill size and the first two are unusually variable, but frequency distributions of each of the bill dimensions of the three species are contiguous and only trivially overlapping (Abbott, Grant and Abbott, in prep.; Lack, 1947).

The rock nuthatches are extremely closely related systematically (Voous and van Marle, 1953), and differ in bill size but not shape. It is noteworthy that in both syntopic and allotopic sympatry they differ in bill length by more than 20%. In syntopic area 9 the mean bill lengths of males only are more than three standard deviations of each apart. In syntopic area 8 the mean bill length of *neumayer* males (the larger sex of the smaller species), and of *tephronota* females (the smaller sex of the larger species), are 2.5 standard deviations of each apart. In allotopic area 7 the comparable inter-sex difference is three standard deviations of each. Thus in both syntopy and allotopy the limiting similarity is clearly exceeded, reinforcing the conclusion reached earlier that the species are ecologically compatible in the allotopic areas. In fact by this criterion, populations as far apart as area 3 (*neumayer*) and 16 (*tephronota*) could coexist. The fragmentary information on the diet of these two species (Tables XVI and XVII) also support the conclusion that the limiting similarity (one standard deviation of resources used) is exceeded in sympatry.

It could be argued that rock nuthatch habitats are poor in resource variety, and the limiting similarity should be larger than four standard deviations of morphological means. But islands, from which the limiting similarity was established, are also low in resource variety (Allan *et al.*, 1973; Janzen, 1973), so the conclusion stands.

SUMMARY

The two species of rock nuthatches, *Sitta tephronota* and *S. neumayer*, provide the classical example of character displacement. The westernmost *neumayer* has a distribution from Yugoslavia to Iran, whereas *tephronota* extends eastwards from Iran to the Tien-Shan mountains of the U.S.S.R. In sympatry, largely Iran, they differ markedly in bill size and eye stripe size, whereas in extreme allopatry they are almost identical. According to theory, the two species were morphologically similar when they first came into contact. Natural selection favored divergent bill sizes and eye stripe sizes in the two populations, as a consequence of which interspecific competition and hybridization were minimized. A morphological study was undertaken to establish more precisely the pattern of geographical variation in bill size and eye stripe size. A field study was undertaken to determine the diet of the species and its relationship with bill size, the role of eye stripe and body size in species discrimination, songs of the species and territorial behavior.

The study confirms that *neumayer* has undergone divergent character displacement in eye stripe size, but in south and west Iran and neighboring Iraq only. *S. tephronota* has probably done so as well, but the interpretation is complicated by the presence of a population with (divergent) large eye stripes in allopatry. Both species may have undergone mutually divergent character displacement in body size as well, but the data are not extensive. Neither species appears to have undergone divergence of bill size in sympatry. The large interspecific difference in bill size in sympatry is attributed partly to allometric consequences of selection for divergent body sizes. These conclusions are drawn from analyses which assume, with field observations supporting, that the two species are fully syntopic in south and west Iran only; analyses based on the simpler assumptions of full syntopy throughout sympatry yield no strong evidence of character displacement in either species.

In northern Iran *neumayer* occurs at higher altitudes than *tephronota*. Their apparent failure to coexist here is attributed to potential reproductive confusion, since their eye stripe sizes are more similar than where they are syntopic. Failure to coexist is not attributed to ecological incompatibility, since their bill sizes differ largely and by the same amount as in syntopy where dietary differences are known to be pronounced. Species discrimination experiments in syntopic synpatry and allopatry support the interpretation involving reproductive confusion. Discrimination is based on eye stripe and body size together. This is shown by the results of experiments in which pairs of models were presented to birds close to their nest, and by observations of the displays of the birds. The potential for reproductive confusion is demonstrated by the strong responses of allopatric *neumayer* to models of sympatric *tephronota,* which contrasts with the avoidance of sympatric *tephronota* models by sympatric *neumayer*.

Adopting the allopatric model of speciation, it is suggested that the two species underwent repeated range shifts and adaptive change in allopatry in relation to the climatic vicissitudes of the Pleistocene. Present day sympatry may represent two separate secondary contacts of the species. Natural selection did not favor convergence of song and the adoption of interspecific territoriality, which is predicted by one theory, because the species were fully ecologically isolated at the time that both secondary contacts were made. The relevance of this to the concept of limiting similarity is briefly discussed.

ACKNOWLEDGMENTS

This study has benefited from the kind help of an unusually large number of people. I gratefully acknowledge all their help. Museum material was supplied by J. C. Barlow (Ontario), M. LeCroy and C. Vaurie (American), G. Mauersberger (Berlin), P. Morgan (Liverpool), F. Roux (Paris), F. Salomonsen (Copenhagen), E. G. F. Sauer and G. Niethammer (Bonn), H. Schifter (Vienna), D. W. Snow (British), L. S. Stepanyan (Moscow), M. A. Traylor (Chicago), J. Wattel (Amsterdam) and J. S. Weske (Smithsonian). Sound recording equipment and a Kay sonagraph were loaned by C. R. Dawkins, G. Hirons and R. E. Lemon. Tape recordings of songs were kindly supplied by E. D. H. Johnson, H. Löhrl and G. Thielcke. R. Lamarche and R. Tanner gave considerable help with photography, and S. Bradbury allowed me (and showed me how) to use a

Quantimet instrument in the Human Anatomy Department, Oxford. M. H. Bokhari helped with the identification of plants. J. Stachiewicz and O. Möhling of the Faculty of Engineering, McGill, determined some of the seed hardnesses with an Instron. The study was initiated and largely conducted at the Animal Ecology Research Group, Oxford, and I thank Prof. J. S. Pringle and J. Phillipson for facilities provided. Fieldwork in Iran was made possible by the assistance of M. A. Taylor and S. S. Jahromi of Pahlevi University, Shiraz. I have corresponded and discussed the subject with many people, including R.-D. Etchecopar, R. F. Porter, C. Vaurie, L. Cornwallis and D. A. Scott. The text bears witness to the large amount of advice and information I have received from the last two in particular. I. Abbott, L. Cornwallis, R. D. Montgomerie, D. A. Scott, and J. N. M. Smith made valuable comments on the manuscript, and my wife made valuable and diverse contributions throughout the study. It was supported by the National Research Council of Canada (A2920).

APPENDIX I

In areas 7 and 11 *neumayer* and *tephronota* occur at different altitudes. S. *neumayer* occurs at the higher altitudes in the Elburz system, above a minimum of 3000 feet (Érard and Etchecopar, 1970) on the northern slopes, and above 5000 feet toward the eastern end of the range (Streseman, 1928), whereas *tephronota* is found at low altitudes, for example at the edges of the Dasht-i-Kavir (Sarudny and Härms, 1923).

Museum specimens of the two species have not been collected from the same locality in either area at the same altitude. But the most compelling evidence of a separation of the species by habitat and altitude is unpublished. Dr. Derek Scott's observations of the two species in 1970-1972 are more extensive than any other made in northern Iran. He writes (*pers. comm.*):

"I have never seen the two species of nuthatch together in the north. *Tephronota* seems to be uncommon in the north and prefers lower, warmer and drier elevations than *neumayer*. My three (locality) records from the north all fall in the range 3000 to 5000 feet above sea level. *Neumayer*, however, is clearly a high mountain species, which I have seen up to 11,000 feet in the Central Alborz (Elburz). Of my 19 records of *neumayer* in the north, no less than 13 fall in the altitudinal range of 7500 to 11,000 feet, whereas only three fall below 5000. Two of these records, at Vinak and Shutlu in the extreme northwestern corner of Azarbaijan (both at 3000 feet) probably fall outside the distribution of *tephronota*. The nearest case of overlap in altitudinal range which I have yet observed is in the Mohammad Reza Shah Wildlife Park in the Eastern Alborz. *Neumayer* was observed at 4000 feet in the central valley during extremely cold weather in February 1972, while *tephronota* was observed at about 3500 feet in the same valley in May 1971." S. *neumayer* descends to 3000 feet on the northern side of the mountains where "*tephronota* seems to be entirely absent."

It is not known whether the altitudinal separation of the species is complete, as suggested by these observations, or partial. In Armenia *neumayer* breeds up to 6000 feet, but descends to the valleys in the winter (Dementiev and Gladkov 1970) where it may coexist with *tephronota* for a time. Buxton (1920) collected specimens of both species between 1000 and 2000 feet in the valley of the Sufed Rud, Elburz, near Menjil (area 11), but did not record whether they were found together or not. Érard and Etchecopar (1970) found the two species nesting on opposite sides of a valley near Bojnurd (area 11). They were only 300 m apart, but in somewhat different habitat (R-D. Etchecopar, *pers. comm.*).

APPENDIX II

Body size estimates can be obtained more easily by measuring the internal diameter of nest entrances than by collecting specimens of birds. Nuthatches construct their nests in such a way that they can just pass through the entrance. The width of the entrance is therefore an indicator of the width of the bird. Mean entrance width of *neumayer* (areas 2 and 9) and *tephronota* (area 9) is positively correlated with mean $\sqrt[3]{}$ body weight; $r = 0.999$, $P< 0.05$. Weights are given in Table VIII, and mean nest entrance widths in millimeters and the standard errors are as follows: *neumayer* area 2, 31.2 $\pm$ 0.3 ($N = 14$); *neumayer* area 9, 25.2 $\pm$ 1.6 ($N = 4$); *tephronota* area 9, 39.0 $\pm$ 0.7 ($N = 11$). Therefore, the question of whether the two species differ as much in body size in areas 7 and 11 as they do in areas 8 and 9 can be investigated by this indirect means, as can the possibility of intraspecific variation in body sizes across the zone boundaries.

REFERENCES

Abbott, I., Abbott, L. K., and Grant, P. R., 1975, Seed selection and handling ability of four species of Darwin's Finches, *Condor* **88:** (In Press).

Allan, J. D., Barnthouse, L. W., Prestbye, R. A., and Strong, D. R., 1973, On foliage arthropod communities of Puerto Rican second growth vegetation, *Ecology* **54:**628–632.

Ashmole, N. P., 1968, Body size, prey size, and ecological segregation in five sympatric tropical terns (Aves: Laridae), *Syst. Zool.* **17:**292–304.

Balát, F., 1962, Contribution to the knowledge of the avifauna of Bulgaria, *Acta Acad. Scient. Cechoslovenicae Basis Brunensis* **10:**445–492.

Blair, W. F., 1974, Character displacement in frogs, *Amer. Zool.* **14:**1119–1125.

Brown, W. L., Jr., 1964, Two evolutionary terms, *Syst. Zool.* **13:**50–52.

Brown, W. L., Jr., and Wilson, E. O., 1956, Character displacement, *Syst. Zool.* **5:**49–64.

Büdel, J., 1951, Die Klimazonen des Eiszeitalters. *Eiszeitalter u Gegenwart, Öhringen* **1:**16–26.

Bulmer, M. G., 1974, Density dependent selection and character displacement, *Am. Nat.* **108:**45–58.

Bush, G. L., 1969, Sympatric host race formation and speciation in frugivorous flies of the genus *Rhagoletis* (Diptera, Tephritidae), *Evolution* **23:**237—251.

Butzer, K. W., 1958, *Quaternary Stratigraphy and Climate in the Near East,* Ferd Dummlers Verlag: Bonn.

Buxton, P. A., 1920, *Bull B.O.C.* **40:**135–139.

Charman, W. N., 1972, Eye marks in vertebrates as aids to vision, *Science* **177:**367.

Clarke, B. C., 1962, Balanced polymorphism and the diversity of sympatric species, *in, Taxonomy and Geography, Syst. Assoc. Publ.* **No. 4:**47–70.

Cody, M. L., 1969, Convergent characteristics in sympatric populations: a possible relation to interspecific territoriality, *Condor* **71:**222–239.

Cody, M. L., 1973, *Competition and the structure of bird communities,* Princeton University Press, Princeton, N.J.

Crozier, R. H., 1974, Niche shape and genetic aspects of character displacement, *Amer. Zool.* **14:**1151–1157.

Dementiev, G. P., and Gladkov, N. A., (eds) 1970, *Birds of the Soviet Union,* Vol. **5:** Israel Program for Sci. Transl., Jerusalem.

Diamond, J. M., 1970, Ecological consequences of island colonization by Southwest Pacific birds, I. Types of niche shifts, *Proc. Natl. Acad. Sci.* **67:**529–536.

Diamond, J. M., 1973, Distributional ecology of New Guinea birds, *Science* **179:**759–769.

Diamond, J. M., 1975, Assembly of species communities, In Cody, M. L., and Diamond, J. M., (eds.) *Ecology and Evolution of Communities,* Harvard University Press, Cambridge (In Press).

Dilger, W. C., 1956, Adaptive modifications and ecological isolating mechanisms in the thrush genera *Catharus* and *Hylocichla, Wilson Bull.* **68:**171–199.

Dumitrashkov, N. V., and Kamanin, L. G., 1946, Paleogeography of Central Siberia and the Baikal Region. *Trans. Inst. Geogr. Acad. Sci. USSR* **37:**132–151.

Dunajewski, A., 1934, Die eurasiatischen Formen der Gattung *Sitta LINN., Acta Ornithol. Musei Zoologici Polonici* **1:**181–251.

Ehrlich, P. R., & Raven, P. H., 1969, Differentiation of populations, *Science* **165:**1228–1232.

Eldredge, N., 1974, Character displacement in evolutionary time, *Amer. Zool.* **14:**1083–1097.

Emberger, L., 1930, Sur une formule climatique applicable en géographie botanique, *C. R. Acad. Sci.* **191:**389–390.

Emberger, L., 1952, Sur le quotient pluviothermique, *C. R. Acad Sci.* **234:**2508–2510.

Emlen, J. M., 1973, *Ecology: an evolutionary approach,* Addison-Wesley, Reading, Mass.

Emlen, S. T., 1972, An experimental analysis of the parameters of bird song eliciting species recognition, *Behavior* **41:**130–171.

Endler, J. A., 1973, Gene flow and population differentiation, *Science* **179:**243–250.

Érard, C., and Etchecopar, R. D., 1970, Contribution à l'étude des oiseaux d'Iran, *Mém. Mus. Nat. d'Histoirie Naturelle* No. 66.

Eriksson, K., 1970, The invasion of *Sitta europaea asiatica* Gould into Fennoscandia in the winters of 1962/63 and 1963/64, *Ann. Zool. Fennici* **7:**121–140.

Ficken, R. W., Matthiae, P. E., and Horowich, R., 1971, Eye marks in vertebrates: aids to vision, *Science* **173:**936–938.

Ford, E. B., 1971, *Ecological Genetics,* 3rd ed. Chapman & Hall, London.

Frenzel, B., 1968, The pleistocene vegetation of northern eurasia, *Science* **161**:637–649.

Fretwell, S. D., 1969, Ecotypic variation in the nonbreeding season in migratory populations: a study of tarsal length in some Fringillidae, *Evolution* **23**:406–420.

Gatter, W., 1974, Beobachtung an Invasionsvögeln des Kleibers (*Sitta europaea caesia*) an Randecher Maar, Schwäbische Alb, *Die Vogelwarte* **27**:203–209.

Gill, F. B., and Lanyon, W. E., 1964, Experiments in species discrimination in Blue-winged Warblers, *Auk* **81**:53–64.

Gingerich, P. D., 1974, Stratigraphic record of early Eocene *Hyopsodus* and the geometry of mammalian phylogeny, *Nature* **248**: 107–109.

Grant, P. R., 1965, The adaptive significance of some size trends in island birds, *Evolution* **19**:355–367.

Grant, P. R., 1966a, The coexistence of two wren species of the genus *Thryothorus*, *Wilson Bull.* **78**:266–278.

Grant, P. R., 1966b, Further information on the relative length of the tarsus in land birds, Yale Peabody Mus. Nat. Hist., *Postilla* **98**:1–13.

Grant, P. R., 1966c, Ecological incompatibility of bird species on islands, *Am. Nat.* **100**:451–462.

Grant, P. R., 1967, Bill length variability in birds of the Tres Marías Islands, Mexico, *Can. J. Zool.* **45**:805–815.

Grant, P. R., 1968, Bill size, body size, and the ecological adaptations of bird species to competitive situations on islands, *Syst. Zool.* **17**:319–333.

Grant, P. R., 1969, Colonization of islands by ecologically dissimilar species of birds, *Can. J. Zool.* **47**:41–43.

Grant, P. R., 1972a, Convergent and divergent character displacement, *Biol. J. Linn. Soc.* **4**:39–68.

Grant, P. R., 1972b, Bill dimensions of the three species of *Zosterops* on Norfolk Island, *Syst. Zool.* **21**:289–291.

Hanson, W. D., 1966, Effect of partial isolation (distance), migration and different fitness requirements among environmental pockets upon steady state gene frequencies, *Biometrics* **22**:453–468.

Hartmann, L. von., 1956, Territory in the Pied Flycather, *Muscicapa hypoleuca, Ibis* **98**:460–475.

Hartmann, L. von., 1957, Adaptation in hole-nesting birds, *Evolution* **11**:339–347.

Hastings, A., and Rohlf, F. J., 1974, Gene flow: effect in stochastic models of differentiation, *Am. Nat.* **108**:701–705.

Hespenheide, H. A., 1971, Food preference and the extent of overlap in some insectivorous birds, with special reference to the Tyrannidae. *Ibis* **113**:59–72.

Hüe, F., and Etchecopar, R.-D., 1970, *Les Oiseaux du proche et du moyen Orient*, N. Boubée, & Cie, Paris.

Huey, R. B., and Pianka, E. R., 1974, Ecological character displacement in a lizard, *Amer. Zool.* **14**:1127–1136.

Huey, R. B., Pianka, E. R., Egan, M. E., and Coons, L. W., 1974, Ecological shifts in sympatry: Kalahari fossorial lizards (*Typhlosaurus*), *Ecology* **55**:304–316.

Huxley, J. S., 1942, *Evolution: the Modern Synthesis*, Allen and Unwin, London.

James, F. C., 1970, Geographic size variation in birds and its relationship to climate, *Ecology* **51**:365–390.

Janzen, D. H., 1973, Sweep samples of tropical foliage insects: effects of seasons, vegetation types, elevation, time of day, and insularity, *Ecology* **54**:687–708.

Jilke, A. and Leisler, B., 1974, Die Einpassung dreier Rohrsängerarten (*Acrocephalus schoenobaenus, A. scirpaceus, A. arundinaceus*) in ihre Lebensräume in bezug auf das Frequenzspectrum ihrer Reviergesange, *J. f. Orn.*, **115**:192–212.

Johnston, R. F., and Selander, R. K., 1964, House Sparrows: rapid evolution of races in North America, *Science* **144**:548–550.

Johnston, R. F., and Selander, R. K., 1971, Evolution in the house sparrow. II. Adaptive differentiation in North American populations, *Evolution* **25**:1–28.

Kilham, L., 1972, Reproductive behavior of White-breasted Nuthatches. II. Courtship. *Auk*. **89**:115–129.

Kilham, L., 1973, Reproductive behavior of the Red-breasted Nuthatch. I. Courtship, *Auk* **90**:597–609.

Kroodsma, R. L., 1974, Species recognition behavior of territorial male Rose-breasted and Black-headed Grosbeaks (*Pheucticus*), *Auk* **91**:54–64.

Lack, D., 1939, The behaviour of the Robin. Part 1. The life-history, with special reference to aggressive behaviour, sexual behaviour, and territory. Part II. A partial analysis of aggressive and recognitional behaviour, *Proc. Zool. Soc. London, Ser. A.* **109**:169–219.

Lack, D., 1944, Ecological aspects of species-formation in passerine birds, *Ibis* **86**:260–286.

Lack, D., 1947, *Darwin's Finches,* Cambridge University, Press.

Lack, D., 1971, *Ecological isolation in birds,* Blackwells: Oxford.

Lanyon, W. E., 1957, The comparative biology of the meadowlarks (*Sturnella*) in Wisconsin, *Nuttall Orinthological Club Publs.,* No. **1.**

Lanyon, W. E., 1960, The Middle American populations of the Crested Flycatcher, *Myiarchus tyrannulus, Condor* **62**:341–350.

Lanyon, W. E., 1963, Experiments on species discrimination in *Myiarchus* flycatchers, *Amer. Mus. Novit.,* No. **2126**:1–16.

LeGrand, H. E., 1973, Hydrological and ecological problems of karst regions, *Science* **179**:859–864.

Levin, D. A., 1970, Reinforcement of reproductive isolation: plants versus animals, *Am. Nat.* **104**:571–581.

Lewis, D. M., 1972, Importance of face-mask in sexual recognition and territorial behavior in the Yellowthroat, *Jack-Pine Warbler* **50**:98–109.

Löhrl, H., 1958, Das Verhalten des Kleibers (*Sitta europaea caesia* Wolf), *Zeits. für Tierpsychol.* **15**:191–252.

MacArthur, R. H., 1972, *Geographical Ecology.* Harper & Row, New York.

MacArthur, R. H., and Levins, R., 1964, Competition, habitat selection and character displacement in a patchy environment, *Proc. Natl. Acad. Sci.* **51**:1207–1210.

MacArthur, R. H., and Levins, R., 1967, The limiting similarity of coexisting species, *Am. Nat.* **101**:377–385.

May, R. M., 1972, *Stability and complexity in model ecosystems,* Princeton University Press, Princeton, N.J.

Mayr, E., 1963, *Animal species and evolution,* Belknap Press, Harvard University, Cambridge.

Mayr, E., 1974, The definition of the term disruptive selection, *Heredity* **32**:404–406.

Mengel, R. M., 1964, The probable history of species formation in some northern Wood Warblers (*Parulidae*), *Living Bird,* **3**:9–43.

Moreau, R. E., 1954, The main vicissitudes of the European avifauna since the Pliocene, *Ibis* **96**:411–431.

Moreau, R. E., 1955, Ecological changes in the Palaeartic region since the Pliocene, *Proc. Zool. Soc. London* **125**:253–295.

Morse, D. H., 1967, Foraging relationships of the Brown-headed Nuthatches and Pine Warblers, *Ecology* **48**:94–103.

Noble, G. K., 1936, Courtship and sexual selection of the Flicker (*Colaptes auratus luteus*), *Auk* **43**:269–283.

Noble, G. K., and Vogt, W., 1935, An experimental study of sex recognition in birds, *Auk* **42**:278–286.

Nørrevang, A., 1959, Double invasions and character displacement, *Vidensk. Meddr. dansk naturh. Foren.* **121**:171-180.

Norris, R. A., 1958, Comparative biosystematics and life history of the nuthatches *Sitta pygmaea* and *Sitta pusilla, Univ. Calif. Publ. Zool.,* **56**:119-300.

Odum, E. P., 1971, *Fundamentals of Ecology* 3rd ed., W. B. Saunders, Philadelphia.

Orians, G. H., 1968, *The study of life,* Allyn and Baker, Boston.

Orians, G. H., and Willson, M. F., 1964, Interspecific territories of birds. *Ecology* **45**:736-745.

Paludan, K., 1940, *Contributions to the ornithology of Iran, Danish Scientific Investigations in Iran,* Part II. 1-54, Munksgaard, Copenhagen.

Paludan, K., 1959, On the birds of Afghanistan, *Vidensk Meddr. Dansk natur. Foren.* **122**:1-332.

Panov, E., 1972, Interspecific hybridization and fate of hybrid populations as exemplified by two species of shrikes: *Lanius collurio* L., and *L. phoenicuroides* Schalow, *Zh. obshchei biol.* **33**:409-426.

Pimentel, D., Smith, G. J. C., and Soans, J., 1967, A population model of sympatric speciation, *Am. Nat.* **101**:493-504.

Pianka, E. R., 1974, Niche overlap and diffuse competition, *Proc. Natl. Acad. Sci.* **71**:2141-2145.

Richardson, F., 1942, Adaptive modifications for tree-trunk foraging in birds, *Univ. Calif. Publ. Zool.* **46**:317-368.

Sarudny, N., and Härms, M., 1923, Bermerkung über einige Vögel Persiens. III, *J.f. Orn.* **71**:398-421.

Schoener, T. W., 1965, The evolution of bill size differences among sympatric congeneric species of birds, *Evolution* **19**:189-213.

Schoener, T. W., 1969, Models of optional size for solitary predators, *Am. Nat.* **103**:277-313.

Schoener, T. W., 1974, Resource partitioning in ecological communities, *Science* **185**;27-39.

Scudder, G. G. E., 1974, Species concepts and speciation, *Can. J. Zool.* **52**:1121-1134.

Selander, R. K., 1971, Systematics and speciation in birds, *In* Farner, D. S., and King, J. R., (eds.) *Avian biology,* Vol. **1,** Academic Press, New York.

Selander, R. K., and Johnston, R. F., 1967, Evolution in the house sparrow. I. Intrapopulation variation in North America, *Condor,* **69**:217-258.

Short, L. L., 1971, The evolution of terrestrial woodpeckers, *Amer. Mus. Novit.* No. **2467**:1-23.

Smith, J. M., 1966, Sympatric speciation, *Am. Nat.* **100**:637-650.

Soans, A. B., Pimentel, D., and Soans, J. S., 1974, Evolution of reproductive isolation in allopatric and sympatric populations, *Am. Nat.* **108**:117-124.

Sokolov, S., Ya., 1954, *Derev'ya i Kustarniki SSSR (Trees and Shrubs of the USSR)* U.S.S.R. Academy of Sciences Press, Moscow.

Stepanyan, L., 1961, The specific independence of the two nuthatches *Sitta neumayer,* Michah, and *Sitta tephronota* Sharpe, *Trudy Inst. Zoologii AN Kaz. SSR.* **15**:139-142.

Störck, H. J., 1968, Morphologische Untersuchungen an Drosseln. Eine Analyse von Anpassungsstrukturen im Körperbau von sechs europäischen Arten der Gattung *Turdus* L., *Zeits. für Wissenschaftliche Zoologie,* Abt. A. **178**:72-185.

Streseman, E., 1920, *Avifauna Macedonica,* Munich.

Streseman, E., 1928, Die Vögel der Elburs-Expedition 1927, *J.f. Orn.* **76**:313-411.

Thielcke, G., 1966, Die Auswertung von Vogelstimmen nach Tonbandaufnahmen, *Die Vogelwelt* **87**:1-14.

Thielcke, G., 1969, Geographic variation in bird vocalizations. *In* Hinde, R. A., *Bird Vocalizations,* University Press, Cambridge.

Thoday, J. M., and Gibson, J. B., 1962, Isolation by disruptive selection, *Nature* **193**:1164-1166.

Trewartha, G. T., 1961, *The earth's problem climates,* Univ. Wisconsin Press, Madison.

Vaurie, C., 1950, Notes on some Asiatic Nuthatches and Creepers, *Amer. Mus. Novit.,* No. **1472**:1–29.

Vaurie, C., 1951, Adaptive differences between two sympatric species of Nuthatches (*Sitta*), *Proc. Tenth Int. Orn. Congr.*:163–166.

Vaurie, C., 1957, Systematic notes on Palearctic Birds, No. 29. The subfamilies Tichodromadinae and Sittinae, *Amer. Mus. Novit.,* No. **1854**:1–26.

Venables, L. S. V., 1938, Nesting of the nuthatch, *British Birds,* **32**:26–33.

Voous, K. H., 1960, *Atlas of European birds,* Thomas Nelson, London.

Voous, K. H., and van Marle, J. G., 1953, The distributional history of the Nuthatch *Sitta europaea* L., *Ardea* **41**:1–67.

Walker, T. J., 1974, Character displacement and acoustic insects. *Amer. Zool.* **14**:1137–1150.

Watson, G. E., 1962, Sympatry in Palearctic *Alectoris* partridges, *Evolution* **16**:11–19.

World Weather Records, 1967a. Vol. 2 Europe, Washington, D.C.

World Weather Records, 1967b. Vol. 4 Asia, Washington, D.C.

Wright, H. E., Jr., 1968, The natural environment of early food production in the Zagros mountains, *Science* **161**:334–339.

Wright, H. E., Jr., McAndrews, J. H., and van Zeist, W., 1967, Modern pollen rain in western Iran, and its relation to plant geography and Quaternary vegetational history, *J. Ecol.* **55**:415–433.

Zohary, M., 1963, On the geobotanical structure of Iran, *Bull Res. Coun. Israel, Sect. D* (*Botany*) 11D, **Suppl.** 113pp.

5

Chromosome Evolution in the Caudate Amphibia

ALESSANDRO MORESCALCHI*

Institute of Histology and Embryology
University of Naples
Naples, Italy 80134

INTRODUCTION

Although chromosome evolution constitutes an aspect of the more general evolution of organisms, exact relationships between the two phenomena, of different levels, are not always easy to assess (White, 1973). A knowledge of this aspect is particularly useful when one is concerned with groups of organisms, such as the present-day amphibians, whose evolutionary history is seen to be especially complex and full of gaps on account of the scarcity of fossils, the high degree of specialization attained by the living species, and the abundance of parallelisms and convergences, tending to mask the course of the phylogeny (Romer, 1966).

Modern techniques of investigation and the findings of molecular biology have enabled comparative karyology not only to obtain information on the evolutionary trends followed by the morphology of the chromosome set (karyotype), but also to assess the quantitative differentiation of the genome (in terms of variations in the nuclear DNA content) and to identify those nucleotide fractions which are most involved in this process.

Various nuclear constituents seem to confer upon the genome functions that are not strictly genetic: apart from the various kinds of proteins, the significance of which is still little known, a large fraction of the chromosome DNA, consisting of equal or similar nucleotide sequences repeated several times, may have structural or functional roles, thus being directly

* Research supported by a grant from C.N.R. (Rome).

subject to natural selection (Walker *et al.,* 1969; Yunis and Yasmineh, 1971). These repetitive fractions constitute most of the genome in certain organisms characterized by quantities of DNA that are decidedly greater than the average (compared with the other members of the same subphylum). This is so with the caudates, some species of which have very high percentages of repetitive DNA (over 80% in *Amphiuma,* according to Britten and Kohne, 1968). It is probable that the genome sizes of these organisms may, taken as a whole, assume adaptive roles; in fact, they are significantly correlated with various cytological, physiological, or ecological factors of this nature (cf. Stebbins, 1966, 1971; Sparrow *et al.,* 1972).

The specific localization on the chromosome of DNA fractions each having peculiar characteristics and functions (now demonstrated in various species with the techniques of chromosome "banding"; cf. Comings, 1974) proposes again on molecular bases the proofs—already obtained on structural and cytogenetic bases—of the particular degree of complexity of the chromosomes of the eukaryotic cells, which have a level of organization decidedly higher than that of a simple set of genes. For these reasons the evolution of the chromosome "phenotype" is largely independent of that of the genotype, whereas it seems to be canalized by various endo- and exocellular factors, probably in response to selective stimuli. The cytogenetic characteristics of living animals, like the somatic, physiological, and behavioral characteristics, are now seen as the result of a long evolutionary history in the course of which selection has favored those chromosome changes that have led to the karyological formulas typical of the living species (White, 1973).

In actual fact, the chromosome differentiation of many groups of organisms proceeds in a nonrandom manner, making use of only some of the possible chromosome rearrangements, which lead to karyotypes having the same characteristics in the whole of the group considered (principle of *karyotypic orthoselection,* after White).

As we shall see, the karyological evolution of the caudates constitutes one of the clearest demonstrations of this principle among the vertebrates. This fact, together with what has been seen regarding the possible adaptive nature of the genome size of these amphibians, may constitute proof of a particular plasticity of the genome of the caudates: in fact, groups of organisms that are less dependent upon the environmental chemicophysical characteristics, e.g., the amniotes, have a karyological situation that is very different and in certain cases, e.g., the mammals, less schematic than that of the caudates as well as of the other amphibians (Morescalchi, 1970, 1973).

The first part of this work will be devoted to a consideration of the differentiation of the karyotype (number, form, and relative size of the chromosomes); the second part to the quantitative differentiation of the genome of the caudates. Perhaps this subdivision of the problem is not a natural one, but it is founded on fairly precise factual data. While the karyotype follows evolutionary lines that are parallel to the phylogeny of the order; the genome size reveals an interspecific variability that is largely independent of this last phenomenon. Moreover, it is probable that the chromosome changes involved in the two processes of differentiation are partially different from one other.

EVOLUTION OF THE KARYOTYPE

Of the three orders of living amphibians, the caudates are the ones that have been studied most thoroughly from a karyological point of view. The chromosomes of over 30% of the species are known (a total of just over 300 species, according to Brame's list, 1967), belonging to 44 of the 54 genera of the order. In the 8 families of the caudates, only 3 genera of Hynobiidae, 1 of Ambystomatidae, 3 of Plethodontidae and 3 of Salamandridae present species that have not yet been studied karyologically (cf. Table I: it is incomplete as regards the Plethodontidae, since Kezer states that he has data on the chromosomes of 20 of the 23 genera of the family; cf. Kezer and Macgregor, 1971).

Statistically, this is a highly significant sample. It is therefore probable that the conclusions that can be drawn from it hold in general for the karyological evolution of the order and that further studies will alter these conclusions only slightly. (This appears possible in the case of certain primitive hynobiids and more likely in the case of tree-dwelling species of the vast and luxuriant radiation of the Bolitoglossini.)

In broad lines we can distinguish two levels of organization among the living caudates: a primitive level, typical of the Hynobiidae and of the related Cryptobranchidae (suborder Cryptobranchoidea), and an advanced level, represented by the remaining families, with the exception of perennibranchiates and perhaps unrelated proteids and sirenids, whose paedogenetic habitus makes it difficult to define their affinity with the other families, but which can generally be considered as intermediate between the above-mentioned levels of organization (Hecht, 1969).

Among the advanced families, Ambystomatidae and Plethodontidae (Ambystomatoidea) probably derive from the same ancestors, perhaps different from those from which the Salamandridae (Salamandroidea) are derived (Wake, 1966); the paedogenetic Amphiumidae seem more akin to

TABLE I.   Chromosome Numbers in the Caudata

Species	$2n$	Reference[a]
HYNOBIIDAE:		
Hynobius keyserlingii	62	1, 2
lichenatus	58	1
leechi	56	3
naevius	56	1, 4
kimurai	56/60	1, 5
okiensis	56	6
stejnegeri	56	6
nebulosus	56	3, 5, 4
tsuensis	56	5, 4
dunni	56	5, 4
nigrescens	56	1, 7
sadoensis	56	6
retardatus	40	1, 7
Pachypalaminus boulengeri	56	5
CRYPTOBRANCHIDAE:		
Cryptobranchus alleganiensis	62	8
Andrias davidianus	60	2
japonicus	64/60	9, 2
AMBYSTOMATIDAE:		
Rhyacotritoninae:		
Rhyacotriton olimpicus	26	10
Dicamptodontinae:		
Dicamptodon ensatus	28	10
Ambystomatinae:		
Ambystoma annulatum	28	11
laterale	28	12
texanum	28	11

[a] *References:* (1) Makino, 1932; (2) This paper; (3) Makino, 1934; (4) Morescalchi, 1973; (5) Sato, 1936; (6) Makino, 1951; (7) Azumi and Sasaki, 1971; (8) Makino, 1935; (9) Iriki, 1932; (10) Humphrey, 1958; (11) Morescalchi *et al.*, 1974; (12) Uzzell, 1963; (13) Wickbom, 1945; (14) Hauschka and Brunst, 1964; (15) Callan, 1966; (16) Uzzell, 1964; (17) Donnelly and Sparrow, 1965; (18) Kezer, unpublished (cf. in the text); (19) Kezer, 1964; (20) Kezer and Macgregor, 1971; (21) Gallien *et al.*, 1965; (22) Lacroix, 1970; (23) Makino, 1962; (24) Ferrier and Beetschen, 1973; (25) Magalhaes, 1942; (26) Kessous *et al.*, 1968; (27) Ullerich, 1970; (28) Mancino and Barsacchi, 1966a; (29) Kawamura and Utsonimiya, 1957; (30) Jaylet, 1966; (31) Mancino and Barsacchi, 1966b; (32) Mancino, 1965; (33) Bhatnagar, unpublished; (34) Callan and Lloyd, 1960b; (35) Mancino and Barsacchi, 1966c; (36) Mancino and Barsacchi, 1969; (37) Nardi and Mancino, 1971; (38) Mancino and Nardi, 1971; (39) Barsacchi *et al.*, 1970; (40) Nardi *et al.*, 1973; (41) Fankhauser, 1938; (42) Gall, 1954; (43) Seto *et al.*, 1964; (44) Kezer *et al.*, 1965; (45) Kezer, 1962; (46) Morescalchi and Olmo, 1974; (47) Morescalchi, 1971; (48) Leon and Kezer, 1974.

[b] Sex chromosomes, male heterogamety.

[c] Heteromorphism in one lampbrush bivalent: female heterogamety.

[d] Supernumerary (or accessory) chromosomes.

TABLE I. (Continued)

Species	$2n$	Reference[a]
maculatum	28	13, 11
macrodactylum	28	10, 4, 11
opacum	28	11
talpoideum	28	11
tigrinum	28	11
gracile	28	10
mexicanum	28	13, 14, 15
jeffersonianum	28	12
platineum	$(3n = 42)$	16
tremblayi	$(3n = 42)$	16
PLETHODONTIDAE:		18
Desmognathinae:		
Desmognathus fuscus	28	19, 11
monticola	28	2
Leurognathus marmoratus	28	2
Plethodontinae:		
Hemidactylini:		
Eurycea bislineata	28	2
longicauda	28	2
lucifuga	28	11
Gyrinophilus porphyriticus	28	11
Pseudotriton ruber	28	11
Plethodontini:		
Aneides aeneus	28	11
ferreus	28	19
lugubris	28	2
Ensatina eschscholtzii	28	11
Plethodon cinereus	28	11, 20
glutinosus	28	2
jordani	28	19
vehiculum	28	19
Bolitoglossini:		18
Batrachoseps attenuatus	26	11, 18
wrighti	26	18
Hydromantes genei	28	18
brunus	28	18
italicus	28	18, 19
platycephalus	28	18
shastae	28	18, 19
Bolitoglossa subpalmata	26	18, 19
Chiropterotriton abscondens	26[b]	18
dimidiatus	26	18
multidentatus	26	18

TABLE I. (Continued)

Species	$2n$	References[a]
Oedipina bonitaensis	26^b	18
poelzi	26^b	18
syndactyla	26^{b+d}	18
uniformis	26^b	18
Pseudoeurycea smithi	26^{+d}	18
werleri	26^{+d}	18
Thorius dubitus	26^b	18
pennatulus	26^b	18
SALAMANDRIDAE:		
Pleurodeles waltlii	24^c	21, 22
poireti	24^c	22
Tylototriton andersoni	24	23
verrucosus	24	24, 4
Chioglossa lusitanica	24	25
Salamandra atra	24	13, 4
salamandra	24	26, 27
Salamandrina terdigitata	24	28
Cynops pyrrhogaster	24	29
Euproctus asper	24	30
montanus	24	31
platycephalus	24	32
Neurergus crocatus	24	33
Paramesotriton hongkongensis	24	2
Triturus alpestris	24	27
cristatus	24^c	13, 34
helveticus	24	35
italicus	24	36
marmoratus	24^c	37, 38
vulgaris	24	13, 27, 39, 40
Notophthalmus viridescens	22	41, 42, 2
Taricha granulosa	22	41, 4
rivularis	22	2
torosa	22	2
AMPHIUMIDAE:		
Amphiuma means	28	11
tridactylum	28	17
PROTEIDAE:		
Necturus maculosus	38	43, 44, 2
Proteus anguinus	38	44, 45, 2
SIRENIDAE:		
Pseudobranchus striatus (4n?)	64	46
Siren intermedia (4n?)	46	4, 46, 47, 48
lacertina (4n?)	52	4, 46

the members of the former than to those of the latter suborder (Regal, 1966) and will be considered here among the Ambystomatoidea (giving importance, as we shall see, to the karyological findings).

Let us look then at the lines followed by the karyological evolution at the level of the three suborders listed below and of the two families of perennibranchiate urodeles.

Regarding the terminology, in a work of a general nature, such as the present report, it does not seem necessary to define the chromosome morphology in detail, according to the classification proposed by Levan *et al.* (1964), used, in fact, by the present author in his work of 1973. Consequently, the term *metacentric* will be used to denote chromosomes with two arms, even if these are of unequal size, and thus the submetacentrics are included, as well as part of the subtelocentrics of the above classification. *Acrocentric* will denote those having only one arm (cf. White, 1973).

Cryptobranchoidea

The most generalized forms among the living caudates belong to the Hynobiidae, at present spread throughout the Palearctic region, and to the Cryptobranchidae, paedogenetic derivatives of the former family, with an Asiatic (*Andrias*) and a Nearctic genus (*Cryptobranchus*) (Regal, 1966).

Among the hynobiids the chromosomes are known in various species of *Hynobius* (perhaps the central genus in the phylogeny of the family) and of the monotypical *Pachypalaminus,* which were studied in the 1930's by Japanese authors (Table I and references). Among the more generalized continental forms, *Hynobius* (*Salamandrella*) *keyserlingii* has $2n = 62$, while *H. leechi* and most of the island species (including *Pachypalaminus boulengeri*) have 56 chromosomes, with the exception of a population (or perhaps another species?) of *H. kimurai* with 60 chromosomes; *H. lichenatus* would appear to have 58 chromosomes, while *H. retardatus* would have $2n = 40$.

The present author (Morescalchi *et al.,* 1970; Morescalchi, 1971, 1973) and Azumi and Sasaki (1971) have restudied the chromosomes of *H. keyserlingii, H. retardatus,* and some species of *Hynobius* having 56 chromosomes, verifying the chromosome numbers reported by previous authors for these species and defining, with the use of modern karyological techniques, the morphology of the individual pairs of homologous chromosomes of each karyotype.

In all the species [comprising *Batrachuperus mustersi,* with more than 62 chromosomes (unpublished) except *H. retardatus*] two categories of chromosomes are distinguished: "normal" ones, comparable in size to the chromosomes of other groups of caudates, metacentric or acrocentric, and of various sizes (large, small, and medium); and microchromosomes, very small dotlike elements, about one micron in length, acrocentric and comparable to the same elements that characterize the karyotype of various sauropsids. According to a calculation made by Olmo (unpublished), the microchromosomes contain about $\frac{1}{25}$ of the DNA of the largest chromosomes of the Cryptobranchoidea.

The number of the microchromosomes is greater in the hynobiids having a higher overall number of chromosomes: there are about 12 pairs of them in *H. keyserlingii* (in some cases the distinction between small chromosomes and microchromosomes is unclear), and 7-8 pairs in the species having 56 chromosomes; *H. retardatus* seems to be lacking in typical microchromosomes (Figs. 2-4).

The interspecific differences in the karyotype of these urodeles concern not only the number of the microchromosomes but also the number of acrocentric chromosomes, which are more numerous in *H. keyserlingii* (12 pairs, as opposed to 7 pairs of large metacentric chromosomes) and reduce to only a few pairs (5-6) in the 56-chromosome species. In *H. retardatus*, the species with the lowest number of chromosomes, the acrocentric chromosomes are reduced to a few pairs, all consisting of small chromosomes (Fig. 10).

The primitive nature of the continental species of *Hynobius* and their greater karyological resemblance to the cryptobranchids (and also, as we shall see, to the Anura and Apoda of the more primitive families) suggest that the formulas with a higher number of chromosomes (such as that of *H. keyserlingii*, with $2n = 62$) are the most generalized ones in the present-day species of the family. It would appear that from these are derived the 56-chromosome formulas and finally the 40-chromosome formula of *H. retardatus* through a progressive numerical decrease in the chromosomes (perhaps due to unequal translocations), especially in the acrocentric and microchromosomes. The evolutionary tendency in the morphology of the karyotype of the family would therefore seem to consist of the acquisition of formulas lacking in microchromosomes and mostly made up of metacentric chromosomes.

As regards the cryptobranchids, *Cryptobranchus alleganiensis* ($2n = 62$) would appear to have a karyological formula that is almost identical to that of *H. keyserlingii* (Makino, 1935). The chromosome number of *Andrias* (*Megalobatrachus*) *japonicus*, calculated as being equal to 64 by Iriki (1932), is in fact 60, both in this species and in the related *A. davidianus* (Morescalchi, 1973; this paper). The karyotype of these last two species also shows various similarities with that of *H. keyserlingii* and of *Batrachuperus*. There are 6 pairs of large or medium metacentric chromosomes, 10 pairs of medium or small acrocentric chromosomes, and 14 pairs of microchromosomes and the set is more or less the same in the two species of *Andrias* (Figs. 1, 10). These data confirm the hypotheses that these caudates may be derived from primitive hynobiid forms, and have preserved the principal karyological characteristics of the latter.

Ambystomatoidea

In this suborder are grouped three families having an essentially Nearctic diffusion and varying complexity: the Ambystomatidae, of direct hynobiid derivation; the Plethodontidae, of probable ambystomatid or protoambystomatid derivation; and the paedogenetic Amphiumidae, possibly derived from some ambystomatid or ambystomatid–plethodontid stock (Regal, 1966; Wake, 1966; Salthe, 1967; Estes, 1969).

According to Tihen (1958) the Ambystomatidae comprise three subfamilies, which have differentiated independently from a protoambysto-matid stock: the monotypical Rhyacotritoninae and Dicamptodontinae, and the Ambystomatinae, with the genera *Ambystoma* (rich in species and including the most generalized forms in the subfamily) and *Rhyacosiredon,* derived from *Ambystoma* and not yet studied karyologically.

With the exception of *Rhyacotriton olympicus* ($2n$ = 26), all the ambystomatids so far studied (*Dicamptodon ensatus* and various species of *Ambystoma*) have $2n$ = 28, with some triploid species of *Ambystoma* with 42 chromosomes (Table I).

The chromosomes of the ambystomatids are all metacentric (with two arms); *Rhyacotriton* seems to have two chromosomes less, among the small ones, compared with the remaining species studied; in no species of the family have microchromosomes been found.

The genus *Ambystoma* has been subdivided by Tihen into three subgenera: *Ambystoma, Linguaelapsus,* and *Bathysiredon,* the first being subdivided, in its turn, into certain groups of species. Figure 7 shows the chromosomes of some species of the first two subgenera and of three groups of species of the subgenus *Ambystoma* (including *A. tigrinum,* perhaps the most generalized species of the subfamily). There exist certain interspecific differences in the morphology of the various pairs of homologous chro-mosomes, especially the smallest ones. However, these differences are also noted within the same group of species (e.g., between *maculatum* and *macrodactylum* of the group *maculatum,* subgenus *Ambystoma*), so that, at present, it is not possible to demonstrate a direct correlation between them and the systematics proposed for these caudates.

Although probably of direct hynobiid derivation (in the past they were included in the same family), the present-day ambystomatids are karyologi-

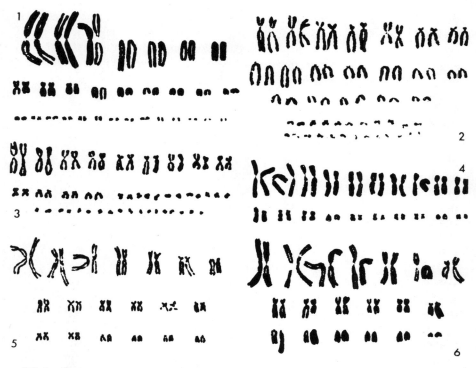

FIG. 1.   Karyotype of *Andrias japonicus* ♀ (2n = 60).
FIG. 2.   Karyotype of *Hynobius keyserlingii* ♀ (2n = 62).
FIG. 3.   Karyotype of *Hynobius nebulosus* ♂ (2n = 56).
FIG. 4.   Karyotype of *Hynobius retardatus* (2n = 40) (from Azumi and Sasaki, 1971).
FIG. 5.   Karyotype of *Proteus anguinus* ♂ (2n = 38).
FIG. 6.   Karyotype of *Necturus maculosus* ♂ (2n = 38).

cally very different from the hynobiids including the species (*H. retardatus*) that is karyologically most differentiated in both the morphology and the number of chromosomes. The ambystomatids seem to have pursued, in the course of their phylogeny, the processes of numerical reduction in the chromosomes of the karyotype and those of elimination of the acrocentric chromosomes that characterize the karyological differentiation of the hynobiids. The 26–chromosome formula of *Rhyacotriton* must be considered as derived with respect to the 28–chromosome formula of the other members of the family, because the latter is (numerically) closer to the formulas of the hynobiids and, moreover, is present in the more generalized forms of the ambystomatids.

The Plethodontidae, with the greatest number of species in the order, are specialized for various types of environment; Wake (1966) subdivides them into two subfamilies of unequal size: the Desmognathinae, with only a few genera, which differentiated at an early date from the central stock of the family; the Plethodontinae, with the most generalized and other highly specialized forms. The second subfamily includes the tribes of Hemidactylini, Plethodontini, and Bolitoglossini (the last with various Neotropical species and one supergenus, *Hydromantes*, with Californian and Mediterranean species).

The karyology of the Plethodontidae has for some years been the subject of research by J. Kezer, who has studied the species of 20 of the 23 genera of the family. Unfortunately, most of this author's results are still unpublished, though they have been the subject of bibliographical references since 1949 (cf. Matthey, 1949; Beatty, 1964; Kezer, 1964; Mancino, 1965; Wake, 1966; Kezer and Macgregor, 1971). According to Kezer, the various species studied have $2n = 28$, with the exception of various Bolitoglossini: in this tribe, all the species of *Hydromantes* would appear to have 28 chromosomes also, whereas those of the other two supergenera (*Batrachoseps* and *Bolitoglossa*) would have 26 chromosomes, two small ones less than the other plethodontids. Various Bolitoglossini of Latin America, moreover, would appear to have markedly heteromorphous sex chromosomes in the male line and, in some populations, accessory or supernumerary chromosomes (Table I).

Studies carried out by other authors are in agreement with these results; the present author too has studied the chromosomes of some species of Plethodontidae belonging to the main subdivisions of the family (Fig. 7). With the exception of the Bolitoglossine *Batrachoseps attenuatus*, with $2n = 26$, the remaining species studied have 28 chromosomes, with elements that are all metacentric, apart from three pairs of homologues in *Aneides aeneus*, consisting of acrocentric chromosomes (Morescalchi *et al.*, 1974). It is probable that in the case of *Aneides* the acrocentrics are of secondary derivation (through pericentromeric inversions in the metacentric chromosomes), since these elements are lacking in the other members of the family (including the primitive Hemidactylini) and are also absent in *Plethodon*, from which *Aneides* is probably derived.

In conclusion, and for the same reasons already seen with regard to the ambystomatids, among the plethodontids the 28–chromosome formula must also be considered the most generalized one, whereas the 26–chromosome formula of advanced groups of the family would appear to be derived from it. The appearance of sex chromosomes and supernumerary chromosomes in certain tree-dwelling Bolitoglossini must be considered as a further karyological specialization of these forms, probably still in the evolutionary stage (Wake, 1966).

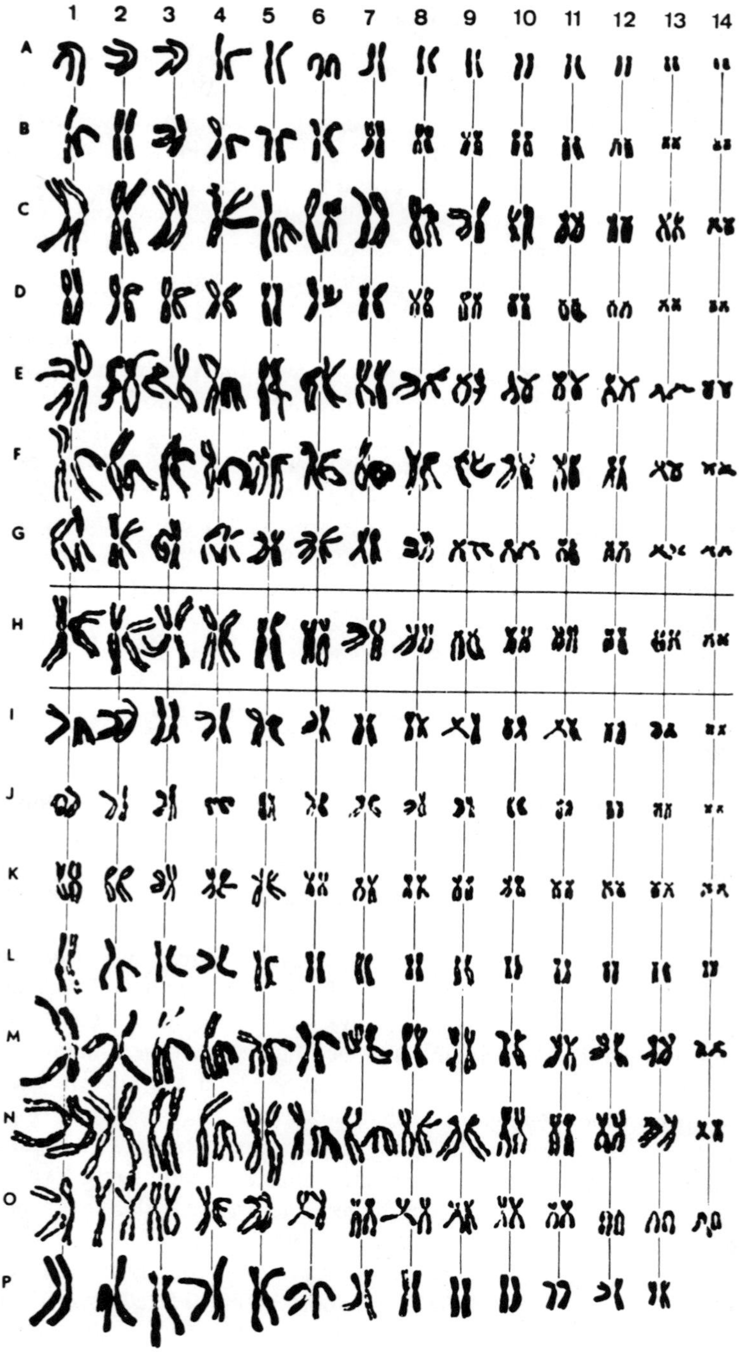

1 2 3 4 5 6 7 8 9 10 11 12 13 14
A
B
C
D
E
F
G
H
I
J
K
L
M
N
O
P

The monogeneric Amphiumidae, once considered akin to the Salamandridae, today seem probably more connected to the other two families of Ambystomatoidea. Two of the three species (or subspecies) of *Amphiuma* (*tridactylum* and *means*) have been studied karyologically (Table I): they have 28 chromosomes, all metacentric and of gradually decreasing length in the various pairs. The karyotype of *Amphiuma*, apart from the quantitative aspects of the DNA, appears more or less equal to that of certain *Ambystoma* and of various plethodontids (Fig. 7) (Morescalchi, 1973; Morescalchi *et al.*, 1974). The karyological data therefore agree with the most recent proposals on the systematics of these caudates, even if there remains the possibility of parallelisms in the karyological differentiation between this family and the others of the suborder.

Salamandroidea

The higher caudates of this suborder, sometimes subdivided into two families, are in most cases included in a single family (Salamandridae), of uncertain origin; according to the opinion most current, it might be derived from some hynobiid stock, perhaps different from the one that has given rise to the Ambystomatidae.

The systematics of the Salamandridae is disputed (cf. von Wahlert, 1953); in general, two subfamilies are distinguished: the Pleurodelinae, more aquatic ("newts"), and the Salamandrinae, more terrestrial (true "salamanders"). However, recent studies have cast doubts on this classification: according to Wake and Özeti (1969), it is possible to distinguish in the family two very primitive genera, which separated independently and at an early date from the rest of the Salamandridae (*Tylototriton* and *Pleurodeles*), and two groups of species of different specialization, one comprising the genera *Chioglossa*, *Salamandra* and *Salamandrina*, the other the remaining genera, including the Nearctic ones.

From the karyological point of view, the Palearctic salamandrids show considerable homogeneity with one another: all the species studied present 24 metacentric chromosomes of regularly decreasing length (with the excep-

FIG. 7. Karyotypes of some Ambystomatidae (A–G), Amphiumidae (H) and Plethodontidae (I–P), suborder Ambystomatoidea. A and B: *Ambystoma texanum* and *A. annulatum* (subgenus *Linguaelapsus*); C–D: *A. maculatum* and *A. macrodactylum* (subgenus *Ambystoma, maculatum* group); E–F: *A. opacum* and *A. talpoideum* (subgenus *Ambystoma, opacum* group); G: *A. tigrinum* (subgenus *Ambystoma; tigrinum* group). H: *Amphiuma* (*means*) *tridactylum*. I: *Desmognathus fuscus* (Desmognathinae); J–K–L: *Gyrinophilus danielsi, Pseudotriton ruber* and *Eurycea lucifuga* (Plethodontinae, Hemidactyliini); M–N–O: *Plethodon cinereus, Ensatina eschscholtzi* and *Aneides aeneus* (Plethodontinae, Plethodontini); P: *Batrachoseps attenuatus* (Plethodontinae, Bolitoglossini) (Morescalchi *et al.*, 1974).

tion of a greater interval between the 8th and 9th pair of homologous chromosomes), with formulas that differ very little even at intergeneric level (Fig. 8; cf. Morescalchi, 1973). The most considerable differences between these karyotypes concern the absolute size of the chromosomes (associated with the quantity of DNA).

The Nearctic genera *Notophthalmus* and *Taricha*, with $2n = 22$, have one pair of chromosomes less than the other salamandrids and a metacentric chromosome set that is even more homogeneous in the relative size of the individual pairs than that of the Palearctic species.

It is interesting to note that viable hybrids, albeit probably sterile, have been obtained between *Pleurodeles waltli* and *Tylototriton verrucosus* (Ferrier and Beetschen, 1973); this fact provides evidence for a considerable affinity between these primitive salamandrids, since, among the caudates, viable hybrids generally occur only between species of the same genus, as in the case of *Hynobius, Triturus* and *Taricha* (cf. Benazzi, 1948; Kawamura, 1953; Twitty, 1964; White, 1973).

The numerous studies so far carried out on the fine morphology of the oocyte lampbrush chromosomes in various species of *Triturus* and *Pleurodeles* suggest that concealed behind an apparent interspecific homogeneity in the morphology of the somatic set lies a certain genetic heterogeneity, since, at the level of loci in homologous position, different species and sometimes subspecies synthesize different gene products (cf. Callan and Lloyd, 1960a; Mancino and Barsacchi, 1966c, 1969; Lacroix, 1968a; more general references in Mancino, 1973; White, 1973).

Moreover, certain of these authors have shown the presence of sections that are heteromorphic as regards the length and shape of the loops in the two homologous chromosomes of a lampbrush bivalent in *Triturus cristatus, T. marmoratus, Pleurodeles waltli,* and *P. poireti*; it is considered that this phenomenon constitutes the proof of the presence of sex chromosomes (female heterogamety) in these species (Callan and Lloyd, 1960b; Lacroix, 1968b, 1970; Mancino and Nardi, 1971).

These last findings, together with those regarding the number of chromosomes in the Salamandridae (the lowest among the caudates), make this family the most specialized one, from the karyological point of view, among the present-day Urodela; within this family, the Nearctic species would appear to be the most differentiated ones.

As regards the sex chromosomes, however, it should be said that, unlike the male line, the female line seems to have been very little studied outside the salamandrids. It is therefore possible that studies on the lampbrush chromosomes of other families may provide fresh information on this problem: in this respect, Leon and Kezer (1974) state that also one of the

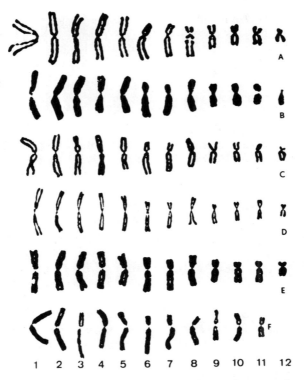

FIG. 8. Haploid karyotypes of species of the main genera of Salamandridae. A: *Salamandra salamandra*; B: *Tylototriton verrucosus*; C: *Pleurodeles waltlii*; D: *Euproctus asper*; E: *Triturus alpestris*; F: *Taricha granulosa*. (Morescalchi, 1973).

oocyte bivalents of *Siren intermedia* (Sirenidae) displays a heteromorphic section. However, it is remarkable that families of the same order, such as Plethodontidae and Salamandridae, have developed chromosome differentiation mechanisms connected with sex in the one case in the male and in the other in the female line [among the tetrapods, a similar case is found in the Sauria (cf. Gorman, 1973)].

Perennibranchiate Families

The two families consisting of permanently larval forms, Proteidae and Sirenidae, are the most problematical in the phylogeny of the Caudata.

The Proteidae, with the genera *Proteus* and *Necturus*, are often associated with the Salamandroidea and are sometimes included, as derived forms, in the same suborder; however, paleontological findings offer evidence that they are very old (Estes, 1969). Also various doubts exist on the real affinity between the two genera of proteids. Hecht (1957) suggests that the similarities between *Proteus* and *Necturus* are due to evolutionary parallelisms; other authors classify them in two different families (Wake, 1966; Salthe, 1967).

Proteus anguinus and *Necturus maculosus* both have $2n = 38$, with chromosomes of different size, mostly metacentric; there are five pairs of acrocentrics in *Proteus* and six in *Necturus* (Figs. 5, 6, and 10). Between the two species there are differences in the chromosome morphology greater than those described by Kezer *et al.* (1965), who have compared the mitotic chromosomes of one species with the meiotic ones of the other; however, from a karyological point of view, the two genera show greater similarities between each other than with other caudates studied so far. While both living proteids seem to have attained similar degrees of karyotype differentiation, this report is not necessarily contrary to Hecht's suggestion: they could be derived from different ancestors having a higher number of chromosomes.

Indeed, the distance between the karyotype of *Necturus* and that of *Proteus* is the same or greater (taking account especially of the genome size and chromosome morphology) than that between *Amphiuma* and some plethodontids (Fig. 10). However, in the actual stage of karyological research it is not possible to make a distinction between true homologies and parallelisms or convergences in the chromosome evolution (to be sure, this is not possible also for many other morphological characters of the amphibians): thus, the inclusion of both *Proteus* and *Necturus* in the same family Proteidae is simply the most parsimonious solution, from a karyological point of view as well.

As regards the number and form of the chromosomes, the two genera of proteids take an intermediate place between the hynobiids and the ambystomatids; they might constitute early derivations from a hynobiid stock that is perhaps older than the one that has given rise to the ambystomatoids (Fig. 14) (Morescalchi, 1973).

The karyological situation of the sirenids is particularly complex. These caudates, about the origins and phyletic affinities of which hardly anything is known, include the monotypic *Pseudobranchus* and *Siren,* with two species. The chromosome numbers of the three species are very different from one another: *P. striatus* has 64 chromosomes, *S. lacertina* and *S. intermedia* have 52 and 46, respectively (Morescalchi, 1971, 1973; Morescalchi *et al.,* 1974; Leon and Kezer, 1974). In spite of the relatively high numbers, none of the three species possesses microchromosomes and only in the first two are there acrocentrics. All the chromosomes of *S. lacertina* and nearly all those of *P. striatus* can be grouped into quartets of elements morphologically very similar to one another, such that it may be suggested that these two species are tetraploid (Fig. 9). In both cases, in the male line (and in the female line in *S. intermedia*) bivalents and not tetravalents are present. This fact may suggest (though the diagnosis on these bases is very imprecise) that the possible tetraploidy may have been acquired through interspecific hybridization (allopolyploidy). In view of the probable antiquity of the present-day sirenids, it is also possible that the presence of bivalents in their germinal line may constitute an indication of the fact that their tetraploid sets are evolving toward diploidy, through the diversification of the single pairs of each quartet (cf. Ohno, 1970).

The hypothesis of tetraploidy appears difficult to apply in the case of *S. intermedia*, which, having 46 chromosomes, remains with one pair of elements left out of a grouping by quartets, although this seems possible for most of the chromosomes (Fig. 9). Leon and Kezer also (1974), while not advancing this hypothesis, offer an idiogram of the chromosomes of *S. intermedia* in which various elements of different pairs seem morphologically very similar to one another (perfect equality is rarely achieved, even among homologues of one and the same pair in these organisms with large chromosomes; this can be seen, for example, in the idiogram of the diploid set of a species of *Salamandra*, presented by Kessous *et al.*, 1968).

Even with the uncertainties involved, the theory of a differentiation of the three species of this family through polyploidy is particularly attractive, because (as we shall see better later) it would enable the karyotype of these caudates to be likened to that of other families of the order. In fact, the theories about an evolution of the sirenids separate from the rest of the caudates (Goin and Goin, 1962) have not so far been confirmed by the systematists (cf. Estes, 1965; Wake, 1966).

Toward Symmetrical Karyotypes

From this initial glance at the data on the morphology of the karyotype of the caudates, and neglecting for the present some quantitative aspects, it is possible to glimpse lines of karyological differentiation that are closely correlated with the behavior of the phylogeny of various families of the order. Applying these concepts to the perennibranchiate families, the karyological findings may contribute to a clarification of the systematic position of these problematical urodeles.

According to a terminology that is most used in plant karyology (cf. Stebbins, 1971) but is particularly effective in the case of the caudates, the

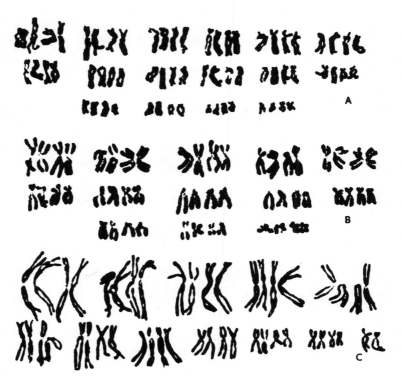

FIG. 9. Karyotypes of the three living Sirenidae, arranged in quartets of chromosomes (save one pair of *Siren intermedia*) following the hypothesis of tetraploidy. A: *Pseudobranchus striatus*; B: *S. lacertina*; C: *S. intermedia*.

karyological formulas of the cryptobranchoids, which are numerically high (2*n* between 62 and 40), they may be defined as *bimodal* (owing to the presence of two categories of chromosomes, the "normal" ones and the microchromosomes) and as *asymmetrical* (owing to the simultaneous presence of metacentric and acrocentric chromosomes, of very different sizes). The degree of asymmetry, which can be estimated from the ratio between acrocentric and metacentric chromosomes, tends to decrease as the total number of chromosomes within the cryptobranchoids is reduced, also in parallel with the reduction in the number of microchromosomes. The forms that are possibly most differentiated from this point of view (*Hynobius retardatus*; cf. see previous discussion on Cryptobranchoidea) have low chromosome numbers (2*n* = 40), are lacking in microchromosomes (*unimodal* karyotypes), and have a minimum ratio between acrocentric and metacentric chromosomes (Fig. 10). Within the suborder, therefore, there exists a tendency toward the production of symmetrical karyotypes.

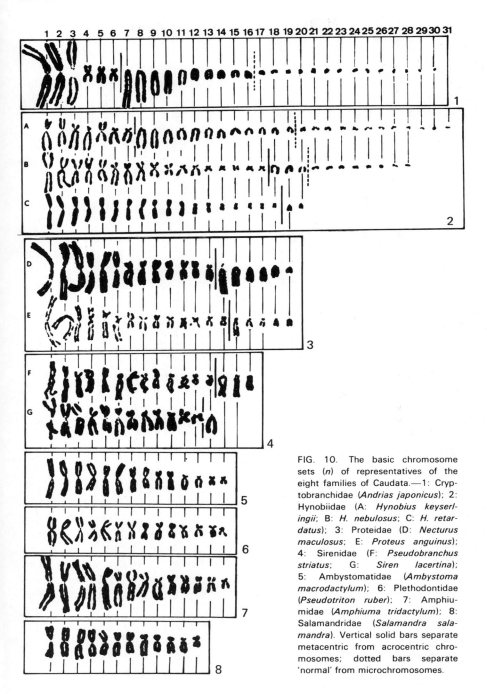

FIG. 10. The basic chromosome sets (n) of representatives of the eight families of Caudata.—1: Cryptobranchidae (Andrias japonicus); 2: Hynobiidae (A: Hynobius keyserlingii; B: H. nebulosus; C: H. retardatus); 3: Proteidae (D: Necturus maculosus; E: Proteus anguinus); 4: Sirenidae (F: Pseudobranchus striatus; G: Siren lacertina); 5: Ambystomatidae (Ambystoma macrodactylum); 6: Plethodontidae (Pseudotriton ruber); 7: Amphiumidae (Amphiuma tridactylum); 8: Salamandridae (Salamandra salamandra). Vertical solid bars separate metacentric from acrocentric chromosomes; dotted bars separate 'normal' from microchromosomes.

At the level of the Ambystomatoidea, this process seems to be largely achieved: the three families have a low number of chromosomes ($2n = 28$), they lack microchromosomes and their chromosomes are all metacentric, decreasing in size in a fairly regular manner (most regular in the pletho-dontids). These karyotypes may be considered *symmetrical* (and hence also unimodal), with a degree of symmetry that is perhaps greater in the pletho-dontids than in the ambystomatids (except for the peculiar and perhaps derived case of *Aneides*, with some acrocentrics). In these two families, a further advance in this tendency has given rise to the 26-chromosome karyotypes of *Rhyacotriton* and many Bolitoglossini.

Among the salamandrids we find, as regards the karyology, the most differentiated forms in the order, with the lowest chromosome numbers and the most symmetrical formulas. From this point of view, the transition between karyotypes similar to those of the Ambystomatoidea and those typical of the salamandrids would require only a few chromosome rear-rangements of the same type as those which, in the suborder, have led from the 28- to the 26-chromosome formulas (Fig. 10). Also in this case the most parsimonious hypothesis would regard the salamandrids as forms derived from some stock of Ambystomatoids. However, this solution should be viewed with great caution, given the existence of parallelisms in the karyo-logical evolution also within the different families of this suborder. Nevertheless, the uncertainty regarding the attribution of certain families (the amphiumids) to one or other grouping of higher caudates suggests that the phyletic distance between ambystomatids and salamandrids may not be so very great.

From the karyological point of view, the interval between Cryp-tobranchoidea and Ambystomatoidea seems greater than that between the latter and the salamandrids. The first interval is partly filled by the karyo-logical situation of the proteids.

These perennibranchiate caudates possess unimodal but still asym-metrical karyotypes, with chromosome numbers intermediate between those of various hynobiids and those of the ambystomatids. The present-day proteids have therefore retained karyological formulas of the type of those that have presumably also characterized the ancestors of the ambysto-matids. However, considering the karyotype of *Hynobius retardatus*, it seems probable that the most important phases in the genesis of a sym-metrical karyotype (such as that of the ambystomatids) have already been effected at the hynobiid level; the proteids might have been differentiated in parallel with the ancestors of the Ambystomatoidea and not necessarily along the same phyletic line.

Also in the morphology of the karyotype, the family that can be placed with the least certainty remains that of the sirenids. Assuming as valid the hypothesis that these caudates have been differentiated from forms that at some moment in their evolution acquired a tetraploid set, it may be noted that the basic set of *Pseudobranchus striatus* (n = 16, $4n$ = 64), unimodal and asymmetrical, closely resembles that of the proteids (n = 19) (Fig. 10). On account of the greater number of chromosomes, *P. striatus* might be the most primitive species among the living sirenids, as regards the karyology. *Siren lacertina* (n = 13, $4n$ = 52), though it has a lower basic number of chromosomes than most of the Ambystomatoidea (n = 14), has a karyotype that is still asymmetrical. Only *S. intermedia,* the species with the least number of chromosomes (46) and the most "diploid" of the family, has a symmetrical karyotype like that of the higher caudates.

In general, the sirenids might be derived from some hynobiid stock that has attained a level of karyological evolution close to that of the present-day proteids but became tetraploid at an early date. The intrafamily differentiation of this tetraploid set would be associated with a reduction in the number of chromosomes and would tend, as in the rest of the caudates, toward the formation of symmetrical karyotypes: the karyotypes of the living sirenids would represent three different stages in this evolutionary process. Hence, also the karyological differentiation of the sirenids may have followed trends of the type already found in the remaining families of the order.

These evolutionary patterns, common to all the caudates, have obviously required the same types of chromosome changes in the course of the phylogeny of the order: the precise manner in which these events have been canalized probably has a selective value (karyotypic orthoselection, after White, 1973).

QUANTITATIVE EVOLUTION OF THE NUCLEAR DNA

DNA and Phylogeny

In cells at the same degree of ploidy and in the presynthetic stage (G_1) the quantity of nuclear DNA ($2c$ in the diploid cells) is constant and is a species-specific characteristic. In some groups of organisms this parameter is relatively stable at interspecific level, in others it is variable. Among the tetrapods, each of the various orders and classes of the Amniota has fairly fixed quantities of nuclear DNA (with the possible exception of Reptiles), whereas the Amphibians sometime display considerable interspecific variations in this quantity (reviews in Goin and Goin, 1968; Morescalchi, 1970; Ohno, 1970; Bachmann *et al.,* 1972).

Among the anurans the quantity of DNA varies from about 2

picograms per nucleus (pg/N) in certain American and Australian deserticolous forms to over 30 pg/N in tropical arboricolous forms. Greater quantities may possibly be expected among various polyploid species, not yet studied with regard to this characteristic, for example, the hexaploid species of *Xenopus* and the tetraploid species of African bufonids and ranids (Tymowska and Fischberg, 1973; Bogart, unpublished). Among the Apoda, few species have been investigated; their DNA ranges from about 7 to 50 pg/N; also in this case it is likely that further studies will extend this range. Among the caudates the smallest quantities of DNA are found in certain hynobiids and plethodontids (about 30 pg/N), while the largest quantities are found in species of paedogenetic families, with values which, according to certain authors, exceed 200 pg (cf. Goin *et al.*, 1968; Sexsmith, 1968; Bachmann, 1970; Beçak *et al.*, 1970; Beçak and Beçak, 1974; Morescalchi, 1970, 1973; Olmo, 1973, 1974).

With the exception of the three existing genera of Dipnoi, which have quantities of DNA ranging between 160 and 284 pg/N (and are therefore similar to those of certain paedogenetic caudates), the other vertebrates do not exceed 10–15 pg of DNA, displaying lower values than most of the living amphibians and a lower interspecific variability (though the latter characterizes the higher Amniota in particular) (besides the above mentioned authors, cf. Mirsky and Ris, 1951; Hinegardner, 1968; Hinegardner and Rosen, 1972; Pedersen, 1971; Thomson *et al.*, 1973).

The greatest quantity of data on the DNA of the caudates have been collected by Sexsmith (1968), Bachmann (1970), and Olmo (1973, 1974, and unpublished); they are shown in Table II. The data of the first two authors incorporate various other data, of the previous literature (in some cases unpublished), some of which are often the average. Other data on the DNA of the caudates, generally in agreement with those given in Table II, can be found in the works of Beçak *et al.*, 1970; Ullerich, 1970; Mizuno and Macgregor, 1974.

It may be noted from Table II that the data relative to certain species studied by several authors differ from one another far more than would be admissible even by taking into consideration instrumental errors or differences in the techniques of study. In some cases this may depend on the use of different subspecies of a given species (for example, fairly extensive differences seem to exist in *Triturus cristatus* between the subspecies *cristatus* and *carnifex*); in other cases, as already mentioned, it may also depend on the fact that the DNA values attributed to certain species are the average of the sometimes heterogeneous data reported in literature. Another reason for these differences consists in the different value attributed by the various authors to the DNA of control species; *Rana pipiens,* often used for this purpose, is stated by Sexsmith to have 10.1 pg/N, whereas according to

TABLE II.　Nuclear DNA Amounts of Caudata (2c, in pg/N)

Species	Olmo, 1973, 1974, unpublished	Sexsmith, 1968	Bachmann, 1970
Hynobiidae			
Hynobius dunni	34		
keyserlingii	43		
naevius	41		
nebulosus	38		
retardatus	38		
tsuensis	33		
Cryptobranchidae			
Andrias davidianus	100		
japonicus	93		
Ambystomatidae			
Ambystoma annulatum	50		
laterale			103
macrodactylum	52		
maculatum	52		88
mexicanum			76
opacum	48	54	
talpoideum	62		
tigrinum	55	57	83
Plethodontidae			
Desmognathus fuscus	30		36
monticola			36
ochrophaeus			36
quadramaculatus			44
Leurognathus marmoratus	33		
Eurycea bislineata	41		71
longicauda	52		
lucifuga	42		
Gyrinophilus danielsi	44		
porphyriticus	50		
Pseudotriton montanus		35	
ruber	49		
Plethodon cinereus	46	39	53
elongatus		51	
glutinosus	54	43	86
jordani			72
Aneides aeneus	86		
lugubris	86		

TABLE II. (Continued)

| Species | Nuclear DNA | | |
	Olmo, 1973, 1974, unpublished	Sexsmith, 1968	Bachmann, 1970
Ensatina eschscholtzii	84	62	
Batrachoseps attenuatus	84		
Amphiumidae			
Amphiuma means	150	130	192
Salamandridae			
Pleurodeles waltlii	39		
Tylototriton verrucosus	49		
Salamandra atra	65		
salamandra		60	
Triturus alpestris		52	
cristatus	44	46	67
vulgaris	48	49	
Taricha granulosa	59		
rivularis	60		
torosa	56		
Notophthalmus viridescens	70	63	86
Paramesotriton hongkongensis	68		
Proteidae			
Necturus lewisi		182	
maculosus	165	136	205
punctatus		180	
Proteus anguinus	97	74	
Sirenidae			
Pseudobranchus striatus	91		
Siren intermedia	108		
lacertina	114		

Bachmann it has 15 pg/N or more (cf. Bachmann, 1970, 1972) and according to Olmo it has 11.6 pg/N. Especially owing to this last fact, Bachmann's data, when they are not "balanced" by the average with the data of other authors, are often found to be the highest (cf. the data relating to *Ambystoma maculatum* and *A. tigrinum, Eurycea bislineata,* and *Plethodon glutinosus*; on this point, see Mizuno and Macgregor, 1974).

In this paper I shall refer particularly to the data reported by Olmo, not because I consider them more accurate but because what is of interest here is not so much the absolute values of the DNA as the extent of the interspecific variations in this characteristic and its behavior at inter family level. These data can be deduced better from Olmo's findings because they are the most consistent and, above all, because they concern species of all the families of the order.

As is already known from the results on other organisms, and as may be deduced from a comparison between Tables I and II, there is no correlation between number of chromosomes and quantity of DNA. Among the caudates, species having the same number of chromosomes with a similar morphology may have very different quantities of DNA, at both intra- and interfamily level.

The considerable intrafamily variability in the DNA makes it difficult to perceive precise relationships between the genome size and the evolutionary level reached. Some authors have advanced the hypothesis that the caudates, like the other amphibians, may be differentiated karyologically through gradual decreases in the quantity of DNA (Goin and Goin, 1968; Bachmann et al., 1972). While admitting that certain species with particular adaptations may be differentiated in parallel with decreases in the DNA, it nevertheless seems to me that the more general tendency in the phylogeny of the caudates (as well as of the other amphibians) has been toward increases in the DNA.

This tendency is seen fairly clearly if the various species of caudates are subdivided into three groups, defined (arbitrarily) on the basis of their quantity of DNA (cf. Olmo's data, given in Table II). The first group comprises the species having less than 50 pg/N of DNA (a value roughly corresponding to the maximum value found in other amphibians), the second comprises the species having values of DNA of between 50 and 90 pg/N, the third, the species having over 90 pg of DNA.

The first group (less than 50 pg of DNA) includes the hynobiids, certain ambystomatids, the Desmognathinae, most of the Hemidactylini, some species of *Plethodon*, the salamandrids *Pleurodeles, Tylototriton,* and some species of *Triturus*. These are groups of caudates comprising the most primitive forms (hynobiids) of the order and some of the oldest or most generalized among various families, subfamilies, or tribes of higher Urodela (cf. Wake, 1966; Wake and Özeti, 1969).

The second group (between 50 and 90 pg of DNA) includes certain ambystomatids that are sometimes neotenic in nature (*Ambystoma maculatum* and *talpoideum*), the advanced Plethodontini, *Aneides* and *Ensatina,* the bolitoglossine, *Batrachoseps,* the salamandrids *Salamandra, Taricha,*

Paramesotriton, Notophthalmus, and *Triturus alpestris.* These are species of phyletically or biologically differentiated groups or else species variously affected by phenomena of paedomorphosis (cf. discussion on DNA and paedogenesis).

The third group (over 90 pg of DNA) includes species of four families belonging to different evolutionary levels but all characterized by a paedogenetic habitus.

This subdivision, even if it appears generally in agreement with the hypothesis of progressive increases in DNA in the phylogeny of the caudates, nevertheless emphasizes the preeminent importance that may be assumed by certain adaptive phenomena in relation to the major interspecific variations in the DNA.

In any case, again in support of the above mentioned hypotheses, two further observations can be made:

1) The caudates generally have higher values of DNA than other amphibians and the rest of the vertebrates (excluding the Dipnoi, the ancestors of which seem to have had far less DNA than the present-day forms; Thomson, 1972). It is therefore probable that the genome of the caudates is differentiated from that of forms that possessed quantities of DNA closer to the average of the other vertebrates (*Latimeria* has only 13 pg/N of DNA; Thomson *et al.,* 1973).

2) None of the four paedogenetic families of caudates (with the highest quantities of DNA in the order) is basic in the phylogeny of these amphibians, so that this peculiar karyological characteristic of theirs must have been acquired in a secondary manner (Morescalchi, 1970, 1973; Thomson, 1972).

DNA and Paedogenesis

Nearly all the families of caudates display phenomena of paedomorphosis (the retention of juvenile characters into adult phenotypes) and neoteny (the attainment of sexual maturity under a larval habitus; when this character is genetically fixed—or at least appears to be so—it is more proper to speak of paedogenesis). Paedomorphosis, in particular, is one of the more important evolutionary patterns for these amphibians (De Beer, 1958), even if it is difficult to assume, as Vandel (1966) did, that the living representatives of the order derive from old neotenic forms. Four of the eight families of Caudata comprise species that are exclusively paedogenetic: Cryptobranchidae and Amphiumidae, which are semilarval since they almost succeed in completing their metamorphosis, Proteidae and Sirenidae, which are permanently larval because they cease to differentiate early in their larval life (Noble, 1931).

Paedogenesis, which confines the animal to an exclusively aquatic environment, often leads to excessive specialization and to degeneration, and leads to evolutionary sterility. Paedomorphosis, on the other hand, often provides an escape from specialization and access to new adaptative zones. The evolutionary role of these phenomena has been investigated in the plethodontids by Wake (1966); as this author points out, there are no recent studies that relate them with modern genetic and evolutionary theories.

A link between the two types of study may be provided by the observed fact that the species of the four paedogenetic families have the highest levels of DNA among those studied. This fact would appear to be of considerable interest, since these species are not phyletically connected with one another (indeed, certain presumed affinities at intrafamily level might derive from evolutionary parallelisms, as in the case of *Proteus* and *Necturus*; cf. Hecht, 1957). Consequently, the karyological characteristics possessed in common by the paedogenetic species assume an adaptive value, since they may be related to the same selective factors that have favored the acquisition of the paedogenetic habitus.

The correlation between quantity of DNA and degree of paedogenesis is not precise; semilarval forms (*Amphiuma*) have far more DNA than other permanently larval forms (*Proteus*, the sirenids). However, we shall see that this fact may be plausibly interpreted in terms of cell metabolism (see discussion; towards a nuclear hypertrophy). Unfortunately there are no data on the DNA of paedogenetic species of other families; those that are sporadically neotenic in nature or, in any case, are involved in extensive phenomena of paedomorphosis (*Ambystoma mexicanum, A. tigrinum*, and *A. talpoideum, Notophthalmus viridescens, Triturus alpestris, Batrachoseps attenuatus*, etc.) generally have levels of DNA of over 50 pg/N. It is therefore possible that the above-mentioned correlation also holds for paedomorphosis in general.

In addition to this, other phenomena of a possibly adaptive nature may be associated with the increases in the DNA (as we shall see, perhaps all those generally favored by decreases in metabolic level). In fact, certain nonpaedomorphic species (*Taricha, Aneides, Ensatina*) possess fairly high quantities of DNA as well.

The distinction made between "evolutionary" and "adaptive" increases in the DNA is purely theoretical; in practice, it is the sum of the two phenomena that has characterized the karyological differentiation of each species of caudates and it is not possible to determine the extent to which each of the two may have contributed to the interspecific variability in the genome size. Since this variability may be very extensive at intrafamily level, it seems probable that the increases of the second type are quantitatively considerable and may in some cases end by masking those having a more strictly evolutionary character; in fact, even paedogenetic species of low evolutionary level have large quantities of DNA (*Andrias*; possibly the proteids).

Whereas the increases correlated with the evolution often seem to be associated with extensive modifications in the morphology and the number of chromosomes (the karyotype is differentiated from the asymmetrical formulas of the Cryptobranchoidea to the symmetrical formulas of the higher families), those having an adaptive character seem to be associated with modifications in the chromosome size alone, since the number and form of the chromosomes are generally unchanged at intrafamily level in spite of the considerable differences in the nuclear DNA. Also species of certain paedogenetic families have similar karyotypes, though with larger chromosomes and with double or triple the quantities of DNA, compared with those of species of allied families, from which they are probably derived (this is the case with *Andrias* compared with *Hynobius* and of *Amphiuma* compared with the ambystomatids). *Proteus* and *Necturus* themselves have karyotypes very similar to one another despite the great quantitative differences in DNA.

Phenomena of this kind are also common in other amphibians. Among the Anura they have been described in detail by Ullerich (1966) in three species of *Bufo* with karyotypes that are identical except for the dimensions of the chromosomes, which are generally larger in the species having the most DNA; the author has been able to exclude the hypothesis of polyteny and has interpreted these differences as the result of local increases in the DNA at interstitial level.

Neither among the caudates are there as yet indications of a plurifibrillar constitution of the chromosomes (Ullerich, 1970; Straus, 1971; White, 1973). The interspecific differences in the chromosome sizes alone—and, in parallel, in the DNA—may be interpreted on the assumption of gene duplications at interstitial level (equivalents in the two chromosome arms) and/or of increases in the centromeric heterochromatin. At interstitial level, in addition to the tandem duplication of preexisting genes, it is also possible that the genesis of new nucleotide sequences may occur (in fact, these differences accompany variations in the unique fractions). Regarding the pericentromeric heterochromatin (which contains much of the highly repetitive DNA), the increases in it would appear to be in agreement with the observation that it is often the redundant fractions that are most involved in the increases in the total DNA (see discussion, the molecular level). Moreover, in *Plethodon* the quantity of this heterochromatin seems to be proportional to the size of each spermatocyte bivalent (Macgregor *et al.*, 1973). In the case of the triploid *Ambystoma* and perhaps of the sirenids, an increase in DNA has occurred through phenomena of polyploidy.

DNA and Cell Size

The nucleus of higher plants and vertebrates contains much more DNA than necessary to code for the required number of functions; thus it seems possible that much of it has no informational role (cf. Strauss, 1974). The studies on these organisms have shown that the interspecific variations in the genome size, often not correlated with evolutionary advancement, are nevertheless not random and probably have an adaptive significance (Stebbins, 1966; Sparrow *et al.*, 1972; White, 1973). In fact, the quantity of DNA seems to be correlated with various factors subject to natural selection, such as the length of the mitotic or meiotic cycle, the basic metabolic rate, the developmental rate, the cell size, certain characteristics of the habitat, etc. (Vialli, 1957; Commoner, 1964a; Van't Hoff and Sparrow, 1963; Stebbins, 1971; Bachmann, 1972; Rees, 1972). The adaptive nature of these correlations is all the more evident since they are encountered in the groups of organisms that are most dependent upon the external environment (such as various plants and, among the vertebrates, the amphibians); in those where this dependence is less (e.g., the homeothermic vertebrates) the variation in nuclear DNA is insignificant.

The correlation that has attracted the greatest attention of research workers is that between the genome and the cell sizes. In fact, the latter is an expression of various factors (rate of oxidative metabolism, level of exchange with the outside, rate of division, etc.) which are more or less directly involved in the adaptive processes (cf. Szarski, 1968).

According to an interesting hypothesis advanced by Szarski (1970) natural selection would act first of all on the cell size, which would vary in adaptive relation to the different environments (among the caudates, big cells are correlated with a low level of metabolism and with the ability of cells to tolerate considerable variations in the composition of the body fluids); variations in DNA would be a consequence of those in cell size. However, certain experiments indicate that the cell volume may be regulated by genes having an intermediate repetitivity, including the ribosomal cistrons (cf. Pedersen, 1971).

Olmo and Morescalchi (1975) have investigated the possible correlations between the quantity of DNA and certain morphometric parameters of the cell in a sample of caudates comprising 39 species of all the families of the order (Table III). Analysis of the results has provided interesting data on the evolution of cell size in these amphibians. A direct correlation has been found between quantity of DNA on the one hand and nuclear volume, cell volume, and cell surface area on the other (Fig. 11). The increase in DNA is therefore associated with increases in cell size, even if the former is proportionally higher than the latter. The surface/volume ratio of the cell

TABLE III. Relations between the Nuclear DNA Amounts and the Main Cell Parameters in Caudates of the Various Families[a]

Species	$2n$	DNA	N_v	C_v	C_s	C_s/C_v
Desmognathus fuscus (P)	28	30.2	122	765	897	1.17
Leurognathus marmoratus (P)	28	33.1	77	417	830	1.99
Hynobius dunni (H)	56	33.8	111	437	799	1.83
Hynobius retardatus (H)	40	38.3	83	413	714	1.73
Hynobius nebulosus (H)	56	38.4	49	380	865	2.28
Hynobius naevius (H)	56	40.9	123	681	1085	1.59
Eurycea bislineata (P)	28	41.5	141	1138	1141	1.00
Eurycea lucifuga (P)	28	42.1	129	557	1003	1.80
Triturus cristatus (S)	24	43.6	91	731	1182	1.62
Gyrinophilus danielsi (P)	28	44.5	182	1079	1351	1.25
Plethodon cinereus (P)	28	46.1	248	1299	1263	0.97
Ambystoma opacum (A)	28	47.7	212	1611	1299	0.81
Ambystoma texanum (A)	28	48.3	287	1462	1250	0.85
Triturus vulgaris (S)	24	48.5	77	449	833	1.85
Pseudotriton ruber (P)	28	48.7	171	1157	1273	1.10
Tylototriton verrucosus (S)	24	49.0	97	524	1157	2.21
Gyrinophilus porphyriticus (P)	28	49.6	359	1664	1171	0.70
Eurycea longicauda (P)	28	52.2	132	704	1061	1.51
Ambystoma macrodactylum (A)	28	52.3	247	1192	883	0.74
Ambystoma maculatum (A)	28	52.4	103	559	1054	1.89
Plethodon glutinosus (P)	28	54.2	326	1791	1287	0.72
Taricha torosa (S)	22	56.0	241	1518	1350	0.89
Taricha granulosa (S)	22	59.1	490	1795	1161	0.65
Taricha rivularis (S)	22	59.8	119	1828	2002	1.09
Ambystoma talpoideum (A)	28	62.2	90	538	1088	2.02
Salamandra atra (S)	24	65.1	407	2649	2308	0.87
Paramesotriton hongkongensis (S)	24	68.4	213	1575	1548	0.98
Notophtalmus viridescens (S)	22	69.6	198	1255	1092	0.87
Batrachoseps attenuatus (P)	26	84.0	228	1233	1092	0.88
Ensatina eschscholtzii (P)	28	84.3	281	1523	1595	1.05
Aneides lugubris (P)	28	85.7	435	1995	1765	0.88
Pseudobranchus striatus (Si)— (4n?)	64	90.8	364	2021	1377	0.68
Andrias japonicus (C)	60	92.9	502	2105	1819	0.86
Proteus anguinus (Pr)	38	96.8	327	1472	1324	0.90
Andrias davidianus (C)	60	100.1	412	1831	1547	0.84
Siren intermedia (Si)—(4n?)	46	107.7	517	1902	1660	0.87
Siren lacertina (Si)—(4n?)	52	114.4	831	2712	1468	0.54
Amphiuma means (Ap)	28	149.9	640	3520	2605	0.74
Necturus maculosus (Pr)	38	165.1	946	3348	2377	0.71

[a] A = Ambystomatidae; Ap = Amphiumidae; C = Cryptobranchidae; H = Hynobiidae; P = Plethodontidae; Pr = Proteidae; S = Salamandridae; Si = Sirenidae. $2n$ = diploid chromosome number; DNA = nuclear DNA amount (pg/N); N_v = nuclear volume (μ^3); C_v = cell volume (μ^3); C_s = cell surface area (μ^2); C_s/C_v = relative cell surface (cf. in the text).

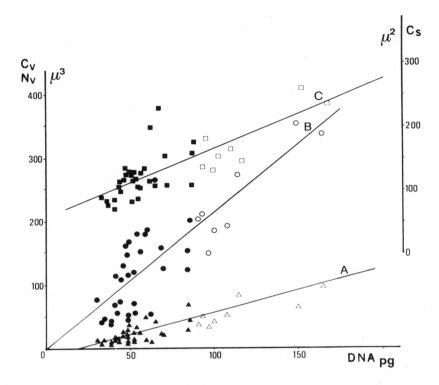

FIG. 11. The nuclear volume (Nv: triangles), cell volume (Cv: circles), cell surface (Cs: squares), on the ordinate, plotted against the nuclear DNA content, on the abscissa, in 39 species of caudates (white symbols for the paedogenetic species; see Table 3). The resulting regression straight lines (A is relative to Nv, B to Cv, and C to Cs) are highly significant statistically (from Olmo and Morescalchi, 1975).

(or *relative surface*), which, as is well known, is directly related to the rate of oxidative metabolism of the cell (Dick, 1959; Holland, 1970), assumes in relation to the quantity of DNA a behavior that approaches that of a branch of an equilateral hyperbola having its focus toward the origin of the axes (Fig. 12). The relative surface tends to decrease rapidly with the increase in nuclear DNA, so that the species with more DNA generally have a lower metabolism (cf. Commoner, 1964b); however, in the species with over 70 pg/N of DNA, further variations in the relative surface are associated with enormous increases in the size of the genome. It is therefore possible that the paedogenetic caudates have metabolic levels that are only slightly lower than those of certain species having a complete metamorphosis, even if they have relatively higher quantities of DNA (see discussion, toward a nuclear hypertrophy).

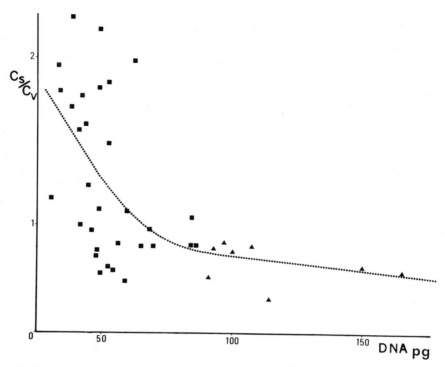

FIG. 12.   The relative cell surface (Cs/Cv, on the ordinate) plotted against the nuclear DNA content (on the abscissa) in the same sample of caudates of Table 3 and Fig. 11 (triangles indicate the paedogenetic species). A well-fitting regression line (F 'Snedecor' test significant at a level of P 0.01) is the hyperbolic curve presented in the figure (from Olmo and Morescalchi, 1975).

The almost geometric behavior of these correlations suggests that among the caudates it is chemicophysical phenomena that regulate the relationships between cell size (or metabolic rate) and quantity of DNA.

The Molecular Level

The increases in DNA, providing the raw material from which new gene complexes can be made, have played important roles in the evolution of the various organisms (Stebbins, 1966; Ohno, 1970). Experiments on the kinetics of renaturation of the DNA have made it possible to identify, in the genome of the eukaryotic cells, three components having different degrees of repetitivity: first, the "unique" nucleotide sequences (repeated less than 100 times), slowly reassociating on hydroxyapatite and probably also containing part of the coding fraction of the genome; second, the sequences repeated even millions of times (fast reassociating), short and tandemly arranged, which do not seem to have the capacity of transcription (this category includes the so-called "satellite" DNA, a highly repetitive fraction separable experimentally from the main band of the DNA and typical of various organisms); and finally, the sequences repeated to an intermediate degree (100–10,000 times), arranged in interstitial regions of the genome, at least part of which are certainly transcriptive (ribosomal RNA, heterogeneous nuclear RNA, etc.).

Different species differ from one another in the proportions of the three fractions; in any case it is generally considered that the major interspecific differences in the DNA depend on the repetitive fractions (especially those having intermediate repetitivity), the differentiation of which might be rapid (saltatory replications) and concomitant with the phenomena of speciation (Britten and Kohne, 1968).

According to certain authors, the evolutionary increase in the DNA of the eukaryotic cells depends not so much on an increase in the structural genes as in that of regulatory genes, capable of perfecting the mechanisms controlling gene activity (Britten and Davidson, 1969; Crick, 1971).

Among the mammals and other organisms the most variable component of the genome is constituted by satellite DNA; on account of its localization in the centromeric heterochromatin, its composition in short nucleotide sequences, and its high specificity, various authors maintain that this fraction of the genome performs functional roles in the nuclear and chromosome physiology, besides having a possible importance in the phenomena of speciation (Walker, 1968; Hennig and Walker, 1970; Yunis and Yasmineh, 1971). It is possible that less repetitive fractions of the genome also perform these roles, especially in connection with the complex meiotic processes (Luykx, 1974).

These theories, however, as well as others on the possible functions of the repetitive DNA (cf. Britten, 1972), do not seem able to provide a convincing interpretation of the significance of the enormous quantities of DNA that are encountered in certain organisms (e.g., in the paedogenetic caudates)—which also have an evolutionary level and a nuclear physiology only slightly different from those of other species of the same taxon—that possess far smaller quantities of nuclear DNA. With regard to this problem, the simplest solution has seemed to be to consider a large part of the DNA of these hypertrophic genomes as *actually unused* or *nonfunctional*; in some cases it has even been termed "junk" DNA (Britten and Davidson, 1971; Comings, 1972; Ohno, 1972).

In the previous chapters we have seen that the quantitative variations in the DNA of the caudates might to a large extent have an adaptive significance. Before returning to this point (see discussion, toward a nuclear hypertrophy), however, it is as well to consider the results of the few studies that have so far been carried out on the molecular biology of the DNA of the caudates.

The presence of satellite fractions, variable between species or subspecies of the same genus, has been described in *Plethodon*; in this genus and in *Notophthalmus* the satellite DNA would seem to have a prevalently centromeric localization (Macgregor and Kezer, 1971; Macgregor *et al.*, 1973; Barsacchi and Gall, 1972).

Britten and Davidson (1971) and Straus (1971) have studied the DNA renaturation kinetics of some anurans and two caudates (*Necturus maculosus* and *Ambystoma tigrinum*) and have pointed out that the quantity of repetitive sequences is greater in the species with a bigger genome.

Morescalchi and Serra (1974) have extended this study to a sample of caudates, comprising various situations in the quantity of nuclear DNA. The study included three paedogenetic species of different families, one species sometimes neotenic in nature, one with the smallest quantities of DNA in the order, and three species of the same family, two congeneric (Table IV).

The results of this study are as follows (Table IV and Fig. 13): the DNA reassociation curves for *Necturus* and *Andrias* indicate a very fast reaction, which would seem to exclude the presence of truly unique sequences. *Amphiuma*, despite the high levels of DNA, has a far slower reaction, which indicates that this paedogenetic urodelan has a genomic "complexity" (the quantity of diverse DNA sequences) greater than that of the more primitive *Necturus* and *Andrias* and comparable with that of *Ambystoma* (phyletically more akin to *Amphiuma*). *Desmognathus*, with very small quantities of DNA, has a very high quantity of highly repetitive sequences; in the percentage of the three categories of sequence considered, this caudate comes closer to certain anurans (e.g., *Rana esculenta*) than to the remaining urodeles (see Serra *et al.*, 1972). As regards the three species of the same family (Salamandridae), the two species of *Taricha* have reassociation curves that are practically coincident, faster than that of *Triturus*, which therefore seems to have greater genetic complexity. From Table IV it may be seen that the quantities of very repetitive sequences are generally higher in the species with more DNA, but that this correlation is imprecise and perhaps insignificant. The interspecific differences seem to concern, in different proportion, all the categories of DNA considered.

These studies therefore highlight the problem of the great interspecific differences also in the slowly reassociating sequences (45 pg in *Amphiuma* and only 6.3 pg in *Desmognathus*, which is possibly of a higher evolutionary level); with others, these differences might be concerned with phenomena of a functional nature, and, in accordance with the fact that in *Amphiuma*, in

TABLE IV.   "Slow" and "Fast" Reassociation Components in the DNA of Eight Species of Caudates

Species	Nuclear DNA (pg/N)	Components with repetition frequency (pg/N)	
		$\leqslant 10^2$	$\geqslant 10^6$
Andrias japonicus (Cryptobranchidae)	93	10.2	26.9
Necturus maculosus (Proteidae)	165	16.5	46.2
Amphiuma means (Amphiumidae)	150	45	25.5
Desmognathus fuscus (Plethodontidae)	30	6.3	10.5
Triturus cristatus (Salamandridae)	44	19.8	5.5
Taricha torosa (Salamandridae)	56	15.1	12.3
Taricha rivularis (Salamandridae)	60	13.2	11.4
Ambystoma tigrinum (Ambystomatidae)	55	22	5.5

spite of the high content of total DNA or unique-sequence DNA, no highly heterogeneous gene products are found (Comings and Berger, 1969). Mizuno and Macgregor (1974) have shown that the slowly reassociating fractions of 15 species of *Plethodon* are different among species but higher levels of homology (40%) are observed than in the case of repetitive sequences (only about 10%). These authors believe that only a small proportion of the unique sequences of *Plethodon* have enough functional significance to justify their conservation in all the species studied. According to Luykx (1974), also the unique sequences are involved in the specific mechanisms of pairing between the homologous chromosomes in meiosis (actually, the chromosomes of *Amphiuma* are far larger than those of *Desmognathus*).

Toward a Nuclear Hypertrophy

In the caudates the tendency toward increases in the DNA has led, in the paedogenetic families, to a sort of nuclear hypertrophy: in these organisms, the giant cells and the enormous quantities of DNA are probably the result of a high degree of specialization at cell level, reached in these proportions only by the Dipnoi among the vertebrates. The possible

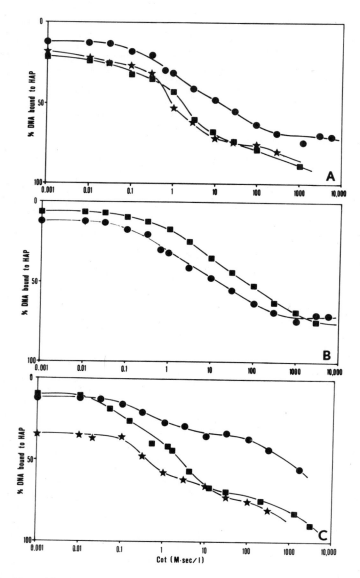

FIG. 13. Plots of DNA renaturation in eight species of Caudates: along the abscissae, the product of DNA concentration (*Co*: Moles of nucleotides/l) and time of incubation (*t*); along the ordinates, the percentage of reassociated DNA (on hydroxiapatite, HAP). A: ● *Amphiuma means*; ■ *Andrias japonicus*; ★ *Necturus maculosus*. B: ● *A. means,* ■ *Ambystoma tigrinum.* C: ● *Triturus cristatus,* ★*Desmognathus fuscus* and ■ *Taricha torosa.* The curve of *Taricha rivularis* is practically coincident with that of *T. torosa* (Morescalchi and Serra, 1974).

adaptive significance of this phenomenon appears clear if we consider as generally valid Szarski's conclusions (1968, 1970) on the "economical" nature of large cells of low metabolism in habitats such as those that characterize the paedogenetic caudates (and the Dipnoi). In these organisms, natural selection might have favored progressive reductions in the cell metabolism; it has been seen (see discussion on DNA and cell size) that these decreases, even if small, beyond certain levels are associated with considerable increases in genome and cell size. This interpretation would justify the discovery of great differences in DNA and in cell volume between some paedogenetic species (such as between *Proteus* and *Necturus,* or *Andrias* and *Amphiuma*) that nevertheless seem to have reached similar levels of specialization: in fact, despite the above-mentioned differences, it is predictable that the metabolic level of these forms may differ little.

This hypothesis therefore lends great importance to the possible adaptive significances of the variations in the DNA of the caudates, which are perhaps the vertebrates that are most dependent upon the outside environment as regards chemicophysical phenomena, but are also among the most plastic in cell physiology. The phenomena of nuclear hypertrophy that characterize the paedogenetic species would appear to represent the equivalent at cell level of the high degree of specialization reached at phenotype level.

Thomson (1972) has studied the pattern of evolution in cell size in fossil Dipnoi; struck by the discovery that the greatest increases in cell size (and perhaps in DNA) occurred after these lungfish had practically ceased to evolve, the author maintains that the enormous increases in DNA have contributed to the cessation of the evolutionary capacities of the Dipnoi (as well as that of the paedogenetic caudates).

The theory that the excessive dimensions of the genome (a sort of nuclear *hypertely*) have been fatal for the further evolution of the caudates is one of the commonest among experts on the karyology of the vertebrates; the fact that the genome of the caudates is far larger than that of other amphibians would also constitute a further proof of a different phyletic origin of the various orders of the class (Ohno, 1967, 1970; Ohno *et al.*, 1968; Beçak *et al.*, 1970; Bachmann *et al.*, 1972).

It has been seen that, in actual fact, only the genome of certain paedognetic species of caudates (*Amphiuma, Necturus*) is of "exceptional" size; that of other paedogenetic forms (*Proteus*, the sirenids, *Andrias*), though enormous, is only slightly larger than that of species having a complete metamorphosis (Table III) and is found to be connected, through a series of genomes of gradually decreasing size, with that of the primitive hynobiids, which, with about 30 pg/N of DNA, have a genome of the same proportions as those of certain Anura and Apoda. Therefore there is no sharp distinction, but merely a gradual passage and perhaps broad areas of "overlapping" in the size of the genome among the three orders of living amphibians. The fact that the Caudata have, on the average, more DNA than the Anura (for the Apoda, this is not yet proved) might depend not so much on a different phyletic origin of them as on the different adaptation generally reached by the species of the two orders (Morescalchi, 1973). It may be significant that the greatest quantities of DNA are reached by species (such as the paedogenetic caudates) characterized by a developmental pattern that is not found among the Anura.

The greatest supporter of the above-mentioned theories is Ohno (1967,1970), according to whom the enormous increases in the DNA of the caudates (and the Dipnoi) have been acquired through tandem duplication of the structural genes alone, which, by synthesizing great quantities of the same proteins, would lead to increases in cell size. In their turn, large cells would require ever greater quantities of the same gene products and hence the continuous activity of the duplicated genes; as a result, it has become impossible for these organisms to eliminate the genetic redundancy. For this reason, the Caudata would be reduced to evolutionary sterile forms, unlike Anura and Amniota, which are capable of further evolution through having maintained genomes of "reasonable" size.

The fact that various interspecific differences in the DNA of the caudates also concern the "unique" fractions is, however, in contrast with this theory; moreover, there are no proofs that the synthetic activity of the large cells of Dipnoi and paedogenetic caudates is directly proportional to the size of their genome, whereas it seems more probable that the opposite is the case (Brown and Dawid, 1968; Pedersen, 1971).

In any case, I maintain that, for the present, it cannot be proved that the great increases in the DNA have led these organisms into an evolutionary *cul-de-sac*. In the caudates, one of the most general evolutionary patterns seems to be paedomorphosis, which appears to be associated with high levels of DNA. The radiation of Ambystomatidae and Salamandridae seems to have occurred at DNA levels that are four to ten times higher than those of the Amniota. The most important radiation in the phylogeny of the order—that of the Bolitoglossini (with about 20% of the species of existing Caudata)—may have occurred at even higher levels, since, of the three supergenera of the tribe, *Batrachoseps* has over 80 pg/N of DNA whereas *Hydromantes* and *Bolitoglossa*, more affected by paedomorphosis than the remaining non-paedogenetic Plethodontidae (Wake, 1966), have very large chromosomes and cells and hence, presumably, high levels of DNA. Indeed, Macgregor and Horner (cited in Mizuno and Macgregor, 1974) have found that species of the supergenus *Bolitoglossa* (genera *Oedipina, Thorius* and *Chiropterotriton*) have the largest genomes of the family. In these urodeles, which are even today in full evolutionary stage in the various ecological niches of the neotropics, the large genomes in no way constitute an obstacle to their differentiation; indeed, they may even be closely correlated with it. As regards the paedogenetic caudates, they may be reduced to brachytelic forms, not so much (or not only) because of the enormous size of their genome as because they are by now forms that are excessively specialized for their environment, by very reason of their paedogenetic habitus.

In conclusion, the presumed exceptional nature of the Caudata as regards their DNA would seem to lie solely in the fact that, in the course of their evolution, they have effected greater improvement in certain mechanisms regulating cellular metabolic activity, which are somehow associated with the size of the genome. In this, though on a smaller scale, they are imitated by the Anura and perhaps by the Apoda. The tendency toward increases in the DNA appears common in the three orders (which, on the average, have far more DNA than the Amniota and a greater interspecific variability). It seems to function through similar methods among the living Amphibia (with the usual reservation for the Apoda, which have been little studied in this respect): by polyploidy, the sporadic appearance of accessory chromosomes and, above all, by the concentration of the genome in a few large metacentric chromosomes (symmetrical karyotypes), the site of large quantitative increases at interspecific level. From the karyological point of view, the characteristics that unite the three orders of Amphibia are certainly more numerous than those that differentiate them from one another, so that it is not necessary—indeed, it appears an improbability—postulate a polyphyletic origin for them (Morescalchi, 1973).

CONCLUSIONS AND PERSPECTIVES

The karyological evolution of the caudates presents an apparent diversity of aspects and problems if it is considered at the level of the morphology of the karyotype (the differentiation of which is closely related to the course of the phylogeny) or at the level of genome size (which seems to be largely associated with adaptive phenomena); this evolution may be considered to be the result of the two processes (Fig. 14). As stated in the introduction, the bulk of the studies carried out in the first field is sufficient to provide a general picture of the karyological differentiation of the living caudates; in the case of quantitative studies, much still remains to be done, particularly on the Plethodontidae and on the paedogenetic species of the higher families.

As regards the karyotype, we find two different situations among the caudates (with various transition formulas): on the one hand, the numerically high, asymmetrical and bimodal karyotypes of most of the Cryp-

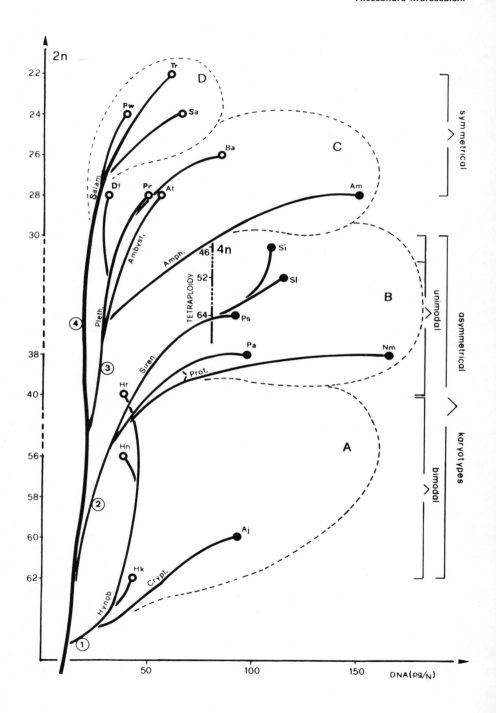

tobranchoidea; on the other hand, the symmetrical karyotypes, with 28 or fewer chromosomes, of the Ambystomatoidea and Salamandroidea (Fig. 10). There is little doubt that the former are older than the latter; in fact, they are typical of the more generalized groups of the order and are the most similar to those of the more primitive families of Anura (Ascaphidae) and Apoda (Ichthyophidae), which are also bimodal and asymmetrical (Morescalchi, 1971, 1973). It is interesting to note that formulas similar to those of the Cryptobranchoidea are found in various aquatic vertebrates (also in the lower Actinopterygii) and in various Sauropsida (including the Chelonia). It has been suggested that they have characterized the stock of vertebrates from which the tetrapods have originated, since, even at the present day, the latter display, in varying degree, karyotypes of this morphology in the oldest groups or in the more conservative groups of various classes (Morescalchi, 1970).

The transition between the formulas of the Cryptobranchoidea and those typical of the higher families, which is above all marked by the apparent loss of the microchromosomes and the acrocentric chromosomes, generally seems to be associated with increases (and at any rate not with reductions) in the nuclear DNA. Consequently, it is likely that the DNA of these peculiar chromosomes has not been lost but has been transferred onto other elements of the set through unequal translocations (including centric fusion, in view of the reduction in the acrocentric/metacentric ratio). The possible significance of this type of phenomenon has been commented upon by various authors (Stebbins, 1966; White, 1973); similar hypotheses, supplemented by others on possible peculiar functions of the microchromosomes in these evolutionary processes, have been extended by the present author (1973) to the three orders of Amphibia.

In the case of the increases in genome size, which in the caudates are relatively independent of the morphological differentiation of the karyotype

---

FIG. 14.   Scheme of the karyological differentiation of the various families of Caudata, following the hypotheses described in the text. Along the abscissae the nuclear DNA amounts (pg/N); along the ordinates the chromosome number (2n, save the case of Sirenidae, possibly polyploids). Differentiation at order level is characterized by progressive decrease in chromosome number and increase in the symmetry of karyotypes; the paedogenetic groups (black circles) tend to have the highest DNA amounts. Encircled numbers and capital letters (A–D) indicate the main branches in Caudate phylogeny of subordinal rank and disposed in four levels of organization characterized by different degrees of symmetry: 1-A: Cryptobranchoidea; 2-B: "Proteoidea" (including the Sirenidae); 3-C: Ambystomatoidea (with the Amphiumidae); 4-D: Salamandroidea.

Small circles indicate species characteristic of each family: Hynobiidae: Hk—*Hynobius keyserlingii*; Hn—*H. nebulosus*; Hr—*H. retardatus.* Cryptobranchidae: Aj—*Andrias japonicus.* Proteidae: Pa—*Proteus anguinus*; Nm—*Necturus maculosus*;. Sirenidae: Ps—*Pseudobranchus striatus*; Sl—*Siren lacertina*; Si—*S. intermedia.* Amphiumidae: Am—*Amphiuma means.* Ambystomatidae: At—*Ambystoma tigrinum.* Plethodontidae: Df—*Desmognathus fuscus*; Ps—*Pseudotriton ruber*; Ba—*Batrachoseps attenuatus.* Salamandridae: Pw—*Pleurodeles waltlii*; Sa—*Salamandra atra*; Tr—*Taricha rivularis.*

(at least as regards number and form of the chromosomes), they may have been acquired with types of chromosome change that are different from the previous ones (such as tandem gene duplications, increases in the pericentromere heterochromatin, etc.), and which have attained their maximum form of expression in the paedogenetic families, independently of their phyletic position.

Among these families, the two semilarval ones have levels of organization comparable with whose of other nonpaedogenetic groups: the Cryptobranchidae are akin to the Hynobiidae, the Amphiumidae to the Ambystomatoidea. From the karyological point of view, these two paedogenetic families differ from those from which they probably originated only in the absolute size of the chromosomes (and hence in the quantity of DNA). On the other hand, the two permanently larval families have different karyotypes from those of other caudates. The Proteidae may be considered intermediate, from this point of view, between the Cryptobranchoidea and the Ambystomatoidea; the Sirenidae occupy a position similar to that of the Proteidae, starting from the assumption that they are derived from tetraploid forms (Fig. 10-14).

Paedogenesis is associated with large quantities of nuclear DNA, generally acquired through interstitial increases in the various families of caudates. The Sirenidae would have achieved the same result by doubling their genome in a single event (tetraploidy) and then continuing to increase it by means of the same mechanisms already found in the other caudates, especially in the 'diploidization' stage that generally follows the stage in which the polyploidy has occurred (cf. Ohno, 1970); in this respect, it may be significant that the species with the greatest number of chromosomes (*Pseudobranchus striatus*) has the lowest amount of DNA in the family.

Proteidae and Sirenidae (in the order) might be independently derived from hynobiid forms at levels of karyological evolution that differ little from one another, perhaps under the stimulus of similar selective factors favourable to the acquisition of a larval habitus. As a working hypothesis, it has been proposed to include the two families in a single suborder (Proteoidea?), having a level of organization intermediate between that of the primitive Cryptobranchoidea and that of the Ambystomatoidea, though probably not on the same phyletic line as the last suborder (Fig. 14).

NOTE ON TECHNIQUES

The modern techniques of chromosome "banding" (Caspersson *et al.*, 1968; Pardue and Gall, 1970; Arrighi and Hsu, 1971), applied especially to the mammals, have allowed a detailed mapping of the chromosomes of various species and a comparison among them on far more radical bases than was possible before with the mere analysis of the general morphology of the chromosomes. The various types of chromosome bands (Q, C, R, G bands), since they are matched by the molecular structure of the chromosome (in the various DNA fractions or, more probably, in the interactions between these fractions and the nonhistonic proteins; cf. Comings,1974), allow a detailed comparison among the individual parts of the various chromosomes, which was originally possible only for the peculiar polytene chromosomes of the Diptera or for the oocyte lampbrush chromosomes of the Amphibia.

In the caudates, whose higher families are characterized by karyological formulas that are morphologically standardized, the new techniques will probably allow the identification of evolutionary trends at intra family and perhaps interspecific levels, which have so far been "masked" by the apparent morphological uniformity of the somatic set, which does not seem to be matched by an equal genetic uniformity, as has already been stated on the basis of the analysis of the lampbrush chromosomes. Results of some importance have already been obtained in this field.

Macgregor and Kezer (1971) have localized the satellite DNA in the centromere heterochromatin of *Plethodon cinereus*, while Macgregor *et al.* (1973) have extended these observations to other species and subspecies of *Plethodon* and have noted differences having a possible systematic significance.

Bailly (1972) has identified the Q bands in *Pleurodeles waltli*, noting a general analogy between them and other bands described by Labrousse *et al.* (1972) in the same species. Nardi *et al.* (1973) have compared the banding patterns obtained with the use of alkali and Giemsa in *Triturus vulgaris* and *T. italicus*, noting interspecific differences of possible systematic significance and analogies in the localization between certain bands and particular pericentromeric loops found on the lampbrush chromosomes of these Salamandridae.

These are obviously the first approaches toward a study at the microsystematic level, which is necessary today in view of the fact that studies on the morphology of the karyotype, by now extended to the more important groups of the caudates, are tending to exhaust the information that can be used at macrosystematic level.

REFERENCES

Arrighi, F. E., and Hsu, T. C., 1971, Localization of heterochromatin in human chromosomes, *Cytogenetics* **10**:81–86.

Azumi, J., and Sasaki, M., 1971, Karyotypes of *Hynobius retardatus* Dunn and *Hynobius nigrescens* Stejneger, *Chrom. Inform. Serv. Hokkaido Univ.* **12**:31–32.

Bachmann, K., 1970, Feulgen slope determinations of urodele nuclear DNA amounts. *Histochemie* **22**:289–293.

Bachmann, K., 1972, Nuclear DNA and developmental rate in frogs, *Quart. J. Fla. Acad. Sci.* **35**:225–231.

Bachmann, K., Goin, O. B., and Goin, C. J., 1972, Nuclear DNA amounts in vertebrates, *Brookhaven Symp. Biol.* **23**:419–450.

Bailly, S., 1972. Etude de la fluorescence des chromosomes mitotiques de *Pleurodeles waltlii* Michah., après coloration par la moutarde de quinacrine, *C. R. Acad. Sci. Paris* **275**:1267–1270.

Barsacchi, G., and Gall, J. G., 1972, Chromosomal localization of repetitive DNA in the newt, *Triturus, J. Cell Biol.* **54**:580–591.

Barsacchi, G., Bussotti, L., Mancino, G., 1970, The maps of the lampbrush chromosomes of *Triturus* (*Amphibia Urodela*). IV. *Triturus vulgaris meridionalis, Chromosoma* **32**:255–279.

Beatty, R. A., 1964, Chromosome deviations and sex in vertebrates. *In:* Armstrong, C. N., and Marshall, A. J., eds.: *Intersexuality in vertebrates including Man,* 17–143, Acad. Press, London and New York.

Beçak, M. L., and Beçak, W., 1974, Diploidization in *Eleutherodactylus* (Leptodactylidae, Amphibia), *Experientia* **30**:624–625.

Beçak, W., Beçak, M. L., Schreiber, G., Lavalle, D., and Amorim, F. O., 1970, Interspecific variability of DNA content in Amphibia. *Experientia* **26**:204–206.

Benazzi, M., 1948, Problemi genetici ed ormonici in rapporto alla ibridazione, *La Ricerca Scientifica* **18**:1–15.

Brame, A. H., Jr., 1967, A list of the world's recent and fossil salamanders, *Herpeton, J. Southwest. Herpetol. Soc.* **2**:1–26.

Britten, R. J., 1972, DNA sequence interspersion and a speculation about evolution, *Brookhaven Symp. Biol.* **23**:80–94.

Britten, R. J., and Davidson, E. H., 1969, Gene regulation for higher cells: a theory , *Science* **165**:349–357.

Britten, R. J., and Davidson, E. H., 1971, Repetitive and non-repetitive DNA sequences and a speculation on the origins of evolutionary novelty, *Quart. Rev. Biol.* **46**:111–138.

Britten, R. J., and Kohne, E. D., 1968, Repeated sequences in DNA, *Science* **161**:529–540.

Brown, D. D. and Dawid, I. B., 1968, Specific gene amplification in oocytes, *Science* **160**:272–279.

Callan, H. G., 1966, Chromosomes and nucleoli of the axolotl *Ambystoma mexicanum, J. Cell Sci.* **1**:85–108.

Callan, H. G., and Lloyd, L., 1960a, Lampbrush chromosomes of crested newts *Triturus cristatus* (Laurenti), *Phil. Trans. R. Soc. London B* **243**:135–219.

Callan, H. G., and Lloyd, L., 1960b, Lampbrush chromosomes. *In:* Walker, P. M. B. ed.: *New Approaches in Cell Biology.* Acad. Press, New York and London, 23–46.

Caspersson, T., Farber, S., Foley, G. E., Kudynowski, J., Modest, E. J., Simonsson, E., Wagh, U., and Zech, L., 1968, Chemical differentiation along metaphase chromosomes, *Exp. Cell Res.* **49**:219–222.

Comings, D. E., 1972, Biochemical mechanisms of chromosome banding and color banding with acridine orange, *In:* Caspersson, T., and Zech, L., eds.: Nobel Symp. 23. *Chromosome identification.* Acad. Press, London and New York, 293–299.

Comings, D. E., 1974, The structure of human chromosomes, *In:* Bush, H., ed. *The Cell Nucleus,* I: 537–563, Acad. Press, London and New York.

Comings, D. E., and Berger, R. O., 1969, Gene products of *Amphiuma:* an amphibian with an excessive amount of DNA, *Biochem. Genet.* **2**:319–333.

Commoner, B., 1964a. Roles of deoxyribonucleic acid in inheritance, *Nature* **202**:960–968.
Commoner, B., 1964b, DNA and the chemistry of inheritance, *Am. Scientist* **52**:365–388.
Crick, F., 1971, General model for the chromosomes of higher organisms, *Nature* **234**:25–27.
De Beer, G., 1958, *Embryos and ancestors,* Oxford Univ. Press, London.
Dick, D.A.T., 1959, Osmotic properties of living cells, *Int. Rev. Cytol.* **8**:387–448.
Donnelly, M. G., and Sparrow, A. H., 1965, Mitotic and meiotic chromosomes of *Amphiuma,* *J. Hered.* **56**:90–98.
Estes, R., 1965, Fossil salamanders and salamander origins, *Amer. Zool.* **5**:319–334.
Estes, R., 1969, Prosirenidae, a new family of fossil salamanders, *Nature* **224**:87–88.
Fankhauser, G., 1938, Triploidy in the newt, *Triturus viridescens, Proc. Amer. Phil. Soc.* **79**:715–739.
Ferrier, V., and Beetschen, J. C., 1973, Etude des chromosomes de *Tylototriton verrucosus* Anderson et de l'hybride viable *Pleurodeles waltlii* ♀ X *Tylototrition verrucosus* ♂ (Amphibiens Urodèles, Salamandridae), *Chromosoma* **42**:57–69.
Gall, J. G., 1954, Lampbrush chromosomes from oocyte nuclei of the newt, *J. Morph.* **94**:283–352.
Gallien, L., Labrousse, M., Picheral, B., and Lacroix, J. C., 1965, Modifications expérimentales du caryotype chez un amphibien urodéle (*Pleurodeles waltlii* Michah.), *Rev. suisse Zool.* **72**:59–85.
Goin, C. J., and Goin, O. B., 1962, *Introduction to herpetology,* W. H. Freeman and Co., London and San Francisco.
Goin, O. B., and Goin, C. J., 1968, DNA and the evolution of Vertebrates. *Am. Midl. Nat.* **80**:289–298.
Goin, O. B., Goin, C. J., and Bachmann, K., 1968, DNA and amphibian life history, *Copeia* **1968**:532–540.
Gorman, G. C., 1973, The chromosomes of the Reptilia, a cytotaxonomic interpretation. *In:* Chiarelli, A. B. and Capanna, E., eds.: *Cytotaxonomy and Vertebrate Evolution.* Acad. Press, London and New York: 349–424.
Hauschka, T. S., and Brunst, V. V., 1964, Sexual dimorphism in the nucleolar autosome of the Axolotl (*Siredon mexicanum*), *Hereditas* **52**:345–356.
Hecht, M. K., 1957, A case of parallel evolution in salamanders, *Proc. Zool. Soc.,* **Mookerjee Mem. Vol.,** Calcutta: 283–292.
Hecht, M. K., 1969, The living tetrapods: their interrelationships and phylogenetic position, *Ann. N.Y. Acad. Sci.* **167**:74–79.
Hennig, W. and Walker, P. M. B., 1970, Variations in the DNA from two rodent families (Cricetidae and Muridae). *Nature* **225**:915–919.
Hinegardner, R. T., 1968, Evolution of cellular DNA content in Teleost fishes, *Amer. Natural.* **102**:517–523.
Hinegardner, R. T., and Rosen, D. E., 1972, Cellular DNA content on the evolution of teleostean fishes, *Amer. Natural.* **106**:621–644.
Holland, R., 1970, Factors determining the velocity of gas uptake by intracellular hemoglobin, *In:* Hershey, D., ed.: *Blood oxygenation,* 1–23, Plenum Press, London and New York.
Humphrey, D. G., 1958. New chromosome number for the order Caudata. *Science* **128**:304.
Iriki, S., 1932, Studies on amphibian chromosomes, VII. On the chromosomes of *Megalobatrachus japonicus, Scient. Rep. Tokyo Bunrika Daigaku B* **1**:91–104.
Jaylet, A., 1966, Le caryotype de l'amphibien Urodéle *Euproctus asper* (Dugés), *Chromosoma* **18**:79–87.
Kawamura, T., 1953, Studies on hybridization in amphibians. V. Physiological isolation among four *Hynobius* species, *J. Sci. Hiroshima Univ. B,* **14**:73–116.
Kawamura, T., and Utsonimiya, Y., 1957, Production of auto- and allotetraploids and

diploid–tetraploid mosaics of newts by a shock of supersonic waves, *J. Sci. Hiroshima Univ. B* **17**:1–12.

Kessous, A., Caussinus, H., Jaylet, A., and Beetschen, J. C., 1967, Essai d'analyse biométrique du caryotype de l'amphibien Urodéle *Salamandra salamandra* L., *Chromosoma* **23**:324–332.

Kezer, J., 1962, The chromosome number of the european cave salamander, *Proteus anguinus* Laurenti, *Bioloski Vestnik* **10**:45–48 (reprinted from Author).

Kezer, J., 1964, Meiosis in salamander spermatocytes, *In:* Stahl, F. W. ed.: *The mechanics of inheritance,* 101–114, Prentice Hall Inc., New Jersey.

Kezer, J., and Macgregor, H. C., 1971, A fresh look at meiosis and centromeric heterochromatin in the red-backed salamander, *Plethodon c. cinereus* (Green), *Chromosoma* **33**:146–166.

Kezer, J., Seto, T., and Pomerat, C. M., 1965, Cytological evidence against parallel evolution of *Necturus* and *Proteus, Amer. Natural.* **99**:153–158.

Labrousse, M., Guillemin, C., and Gallien, L., 1972, Mise en évidence sur les chromosomes de l'Amphibien *Pleurodeles waltlii* Michah., de secteurs d'affinité différente pour le colorant de Giemsa à PH 9, *C.R. Acad. Sci. Paris* **274**:1063–1065.

Lacroix, J. C., 1968*a*, Etude descriptive des chromosomes en écouvillon dans le genre Pleurodeles (amphibien, urodéle), *Ann. Embr. Morph.* **1**:179–202.

Lacroix, J. C., 1968*b*, Variations expérimentales ou spontanées de la morphologie et de l'organisation des chromosomes en écouvillon dans le genre *Pleurodeles* (amphibien, urodéle), *Ann. Embr. Morph.* **1**:205–248.

Lacroix, J. C., 1970, Mise en évidence sur les chromosomes en écouvillon de *Pleurodeles poireti* Gervais, amphibien urodéle, d'une structure liée au sexe, identifiant le bivalent sexuel et marquant le chromosome W, *C. R. Acad. Sci. Paris* **271**:102–104.

Leon, P. and Kezer, J., 1974, The chromosomes of *Siren intermedia nettingi* (Goin) and their significance to comparative salamander karyology, *Herpetologica* **30**:1–11.

Levan, A., Fredga, K., and Sandberg, A. A., 1964, Nomenclature for centromeric position on chromosomes, *Hereditas* **52**:201–220.

Luykx, P., 1974, The organization of meiotic chromosomes, *In:* Busch, H. ed. *The cell nucleus* **2**:163–207, Acad. Press, London and New York.

Macgregor, H. C., and Kezer, J., 1971, The chromosomal localization of heavy satellite DNA in the testis of *Plethodon c. cinereus, Chromosoma* **33**:167–182.

Macgregor, H. C., Horner, H., Owen, C. A., and Parker, I., 1973, Observations on centromeric heterochromatin and satellite DNA in salamandes of the genus *Plethodon, Chromosoma* **43**:329–348.

Magalhaes, de M. A., 1942, Contribution a l'etude des chromosomes de la *Chioglossa lusitanica* Boc. II. Chromosomes sexuels, *Publ. Inst. Zool. Dr. A. Nobre* (Port.) **25**:5–7.

Makino, S., 1932, The chromosome number in some salamanders from Northern Japan, *J. Fac. Sci. Hokkaido Univ.* **2**:97–108.

Makino, S., 1934, The chromosomes of *Hynobius leechi* and *H. nebulosus, Trans. Sapporo Nat. Hist. Soc.* **13**:351–354.

Makino, S., 1935, The chromosomes of *Cryptobranchus allegheniensis, J. Morph.* **58**:573–580.

Makino, S., 1951, An atlas of the chromosome numbers in animals, 2nd Ed. *Iowa State College Press,* Ames.

Makino, S., 1962, Chromosome numbers: Vertebrates. *In:* Altman, P. C. and Dittmer, D. S., eds., *Growth,* 1–7, *F.A.S.E.B.,* Wash., D.C.

Mancino, G., 1965, Osservazioni cariologiche sull'Urodelo della Sardegna *Euproctus platycephalus:* morfologia dei bivalenti meiotici e dei lampbrush chromosomes, *Rend. Acc. Naz. Lincei* **39**:540–548.

Mancino, G., 1973, Lampbrush chromosomes in cytotaxonomic and cytogenetic studies. *In:* Chiarelli, A. B., and Capanna, E., eds., *Cytotaxonomy and Vertebrate Evolution,* 83–92, Acad. Press, London and New York.

Mancino, G. S., and Barsacchi, G., 1966a, Cariologia di *Salamandrina perspeicillata* (Anfibi Urodeli), *Boll. Zool.* **33**:251–267.

Mancino, G. S., and Barsacchi, G., 1966b, Il corredo cromosomico di *Euproctus montanus,* Urodelo della Corsica, *Rend. Acc. Naz. Lincei* **1**:396–401.

Mancino, G., and Barsacchi, G., 1966c, The maps of the lampbrush chromosomes of *Triturus* (Amphibia, Urodela). II. *Triturus helveticus helveticus, Riv. Biol.* **59**:311–351.

Mancino, G., and Barsacchi, G., 1969, The maps of the lampbrush chromosomes of *Triturus.* III. *Triturus italicus,* Ann. Embr. Morph. **2**:355–377.

Marcino, G. and Nardi, I., 1971, Chromosome heteromorphism and female heterogamety in the marbled newt *Triturus marmoratus* (Latreille), *Experientia* **27**:821–822.

Matthey, R., 1949, Les chromosomes des Vertébrés, F. Rouge, Lausanne.

Mirsky, A. E., and Ris, H., 1951, The deoxyribonucleic acid content of animal cells and its evolutionary significance, *J. Gen. Physiol.* **34**:451–462.

Mizuno, S. and Macgregor, H. C., 1974, Chromosomes, DNA sequences, and evolution in salamanders of the genus *Plethodon. Chromosoma,* **48**:239–296.

Morescalchi, A., 1970, Karyology and vertebrate phylogeny, *Boll. Zool.* **37**:1–28.

Morescalchi, A., 1973, Amphibia, *In:* Chiarelli, A. B., and Capanna, E., eds. *Cytotaxonomy and Vertebrate Evolution,* 233–348, Acad. Press, London and New York.

Morescalchi, A., and Olmo, E., 1974, Sirenids: a family of polyploid Urodeles? Experientia **30**:491–492.

Morescalchi, A., and Serra, V., 1974, DNA renaturation kinetics in some paedogenetic Urodeles, *Experientia* **30**:487–489.

Morescalchi, A., Gargiulo, G., and Olmo, E., 1970, Notes on the chromosomes of some Amphibia, *J. Herpetol.* **4**:77–79.

Morescalchi, A., Olmo, E., and Serra, V., 1974, Chromosomes and DNA of the Ambystomatoid salamanders, *Experientia* **30**:619–620.

Nardi, I. and Mancino, G., 1971, Mitotic karyotype and nucleoli in the marbled newt *Triturus marmoratus* (Latreille), *Experientia* **27**:424–427.

Nardi, I., Ragghianti, M., and Mancino, G., 1972, Characterization of the lampbrush chromosomes of the marbled newt *Triturus marmoratus* (Latreille, 1800), *Chromosoma* **37**:1–22.

Noble, G. K., 1931, *The biology of the Amphibia,* McGraw-Hill Book Co., New York.

Ohno, S., 1967, *Sex chromosomes and sex-linked genes,* Monogr. Endocr., Springer-Verlag, Berlin-Heidelberg-New York.

Ohno, S., 1970, *Evolution by gene duplication,* Springer-Verlag, Berlin-Heidelberg-New York.

Ohno, S., 1972, So much "junk" DNA in our genome, *Brookhaven Symp. Biol.* **23**:366–370.

Ohno, S., Wolf, U., and Atkin, N. B., 1968, Evolution from fish to mammals by gene duplication, *Hereditas* **59**:169–187.

Olmo, E., 1973, Quantitative variations in the nuclear DNA and phylogenesis of the Amphibia, *Caryologia* **26**:43–68.

Olmo, E., 1974, Further data on the genome size in the Urodeles, *Boll. Zool.* **41**:29–33.

Olmo, E., and Morescalchi, A., 1975, Evolution of the genome and cell size in Salamanders, *Experientia,* **31**: in press.

Pardue, M. L., and Gall, J. G., 1970, Chromosomal localization of mouse satellite DNA. *Science* **168**:1356–1358.

Pedersen, R. A., 1971, DNA content, ribosomal gene multiplicity, and cell size in Fish, *J. Exp. Zool.* **177**:65–78.

Rees, H., 1972, DNA in higher plants, *Brookhaven Symp. Biol.* **3**:394–417.

Regal, P. J., 1966, Feeding specialization and the classification of terrestrial salamanders, *Evolution* **20**:392–407.

Romer, A. S., 1966, *Vertebrate paleontology*, 3rd Ed., Univ. Chicago Press.

Salthe, S. N., 1967, Courtship patterns and the phylogeny of the Urodeles, Copeia **1967**:100–117.

Sato, I., 1936. On the chromosomes in some hynobiid salamanders from S. Japan. *J. Sci. Hiroshima Univ. B* **4**:143–154.

Serra, V., Olmo, E., and Morescalchi, A., 1972, Cinetica di riassociazione del DNA di alcuni Anfibi, *Boll. Zool.* **39**:662–663.

Seto, T., Pomerat, C. M., and Kezer, J., 1964, The chromosomes of *Necturus maculosus* as revealed in cultures of leukocytes, *Amer. Natur.* **98**:71–78.

Sexsmith, E., 1968, DNA values and karyotypes of Amphibia, Ph.D. Thesis, Univ. Toronto.

Sparrow, A. H., Price, H. J., and Underbrink, A. G., 1972, A survey of DNA content per cell and per chromosome of prokaryotic and eukaryotic organisms: some evolutionary considerations, *Brookhaven Symp. Biol.* **23**:451–492.

Stebbins, G. L., 1966, Chromosomal variation and evolution, *Science* **152**:1463–1469.

Stebbins, G. L., 1971, *Chromosomal evolution in higher plants*, Edward Arnold (Publ.s) Ltd., London.

Straus, N. A., 1971, Comparative DNA renaturation kinetics in Amphibians, *Proc. Nat. Acad. Sci.* **68**:799–802.

Strauss, S., 1974, Nuclear DNA, *In:* Bush, H. ed., *The cell nucleus*, 3:3–35. Acad. Press, London and New York.

Szarski, H., 1968, Evolution of cell size in lower vertebrates, *In:* Tor Örvig, ed. Nobel Symposium 4: *Current problems of lower vertebrate phylogeny.* 445–453, Almqvist and Wicksell, Stockholm.

Szarski, H., 1970, Changes in the amount of DNA in cell nuclei during vertebrate evolution, *Nature* **226**:651–652.

Thomson, K. S., 1972, An attempt to reconstruct evolutionary changes in the cellular DNA content of lungfish, *J. Exp. Zool.* **180**:363–372.

Thomson, K. S., Gall, J. G., and Coggins, L. W., 1973, Nuclear DNA contents of coelacanth erythrocytes, *Nature* **241**:126.

Tihen, J. A., 1958, Comments on the osteology and phylogeny of Ambystomatoid salamanders, *Bull. Fla. St. Mus. Biol. Sci.* **3**:1–50.

Twitty, V. C., 1964, Fertility of *Taricha* species hybrids and viability of their offspring, *Proc. Nat. Acad. Sci.* **51**:156–161.

Tymowska, J., and Fischberg, M., 1973, Chromosome complements of the genus *Xenopus*, *Chromosoma* **44**:335–342.

Ullerich, F. H., 1966, Karyotyp and DNS-Gehalt von *Bufo bufo, B. viridis, B. bufo* × *B viridis* and *B. calamita* (Amphibia, Anura), *Chromosoma* **18**:316–342.

Ullerich, F. H., 1970, DNA content and chromosome structure in amphibians, *Chromosoma* **21**:343–368.

Uzzell, T. M. Jr., 1963, Natural triploidy in salamanders related to *Ambystoma jeffersonianum*, *Science* **139**:113–115.

Uzzell, T. M., Jr., 1964, Relations of the diploid and triploid species of the *Ambystoma jeffersonianum* complex (Amphibia Caudata), *Copeia* **1964**:357:300.

Vandel, A., 1966, Le Protée et sa place dans l'embranchement des Vertébrés, *Bull. Soc. Zool. France* **91**:171–178.

Van't Hoff, J. and Sparrow, A. H., 1963, A relationship between DNA content, nuclear volume, and minimum mitotic cycle time, *Proc. Nat. Acad. Sci.* **49**:897–902.

Vialli, M., 1957, Volume et contenu en ADN par noyau, *Exp. Cell Res. Suppl.* **4**:284–293.

Wahlert, G. von, 1953, Eileiter, Laich und Kloake der Salamandriden, *Zool. Jahrb. Anat.* **73**:276–324.

Wake, D. B., 1966, Comparative osteology and evolution of the lungless salamanders, family *Plethodontidae, Mem. Southern Calif. Acad. Sci.* **4**:1–111.

Wake, D. B., and Ozeti, N., 1969, Evolutionary relationships in the family *Salamandridae, Copeia* **1969**:124–137.

Walker, P. M. B., 1968, How different are the DNAs from related animals? *Nature* **219**:228–232.

Walker, P. M. B., Flamm, W. G., and McLaren, A., 1969, Highly repetitive DNA in Rodents, *In:* Lima-de-Faria, A. ed. *Handbook of molecular biology,* 52–66, North Holland Publ. Co., Amsterdam-London.

White, M. J. D., 1973, *Animal cytology and evolution,* 3rd Ed., Cambridge Univ. Press, London.

Wickbom, T., 1945, Cytological studies on Dipnoi, Urodela, Anura and *Emys, Hereditas* **31**:241–346.

Yunis, J. J., and Yasmineh, W. G., 1971, Heterochromatin, satellite DNA, and cell functions, *Science* **174**:1200–1209.

Index

Vialli, M., 1957, Volume et contenu en ADN par noyau, *Exp. Cell Res. Suppl.* **4**:284–293.

Wahlert, G. von, 1953, Eileiter, Laich und Kloake der Salamandriden, *Zool. Jahrb. Anat.* **73**:276–324.

Wake, D. B., 1966, Comparative osteology and evolution of the lungless salamanders, family *Plethodontidae, Mem. Southern Calif. Acad. Sci.* **4**:1–111.

Wake, D. B., and Ozeti, N., 1969, Evolutionary relationships in the family *Salamandridae, Copeia* **1969**:124–137.

Walker, P. M. B., 1968, How different are the DNAs from related animals? *Nature* **219**:228–232.

Walker, P. M. B., Flamm, W. G., and McLaren, A., 1969, Highly repetitive DNA in Rodents, *In:* Lima-de-Faria, A. ed. *Handbook of molecular biology,* 52–66, North Holland Publ. Co., Amsterdam-London.

White, M. J. D., 1973, *Animal cytology and evolution,* 3rd Ed., Cambridge Univ. Press, London.

Wickbom, T., 1945, Cytological studies on Dipnoi, Urodela, Anura and *Emys, Hereditas* **31**:241–346.

Yunis, J. J., and Yasmineh, W. G., 1971, Heterochromatin, satellite DNA, and cell functions, *Science* **174**:1200–1209.

Index